Lecture Notes in Computer Science 11872

More information about this series at http://www.springer.com/series/7409

Hujun Yin · David Camacho ·
Peter Tino · Antonio J. Tallón-Ballesteros ·
Ronaldo Menezes · Richard Allmendinger (Eds.)

Intelligent Data Engineering and Automated Learning – IDEAL 2019

20th International Conference
Manchester, UK, November 14–16, 2019
Proceedings, Part II

 Springer

Editors
Hujun Yin (iD)
University of Manchester
Manchester, UK

Peter Tino (iD)
University of Birmingham
Birmingham, UK

Ronaldo Menezes
University of Exeter
Exeter, UK

David Camacho (iD)
Technical University of Madrid
Madrid, Spain

Antonio J. Tallón-Ballesteros (iD)
University of Huelva
Huelva, Spain

Richard Allmendinger (iD)
University of Manchester
Manchester, UK

ISSN 0302-9743   ISSN 1611-3349   (electronic)
Lecture Notes in Computer Science
ISBN 978-3-030-33616-5   ISBN 978-3-030-33617-2   (eBook)
https://doi.org/10.1007/978-3-030-33617-2

LNCS Sublibrary: SL3 – Information Systems and Applications, incl. Internet/Web, and HCI

This Springer imprint is published by the registered company Springer Nature Switzerland AG
The registered company address is: Gewerbestrasse 11, 6330 Cham, Switzerland

Preface

This year saw the 20th edition of the International Conference on Intelligent Data Engineering and Automated Learning (IDEAL 2019), held for the second time in Manchester, UK – the birthplace of one of the world's first electronic computers as well as artificial intelligence (AI) marked by Alan Turing's seminal and pioneering work. The IDEAL conference has been serving its unwavering role in data analytics and machine learning for the last 20 years. It strives to provide an *ideal* platform for the scientific communities and researchers from near and far to exchange latest findings, disseminate cutting-edge results, and to forge alliances on tackling many real-world challenging problems. The core themes of the IDEAL 2019 include big data challenges, machine learning, data mining, information retrieval and management, bio-/neuro-informatics, bio-inspired models (including neural networks, evolutionary computation, and swarm intelligence), agents and hybrid intelligent systems, real-world applications of intelligent techniques, and AI.

In total, 149 submissions were received and subsequently underwent rigorous peer reviews by the Program Committee members and experts. Only the papers judged to be of the highest quality and novelty were accepted and included in the proceedings. These volumes contain 94 papers (58 for the main track and 36 for special sessions) accepted and presented at IDEAL 2019, held during November 14–16, 2019, at the University of Manchester, Manchester, UK. These papers provided a timely snapshot of the latest topics and advances in data analytics and machine learning, from methodologies, frameworks, and algorithms to applications. IDEAL 2019 enjoyed outstanding keynotes from leaders in the field, Thomas Bäck of Leiden University and Damien Coyle of University of Ulster, and an inspiring tutorial from Peter Tino of University of Birmingham.

IDEAL 2019 was hosted by the University Manchester and was co-sponsored by the Alan Turing Institute and Manchester City Council. It was also technically co-sponsored by the IEEE Computational Intelligence Society UK and Ireland Chapter.

We would like to thank our sponsors for their financial and technical support. We would also like to thank all the people who devoted so much time and effort to the successful running of the conference, in particular the members of the Program Committee and reviewers, organizers of the special sessions, as well as the authors who contributed to the conference. We are also very grateful to the hard work by the local Organizing Committee at the University of Manchester, in particular, Yao Peng and

Jingwen Su for checking through all the camera-ready files. Continued support, sponsorship, and collaboration from Springer LNCS are also greatly appreciated.

September 2019   Hujun Yin
David Camacho
Peter Tino
Antonio J. Tallón-Ballesteros
Ronaldo Menezes
Richard Allmendinger

Organization

General Chairs

Hujun Yin	University of Manchester, UK
David Camacho	Universidad Politecnica de Madrid, Spain
Peter Tino	University of Birmingham, UK

Programme Co-chairs

Hujun Yin	University of Manchester, UK
David Camacho	Universidad Politecnica de Madrid, Spain
Antonio J. Tallón-Ballesteros	University of Huelva, Spain
Peter Tino	University of Birmingham, UK
Ronaldo Menezes	University of Exeter, UK
Richard Allmendinger	University of Manchester, UK

International Advisory Committee

Lei Xu	Chinese University of Hong Kong, Hong Kong, China, and Shanghai Jiaotong University, China
Yaser Abu-Mostafa	CALTECH, USA
Shun-ichi Amari	RIKEN, Japan
Michael Dempster	University of Cambridge, UK
Francisco Herrera	Autonomous University of Madrid, Spain
Nick Jennings	University of Southampton, UK
Soo-Young Lee	KAIST, South Korea
Erkki Oja	Helsinki University of Technology, Finland
Latit M. Patnaik	Indian Institute of Science, India
Burkhard Rost	Columbia University, USA
Xin Yao	Southern University of Science and Technology, China, and University of Birmingham, UK

Steering Committee

Hujun Yin (Chair)	University of Manchester, UK
Laiwan Chan (Chair)	Chinese University of Hong Kong, Hong Kong, China
Guilherme Barreto	Federal University of Ceará, Brazil
Yiu-ming Cheung	Hong Kong Baptist University, Hong Kong, China
Emilio Corchado	University of Salamanca, Spain
Jose A. Costa	Federal University of Rio Grande do Norte, Brazil
Marc van Hulle	KU Leuven, Belgium

Samuel Kaski	Aalto University, Finland
John Keane	University of Manchester, UK
Jimmy Lee	Chinese University of Hong Kong, Hong Kong, China
Malik Magdon-Ismail	Rensselaer Polytechnic Institute, USA
Peter Tino	University of Birmingham, UK
Zheng Rong Yang	University of Exeter, UK
Ning Zhong	Maebashi Institute of Technology, Japan

Publicity Co-chairs/Liaisons

Jose A. Costa	Federal University of Rio Grande do Norte, Brazil
Bin Li	University of Science and Technology of China, China
Yimin Wen	Guilin University of Electronic Technology, China

Local Organizing Committee

Hujun Yin	Richard Allmendinger
Richard Hankins	Yao Peng
Ananya Gupta	Jingwen Su
Mengyu Liu	

Program Committee

Ajith Abraham	Josep Carmona
Jesus Alcala-Fdez	Mercedes Carnero
Richardo Aler	Carlos Carrascosa
Davide Anguita	Andre de Carvalho
Anastassia Angelopoulou	Joao Carvalho
Ángel Arcos-Vargas	Pedro Castillo
Romis Attux	Luís Cavique
Martin Atzmueller	Darryl Charles
Dariusz Barbucha	Richard Chbeir
Mahmoud Barhamgi	Songcan Chen
Bruno Baruque	Xiaohong Chen
Carmelo Bastos Filho	Sung-Bae Cho
Lordes Borrajo	Stelvio Cimato
Zoran Bosnic	Manuel Jesus Cobo Martin
Vicent Botti	Leandro Coelho
Edyta Brzychczy	Carlos Coello Coello
Fernando Buarque	Roberto Confalonieri
Andrea Burattin	Rafael Corchuelo
Robert Burduk	Francesco Corona
Aleksander Byrski	Nuno Correia
Heloisa Camargo	Luís Correia

Cristian Ramírez-Atencia
Ajalmar Rêgo Da Rocha Neto
Izabela Rejer
Victor Rodriguez Fernandez
Matilde Santos
Pedro Santos
Jose Santos
Rafal Scherer
Ivan Silva
Leandro A. Silva
Dragan Simic
Anabela Simões
Marcin Szpyrka
Jesús Sánchez-Oro
Ying Tan
Qing Tian
Renato Tinós
Stefania Tomasiello

Pawel Trajdos
Carlos M. Travieso-González
Bogdan Trawinski
Milan Tuba
Turki Turki
Eiji Uchino
Carlos Usabiaga Ibáñez
José Valente de Oliveira
Alfredo Vellido
Juan G. Victores
José R. Villar
Lipo Wang
Tzai-Der Wang
Dongqing Wei
Raymond Kwok-Kay Wong
Michal Wozniak
Xin-She Yang
Huiyu Zhou

Additional Reviewers

Sabri Allani
Ray Baishkhi
Samik Banerjee
Avishek Bhattacharjee
Nurul E'zzati Binti Md Isa
Daniele Bortoluzzi
Anna Burduk
Jose Luis Calvo Rolle
Walmir Caminnhas
Meng Cao
Hugo Carneiro
Giovanna Castellano
Hubert Cecotti
Carl Chalmers
Lei Chen
Zonghai Chen
Antonio Maria Chiarelli
Bernat Coma-Puig
Gabriela Czanner
Mauro Da Lio
Sukhendu Das
Klaus Dietmayer
Paul Fergus
Róża Goścień
Ugur Halici

Hongmei He
Eloy Irigoyen
Suman Jana
Ian Jarman
Jörg Keller
Bartosz Krawczyk
Weikai Li
Mengyu Liu
Nicolás Marichal
Wojciech Mazurczyk
Philippa Grace McCabe
Eneko Osaba
Kexin Pei
Mark Pieroni
Alice Plebe
Girijesh Prasad
Sergey Prokudin
Yu Qiao
Juan Rada-Vilela
Dheeraj Rathee
Haider Raza
Yousef Rezaei Tabar
Patrick Riley
Gastone Pietro Rosati Papini
Salma Sassi

Ivo Siekmann
Mônica Ferreira Da Silva
Kacper Sumera
Yuchi Tian
Mariusz Topolski

Marley Vellasco
Yu Yu
Rita Zgheib
Shang-Ming Zhou

Special Session on Fuzzy Systems and Intelligent Data Analysis

Organizers

Susana Nascimento	NOVA University, Portugal
Antonio J. Tallón-Ballesteros	University of Huelva, Spain
José Valente de Oliveira	University of Algarve, Portugal
Boris Mirkin	National Research University Moscow, Russian

Special Session on Machine Learning towards Smarter Multimodal Systems

Organizers

Nuno Correia	NOVA University of Lisboa, Portugal
Rui Neves Madeira	NOVA University of Lisboa, Polytechnic Institute of Setúbal, Portugal
Susana Nascimento	NOVA University, Portugal

Special Session on Data Selection in Machine Learning

Organizers

Antonio J. Tallón-Ballesteros	University of Huelva, Spain
Raymond Kwok-Kay Wong	University of New South Wales, Australia
Ireneusz Czarnowski	University of Novi Sad, Serbia
Simon James Fong	University of Macau, Macau, China

Special Session on Machine Learning in Healthcare

Organizers

Ivan Olier	Liverpool John Moores University, UK
Sandra Ortega-Martorell	Liverpool John Moores University, UK
Paulo Lisboa	Liverpool John Moores University, UK
Alfredo Vellido	Universitat Politècnica de Catalunya, Spain

Special Session on Machine Learning in Automatic Control

Organizers

Matilde Santos	University Complutense of Madrid, Spain
Juan G. Victores	University Carlos III of Madrid, Spain

Special Session on Finance and Data Mining

Organizers

Fernando Núñez Hernández	University of Seville, Spain
Antonio J. Tallón-Ballesteros	University of Huelva, Spain
Paulo Vasconcelos	University of Porto, Portugal
Ángel Arcos-Vargas	University of Seville, Spain

Special Session on Knowledge Discovery from Data

Organizers

Barbara Pes	University of Cagliari, Italy
Antonio J. Tallón-Ballesteros	University of Huelva, Spain
Julia Handl	University of Manchester, UK

Special Session on Machine Learning Algorithms for Hard Problems

Organizers

Pawel Ksieniewicz	Wroclaw University of Science and Technology, Poland
Robert Burduk	Wroclaw University of Science and Technology, Poland

Contents – Part II

Special Session on Machine Learning in Healthcare

Special Session on Machine Learning in Automatic Control

Special Session on Finance and Data Mining

Special Session on Knowledge Discovery from Data

Special Session on Machine Learning Algorithms for Hard Problems

Contents – Part I

Special Session on Fuzzy Systems and Intelligent Data Analysis

Computational Generalization in Taxonomies Applied to: (1) Analyze Tendencies of Research and (2) Extend User Audiences

Dmitry Frolov[1,5(✉)], Susana Nascimento[2], Trevor Fenner[3],
Zina Taran[4], and Boris Mirkin[1,3]

[1] National Research University Higher School of Economics, Moscow, Russia
dfrolov@hse.ru
[2] Universidade Nova de Lisboa, Caparica, Portugal
[3] Birkbeck, University of London, London, UK
[4] Delta State University, Cleveland, MS, USA
[5] Natimatica, Ltd., Moscow, Russia

Abstract. We define a most specific generalization of a fuzzy set of topics assigned to leaves of the rooted tree of a domain taxonomy. This generalization lifts the set to its "head subject" node in the higher ranks of the taxonomy tree. The head subject is supposed to "tightly" cover the query set, possibly bringing in some errors referred to as "gaps" and "offshoots". Our method, ParGenFS, globally minimizes a penalty function combining the numbers of head subjects and gaps and offshoots, differently weighted. Two applications are considered: (1) analysis of tendencies of research in Data Science; (2) audience extending for programmatic targeted advertising online. The former involves a taxonomy of Data Science derived from the celebrated ACM Computing Classification System 2012. Based on a collection of research papers published by Springer 1998–2017, and applying in-house methods for text analysis and fuzzy clustering, we derive fuzzy clusters of leaf topics in learning, retrieval and clustering. The head subjects of these clusters inform us of some general tendencies of the research. The latter involves publicly available IAB Tech Lab Content Taxonomy. Each of about 25 mln users is assigned with a fuzzy profile within this taxonomy, which is generalized offline using ParGenFS. Our experiments show that these head subjects effectively extend the size of targeted audiences at least twice without loosing quality.

Keywords: Generalization · Fuzzy thematic cluster · Annotated suffix tree · Research tendencies · Targeted advertising

1 Introduction

The notion of generalization is not absent from the current developments in knowledge engineering and artificial intelligence. Just the opposite. For example,

© Springer Nature Switzerland AG 2019
H. Yin et al. (Eds.): IDEAL 2019, LNCS 11872, pp. 3–11, 2019.
https://doi.org/10.1007/978-3-030-33617-2_1

building a supervised classifier fits exactly into the concept of generalization: a classifier generalizes given instances of "yes"-objects into a decision rule to separate the "yes"-class from the rest. This, however, relates to a very special case at which all the objects are elements of the same variable space. We are going to tackle the case at which we are presented with a crisp or fuzzy subset of different concepts, and one wishes to generalize this subset into a coarser concept tightly embracing the subset. This is, partly, the meaning of the term "generalization" which, according to the Merriam-Webster dictionary, refers to deriving a general conception from particulars. We assume that a most straightforward medium for such a derivation, a taxonomy of the field, is given to us.

Currently, taxonomic constructions mostly concentrate on develop-ing taxonomies, especially those involving referred to in linguistics as hyponymic/hypernymic relations (see, for example, [6,8]) Also, some activities go in the direction of "operational" generalization: generalized case descriptions involving taxonomic relations between generalized states and their parts are used to achieve a tangible goal such as improving characteristics of text retrieval [7].

This paper does not attempt to develop or change any taxonomy, but rather uses an existing taxonomy. The situation of our concern is a case at which we are to generalize a fuzzy set of taxonomy leaves representing the essence of an empirically observed phenomenon. The rest of the paper is organized accordingly. Section 2 presents a mathematical formalization of the generalization problem as of parsimoniously lifting of a given fuzzy leaf set to nodes in higher ranks of the taxonomy and provides a recursive algorithm leading to a globally optimal solution to the problem. Section 3 describes an application of this approach to deriving tendencies in development of the data science, that are discerned from a set of about 18,000 research papers published by the Springer Publishers in 17 journals related to Data Science for the past 20 years. Its subsections describe our approach to finding and generalizing fuzzy clusters of research topics. In the end, we point to tendencies in the development of the corresponding parts of Data Science, as drawn from the lifting results. Section 4 describes an application of the parsimonious generalization method to efficiently extend the audience of targeted advertising over the Internet. More detailed description can be found in [3].

2   Generalization: Parsimoniously Lifting a Fuzzy Thematic Set in Taxonomy

We consider the following problem. Given a rooted taxonomy tree and fuzzy set S of taxonomy leaves, find a node $h(S)$ of higher rank in the taxonomy, that tightly covers the set S.

The problem is not as simple as it may seem to be. Consider, for the sake of simplicity, a hard set S shown with five black leaf boxes on a fragment of a tree in Fig. 1 illustrating the situation at which the set of black boxes is lifted to the root. If we accept that set S may be generalized by the root, this would lead to a number, four, of white boxes to be covered by the root and, thus, in this way, falling in the same concept as S even as they do not belong in S. Such a situation will be referred to as a gap. Lifting with gaps should be penalized. Altogether, the number of conceptual elements introduced to generalize S here is 1 head subject, that is, the root to which we have assigned S, and the 4 gaps occurred just because of the topology of the tree, which imposes this penalty. Another lifting decision is illustrated in Fig. 2: here the set is lifted just to the root of the left branch of the tree. The number of gaps here has decreased, to just 1. However, another oddity emerged: a black box on the right, belonging to S but not covered by the node at which the set S is mapped. This type of error will be referred to as an offshoot. At this lifting, three new items emerge: one head subject, one offshoot, and one gap. This is less than the number of items emerged at lifting the set to the root (one head subject and four gaps, that is, five), which makes it more preferable. Of course, this conclusion holds only if the relative weight of an offshoot is less than the total relative weight of three gaps.

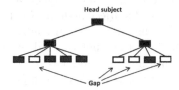

Fig. 1. Generalization of the black box query set by mapping it to the root, with the price of four gaps emerged at the lift.

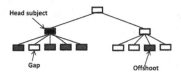

Fig. 2. Generalization of the black box query set by mapping it to the root of the left branch, with the price of one gap and one offshoot emerged at this lift.

We are interested to see whether a fuzzy set S can be generalized by a node h from higher ranks of the taxonomy, so that S can be thought of as falling within the framework covered by the node h. The goal of finding an interpretable pigeon-hole for S within the taxonomy can be formalized as that of finding one or more "head subjects" h to cover S with the minimum number of all the elements introduced at the generalization: head subjects, gaps, and offshoots. This goal realizes the principle of Maximum Parsimony.

Consider a rooted tree T representing a hierarchical taxonomy so that its nodes are annotated with key phrases signifying various concepts. We denote the set of all its *leaves* by I. The relationship between nodes in the hierarchy is conventionally expressed using genealogical terms: each node $t \in T$ is said to be the *parent* of the nodes immediately descending from t in T, its *children*. We use $\chi(t)$ to denote the set of children of t. Each *interior* node $t \in T - I$ is assumed to correspond to a concept that generalizes the topics corresponding to the leaves $I(t)$ descending from t, viz. the leaves of the subtree $T(t)$ rooted at t, which is conventionally referred to as the *leaf cluster of* t.

A *fuzzy set* on I is a mapping u of I to the non-negative real numbers that assigns a membership value $u(i) \geq 0$ to each $i \in I$. We refer to the set $S_u \subset I$, where $S_u = \{i \in I : u(i) > 0\}$, as the *base* of u.

Given a fuzzy set u defined on the leaves I of the tree T, one can consider u to be a (possibly noisy) projection of a higher rank concept, u's "head subject", onto the corresponding leaf cluster. Under this assumption, there should exist a head subject node h among the interior nodes of the tree T such that its leaf cluster $I(h)$ more or less coincides (up to small errors) with S_u. This head subject is the generalization of u to be found. The two types of possible errors associated with the head subject, if it does not cover the base precisely, are false positives and false negatives, referred to in this paper, as *gaps* and *offshoots*, respectively (see Figs. 1 and 2). Given a head subject node h, a gap is a node t covered by h but not belonging to u, so that $u(t) = 0$. In contrast, an offshoot is a node t belonging to u so that $u(t) > 0$ but not covered by h. Altogether, the total number of head subjects, gaps, and offshoots has to be as small as possible. For each of these elements a penalty is defined: 1 is the penalty for a head subject, γ, the penalty for a gap, and λ is the penalty for an offshoot.

Consider a candidate node h in T and its meaning relative to fuzzy set u. An *h-gap* is a node g of $T(h)$, other than h, at which a *loss* of the meaning has occurred, that is, g is a maximal u-irrelevant node in the sense that its parent is not u-irrelevant. Conversely, establishing a node h as a head subject can be considered as a *gain* of the meaning of u at the node. The set of all h-gaps will be denoted by $G(h)$. A node $t \in T$ is referred to as *u-irrelevant* if its leaf-cluster $I(t)$ is disjoint from the base S_u. Obviously, if a node is u-irrelevant, all of its descendants are also u-irrelevant.

An *h-offshoot* is a leaf $i \in S_u$ which is not covered by h, i.e., $i \notin I(h)$. The set of all h-offshoots is $S_u - I(h)$. Given a fuzzy topic set u over I, a set of nodes H will be referred to as a *u-cover* if: (a) H covers S_u, that is, $S_u \subseteq \bigcup_{h \in H} I(h)$, and (b) the nodes in H are unrelated, i.e. $I(h) \cap I(h') = \emptyset$ for all $h, h' \in H$ such that $h \neq h'$. The interior nodes of H will be referred to as *head subjects* and the leaf nodes as *offshoots*, so the set of offshoots in H is $H \cap I$. The set of *gaps* in H is the union of $G(h)$ over all head subjects $h \in H - I$.

The penalty function $p(H)$ for a u-cover H is:

$$p(H) = \sum_{h \in H-I} u(h) + \sum_{h \in H-I} \sum_{g \in G(h)} \lambda v(g) + \sum_{h \in H \cap I} \gamma u(h), \qquad (1)$$

and we are to find a u-cover H that globally minimizes the penalty $p(H)$. Such a u-cover is the parsimonious generalization of the set u.

First, the tree is pruned from all the non-maximal u-irrelevant nodes. Simultaneously, the sets of gaps $G(t)$ and the internal summary gap importance

$V(t) = \sum_{g \in G(t)} v(g)$ in Eq. (1) are computed for each interior node t. After this, our lifting algorithm ParGenFS applies. For each node t, the algorithm ParGenFS computes two sets, $H(t)$ and $L(t)$, containing those nodes in $T(t)$ at which respectively gains and losses of head subjects occur (including offshoots). The associated penalty $p(t)$ is computed too.

Sets $H(t)$ and $L(t)$ are defined assuming that the head subject has not been gained (nor therefore lost) at any of t's ancestors. The algorithm ParGenFS recursively computes $H(t)$, $L(t)$ and $p(t)$ from the corresponding values for the child nodes in $\chi(t)$.

Specifically, for each leaf node that is not in S_u, we set both $L(\cdot)$ and $H(\cdot)$ to be empty and the penalty to be zero. For each leaf node that is in S_u, $L(\cdot)$ is set to be empty, whereas $H(\cdot)$, to contain just the leaf node, and the penalty is defined as its membership value multiplied by the offshoot penalty weight γ. To compute $L(t)$ and $H(t)$ for any interior node t, we analyze two possible cases: (a) when the head subject has been gained at t and (b) when the head subject has not been gained at t.

In case (a), the sets $H(\cdot)$ and $L(\cdot)$ at its children are not needed. In this case, $H(t)$, $L(t)$ and $p(t)$ are defined by:

$$H(t) = \{t\}, \quad L(t) = G(t), \quad p(t) = u(t) + \lambda V(t). \tag{2}$$

In case (b), the sets $H(t)$ and $L(t)$ are the unions of those of its children, and $p(t)$ is the sum of their penalties. To obtain a parsimonious lift, whichever case gives the smaller value of $p(t)$ is chosen.

The output of the algorithm consists of the values at the root, namely, H – the set of head subjects and offshoots, L – the set of gaps, and p – the associated penalty.

The algorithm ParGenFS is proven to lead to an optimal lifting indeed [3].

3   Highlighting Tendencies in Research

Being confronted with the problem of structuring and interpreting a set of research publications in a domain, one can think of either of the following two pathways to take. The first pathway tries to discern main categories from the texts, the other, from knowledge of the domain. The first approach is exemplified by clustering and topic modeling; the second approach, by using an expert-driven taxonomy. The main difference between these approaches lies in the level of granularity: the former pathway uses concepts of the same level of granularity as those in texts, whereas the latter approach may bring forth coarser concepts from the higher ranks of a taxonomy.

This paper follows the second pathway by moving, in sequence, through the stages covered in separate Subsects. 3.1 to 3.6.

3.1   Scholarly Text Collection

We downloaded a collection of 17685 research papers together with their abstracts published in 17 journals related to Data Science for 20 years from 1998–2017. We take the abstracts to these papers as a representative collection.

3.2   DST Taxonomy

The subdomain of our choice is Data Science, comprising such areas as machine learning, data mining, data analysis, etc. We take that part of the six-layer ACM-CCS 2012 taxonomy of computing subjects [1], which is related to Data Science, and add a few leaves related to more recent Data Science developments The taxonomy itself, with all its 317 leaves, can be found in [3].

3.3   Relevance Topic-to Text Score and Co-relevance Topic-to-topic Similarity Index

We first obtain a keyphrase-to-document matrix R of relevance scores by using the Annotated Suffix Tree approach [2]. This matrix R is converted to a keyphrase-to-keyphrase similarity matrix A for scoring the "co-relevance" of keyphrases according to the text collection structure. The similarity score $a_{ii'}$ between topics i and i' is computed as the inner product of vectors of scores $r_i = (r_{iv})$ and $r_{i'} = (r_{i'v})$. The inner product is moderated by a natural weighting factor assigned to texts in the collection. The weight of text v is defined as the ratio of the number of topics n_v relevant to it and n_{max}, the maximum n_v over all $v = 1, 2, \ldots, V$. A topic is considered relevant to v if its relevance score is greater than 0.2 (a threshold found experimentally, see [2]).

3.4   Fuzzy Thematic Clusters of Taxonomy Topics

Clusters of topics should reflect co-occurrence of topics: the greater the number of texts to which both t and t' topics are relevant, the greater the interrelation between t and t', the greater the chance for topics t and t' to fall in the same cluster. We have tried several popular clustering algorithms at our data. Unfortunately, no satisfactory results have been found. Therefore, we present here results obtained with the Fuzzy ADDItive Spectral (FADDIS) clustering algorithm developed in [5] specifically for finding thematic clusters.

After computing the 317×317 topic-to-topic co-relevance matrix, converting in to a topic-to-topic Laplace–transformed similarity matrix [5], and applying FADDIS clustering, we sequentially obtained 6 clusters, of which three clusters are obviously homogeneous. They relate to "Learning", "Retrieval", and "Clustering".

3.5   Lifting the Clusters

The three clusters mentioned above are lifted in the DST taxonomy using Par-GenFS algorithm with the gap penalty $\lambda = 0.1$ and off-shoot penalty $\gamma = 0.9$.

Lifting Cluster L brings three head subjects: Machine Learning, Machine Learning Theory, and Learning to Rank. Lifting of Cluster R: Retrieval leads to two head subjects: Information Systems and Computer Vision. For Cluster C, 16 (!) head subjects were obtained.

3.6   Drawing Conclusions

The "Learning" head subjects show that main work here still concentrates on theory and method rather than applications. A good news is that the field of learning, formerly focused mostly on tasks of learning subsets and partitions, is expanding towards learning of ranks and rankings.

Lifting results for the information retrieval cluster R, clearly show: Rather than relating the term "information" to texts only, as it was in the previous stages of the process of digitalization, visuals are becoming parts of the concept of information. There is a catch, however. Unlike the multilevel granularity of meanings in texts, developed during millennia of the process of communication via languages, there is no comparable hierarchy of meanings for images. One may only guess that the elements of the R cluster related to segmentation of images and videos, as well as those related to data management systems, are those that are going to be put in the base of a future multilevel system of meanings for images and videos. This is a direction for future developments clearly seen from lifting results.

Regarding the "clustering" cluster C with its 16 (!) head subjects, one may conclude that, perhaps, a time moment has come or is to come real soon, when the subject of clustering must be raised to a higher level in the taxonomy to embrace all these "heads". Currently, clustering is not just an auxiliary instrument but rather a model of empirical classification, a big part of the knowledge engineering.

4   Efficient Audience Extending in Targeted Advertising

We consider a company, such as start-up Natimatica, Ltd. (see https:// natimatica.com) associated with us, that supports a service of native advertising. This service follows millions of individual users visiting popular sites providing news, shopping, specific contents, etc. Information of these users is stored in a special system, Data Management Platform (DMP). Each individual user in the DMP is assigned with a subset of the IAB taxonomy of goods and services [4] segments (leaves) relevant to their visits. Each of the segments i is assigned with a real number a_i, a fraction of unity, according to the history of the user's visits to the sites under our observation. The totality of the taxonomy segments and membership values assigned to them constitute what can be referred to as the user's profile. An advertiser formulates their advert needs as a set of IAB leaves (segments) relevant to the advert. In practice, such a formulation is produced manually by an employee, after a detailed discussion of the advert with the advertiser. A conventional, currently most popular, approach (CAS) requires to pre-specify a threshold t (usually, $t = 0.3$), so that a condition $A(t)$ can be applied: Given a set S of taxonomy segments and a user's profile, check if the profile has at least one of the S segments with the value a_i assigned to them so that $a_i > t$. Then the CAS rule requires to expose the advert to all those users for whom condition $A(t)$ holds. An issue with CAS is that the number of users satisfying condition $A(t)$ may be less than the number specified in the advertisement order. In this case, a conventional strategy is to have CAS extended by lessening t to $t^0 < t$, so that more users satisfy condition $A(t^0)$ than $A(t)$ (CASE). In contrast, our strategy, Generalization of User Segments (GUS), is based on the optimal generalizations of user profiles in the IAB taxonomy made off-line. GUS tests condition $A(t)$ by applying it not to the segment concerned but rather to the head subjects.

Table 1 presents comparative results of testing our GUS method at real life advertising campaigns in Natimatica, Ltd. The comparison criteria are: (a) advertising impressions (Imprs in the Table 1) obtained, (b) numbers of clicks, and (3) click through rates (CTR, (b)/(a)). Our method GUS significantly outperforms the conventional strategies.

Table 1. Advertising campaign results at different targeting methods

Campaign	IAB segments	Metric	CAS	CASE	GUS
Software for parental control of children. Dur. 10 days	Daycare and Pre-School, Internet Safety, Parenting Children Aged 4–11, Parenting Teens, Antivirus Software	Imprs	378933	1017598 (+168.5%)	942104 (+148.6%)
		Clicks	1061	1526 (+43.8%)	2544 (+139.8%)
		CTR,%	0.28	0.15 (−46.4%)	0.27 (−3.6%)
Mortgage at a major Russian bank Dur. 10 days	Home Financing, Personal Loans	Imprs	159342	275035 (+72.6%)	289308 (+81.6%)
		Clicks	749	853 (+13.9%)	1302 (+73.9%)
		CTR,%	0.47	0.31 (−34.0%)	0.45 (−4.3%)

Acknowledgment. D.F. and B.M. acknowledge continuing support by the Academic Fund Program at the NRU HSE (grant -19-04-019 in 2018–2019) and by the DECAN Lab NRU HSE, in the framework of a subsidy granted to the HSE by the Government of the Russian Federation for the implementation of the Russian Academic Excellence Project "5-100". S.N. acknowledges the support by FCT/MCTES, NOVA LINCS (UID/CEC/04516/2019).

References

1. The 2012 ACM Computing Classification System. http://www.acm.org/about/class/2012. Accessed 30 Apr 2018
2. Chernyak, E.: An approach to the problem of annotation of research publications. In: Proceedings of the 8th ACM WSDM, vol. 429–434. ACM (2015)
3. Frolov, D., Mirkin, B., Nascimento, S., Fenner, T.: Finding an appropriate generalization for a fuzzy thematic set in taxonomy. Working paper WP7/2018/04, Moscow, Higher School of Economics Publ. House (2018)
4. IAB Tex Lab Content Taxonomy (2017). https://www.iab.com/guidelines/iab-quality-assurance-guidelines-qag-taxonomy/. Accessed 5 July 2019
5. Mirkin, B., Nascimento, S.: Additive spectral method for fuzzy cluster analysis of similarity data including community structure and affinity matrices. Inf. Sci. **183**(1), 16–34 (2012)
6. Vedula, N., Nicholson, P.K., Ajwani, D., Dutta, S., Sala, A., Parthasarathy, S.: Enriching taxonomies with functional domain knowledge. In: The 41st International ACM SIGIR Conference on R&D in Information Retrieval, pp. 745–754. ACM (2018)
7. Waitelonis, J., Exeler, C., Sack, H.: Linked data enabled generalized vector space model to improve document retrieval. In: Proceedings of NLP & DBpedia 2015 Workshop and 14th ISW Conference, CEUR-WS, vol. 1486 (2015)
8. Wang, C., He, X., Zhou, A.: A short survey on taxonomy learning from text corpora: issues, resources and recent advances. In: Proceedings of the 2017 Conference on Empirical Methods in Natural Language Processing, pp. 1190–1203 (2017)

Unsupervised Initialization of Archetypal Analysis and Proportional Membership Fuzzy Clustering

Susana Nascimento$^{(\boxtimes)}$ (ID) and Nuno Madaleno

Computer Science Department and NOVA LINCS,
Faculdade de Ciências e Tecnologia, Universidade Nova de Lisboa,
Lisbon, Portugal
snt@fct.unl.pt

Abstract. This paper further investigates and compares a method for fuzzy clustering which retrieves *pure* individual types from data, known as the fuzzy clustering with proportional membership (FCPM), with the FurthestSum Archetypal Analysis algorithm (FS-AA). The Anomalous Pattern (AP) initialization algorithm, an algorithm that sequentially extracts clusters one by one in a manner similar to principal component analysis, is shown to outperform the FurthestSum not only by improving the convergence of FCPM and AA algorithms but also to be able to model the number of clusters to extract from data.

A study comparing nine information-theoretic validity indices and the soft ARI has shown that the soft Normalized Mutual Information max (NMI_{sM}) and the Adjusted Mutual Information (AMI) indices are more adequate to access the quality of FCPM and AA partitions than soft internal validity indices. The experimental study was conducted exploring a collection of 99 synthetic data sets generated from a proper data generator, the FCPM-DG, covering various dimensionalities as well as 18 benchmark data sets from machine learning.

Keywords: Archetypal analysis · Fuzzy clustering ·
Number of clusters · Information-Theoretic validity indices

1 Introduction

Archetypal Analysis [3] is a statistical method that synthesizes a set of multivariate data points through a few representatives, called *archetypes*, which capture the most discriminating features of the data and lie on the convex hull of the data set. Archetypal Analysis (AA) has been applied to various impact real world problems (e.g. [2,5,6,15,18].

An effective AA algorithm was developed in the framework of unsupervised machine learning [11] to guarantee a fast convergence of AA on finding a predefined number of archetypes. The alternating optimization (AO) algorithm is initialized with the K data points furthest way from the centre of the data

© Springer Nature Switzerland AG 2019
H. Yin et al. (Eds.): IDEAL 2019, LNCS 11872, pp. 12–20, 2019.
https://doi.org/10.1007/978-3-030-33617-2_2

set by the so-called FurthestSum (FS), and uses a simple gradient projection method. Despite its success, the FS-AA algorithm suffers from the problem that the number of archetypes, K, has to be prespecified. On the other hand, to determine the best number of archetypes the authors claim that the analysis of scree plot establishes a reasonable criterion to select K. However, this is not the case in many real world data, mainly in those ones not showing a clear cut cluster structure.

The Fuzzy Clustering with Proportional Membership (FCPM) [12,13] is a clustering algorithm whose model assumes the existence of some prototypes serving as *ideal* patterns underlying the data. To relate the observed data to the prototypes, the FCPM model considers any entity to share parts of the prototypes, such that an entity may bear 60% of a prototype $C1$ and 40% of prototype $C2$, which simultaneously expresses the entity's membership to the respective clusters. The baseline clustering criterion of the model, FCPM0, leads to cluster structures with central prototypes, like the popular fuzzy c-means (FCM) [1], while its smoother version, FCPM2, leads to cluster structures with extremal prototypes that are similar to archetypes.

In this article we focus on two important issues of partitional clustering, specifically, for archetypal analysis and fuzzy proportional membership: (i) strategy to initialize the algorithms; and (ii) assessment of the clustering partitions. Concerning (i) we explore a powerful initialization clustering method, the Incremental Anomalous Pattern (AP) [9,10], that sequentially extracts clusters one by one in a manner similar to principal component analysis, providing the seed points of the multidimensional space of the data and, simultaneously, allows to determine the number, K, of seeds. Taking advantage of this, the AP algorithm had been successfully applied with FCM to the problem of unsupervised segmentation of Sea Surface Temperature (SST) images [14]. Problem (ii), designated as *cluster validity*, is a wide topic in cluster analysis. It comprises computational methods able to identify the "best" K-partition among a set of candidate partitions from $K = K_{min}$ to $K = K_{max}$. This is typically performed by the use of scalar measures, the cluster validity indices (CVIs). We explore a collection of nine information-theoretic cluster validity indices as well as the soft version of the ARI index [7,17] and compare them against five premier internal indices explored in a previous study by one of the authors [8].

The remainder of the paper is organized as follows. The next section briefly describes the AA and FCPM models as well as the corresponding AO algorithms. Section 3 presents the Anomalous Pattern (AP) initialization method. In Sect. 4 we present and discuss the main experimental results comparing nine information-theoretic validity indices to assess the quality of AA and FCPM partitions, the fine-tuning of AP stop condition to model the number of clusters, and the comparison of AP against FS initialization honoring the number of clusters, applied to synthetic and real data. Section 5 concludes the paper.

2 Furthest Sum Archetypal Analysis and Fuzzy Clustering with Proportional Membership

Archetypal analysis aims to represent the observations $\mathbf{x}_i$ $(i = 1, ..., N)$ with $\mathbf{x}_i \in \Re^D$ of a given data set $X = \{\mathbf{x}_1, \mathbf{x}_2, ..., \mathbf{x}_N\}$ as a convex combination of the most extremal K data points, the archetypes $A = \{\mathbf{a}_1, \mathbf{a}_2, ..., \mathbf{a}_K\}$. The K archetypes that better fits the data X are found by minimizing the residual sum of squares (RSS) criterion:

$$RSS = \min_{\delta, \beta} \sum_{i=1}^{N} ||\mathbf{x}_i - \sum_{k=1}^{K} \mathbf{a}_k.\delta_{ki}||^2 = \min_{\delta, \beta} \sum_{i=1}^{N} ||\mathbf{x}_i - \sum_{k=1}^{K} \sum_{j=1}^{N} \mathbf{x}_j.\beta_{jk}.\delta_{ki}||^2, \quad (1)$$

where the data points are convex combinations of the archetypes,

$$\delta_{ki} \geq 0, \qquad \sum_{k=1}^{K} \delta_{ki} = 1, \qquad (2)$$

and each archetype $\mathbf{a}_k$ is expressed as a convex combinations of the data points:

$$\mathbf{a}_k = \sum_{i=1}^{N} \beta_{ik}\mathbf{x}_i, \qquad \text{with} \qquad \beta_{ik} \geq 0, \qquad \sum_{i=1}^{N} \beta_{ik} = 1. \qquad (3)$$

As shown in [3], for a space dimensionality $D > 1$, the archetypes that minimize RSS criterion (1) fall on the convex hull of the data, making them extreme data-values, while for $D = 1$, the sample mean minimizes the RSS. To minimize the RSS criterion (1), it is then needed to find both the $K \times N$ membership matrix $\Delta = [\delta_{ki}]$ and $N \times K$ matrix $B = [\beta_{ik}]$ of β's coefficients, which requires an AO algorithm. Several extensions of the original AO algorithm have been developed to optimize the AA clustering criterion (1) (e.g [2,4,11]).

We explore a recent AA algorithm introduced in [11] which takes advantage of a simple gradient project method and initializes with K data points furthest away from the center of the data set, by the so called Furthest Sum (FS) method. Starting from a randomly selected data observation, the algorithm selects the seeding archetypes as the K data points that are at the maximum distance from the archetypes already selected, $\mathbf{x}_j$. In what follows this algorithm is designated as FS-AA algorithm.

The model of Fuzzy Clustering with Proportional Membership (FCPM) [12,13] assumes that observed data are generated according to a cluster model underlying the data: *observed data = model data + noise.*

Consider an entity-to-feature data matrix Y, preprocessed from X by shifting the origin to the center of the data, and rescaling the features by their ranges. The generic proportional fuzzy membership model takes each membership value u_{ki} not just as a weight but an expression of the proportion of prototype $\mathbf{v}_k$ which is present in the observation $\mathbf{y}_i$. Formally, the FCPM model is defined as:

$$y_{ih} = u_{ki}v_{kh} + e_{kih}, \qquad (4)$$

with e_{kih} as the residual values.

To minimize the residual values e_{kih} a square-error clustering criterion is defined by fitting each data point sharing a proportion of each of the prototypes, represented by the degree of membership, as follows:

$$E_m(U,V) = \sum_{k=1}^{K}\sum_{i=1}^{N}\sum_{h=1}^{D} u_{ki}^m(y_{ih} - u_{ki}v_{kh})^2, \tag{5}$$

with regard to the fuzzy constraints:

$$0 \le u_{ki} \le 1, \text{ for all } k = 1,\ldots,K, \, i = 1,\ldots,N \tag{6}$$

$$\sum_{k=1}^{K} u_{ki} = 1, \text{ for all } i = 1,\ldots,N. \tag{7}$$

The role of membership coefficient u_{ki}^m with integer exponent $m = 0, 1, 2$ in criterion (5) is to smooth the influence of high residual values at small memberships u_{ki}, that is, only meaningful proportions, those with high membership values, are to be taken into account.

The minimization of clustering criterion E_m (5) considers an iterative AO algorithm as follows: (i) initialize the set of prototypes V with random values selected from the data space; (ii) update the fuzzy membership matrix U from this initial V using a gradient project method; (iii) alternate between minimizing the weights, U, given the centroids, $\hat{V}$, and minimizing V, given the updated $\hat{U}$; (iv) stop when the AO algorithm converges. A detailed description of the FCPM-AO algorithm and its foundation is in [12].

In what follows we only consider the models with parameter $m = 0$ and $m = 2$ designated as FCPM0 and FCPM2, respectively.

3 Anomalous Pattern Algorithm

The Incremental Anomalous Pattern (AP) algorithm [9,10] is an alternative initialization to the FurthestSum (FS), that is deterministic and, by modeling its stop condition, automatically determines the number of clusters to be retrieved from data.

The AP algorithm sequentially extracts clusters one by one from the data set as follows. Let Y denote the standardized data set, by shifting the origin of the original data X to the grand mean, $\bar{x}$. The feature vector $\bar{x}$ is taken as the reference point unvaried all over the sequential process, and take as seed point the data point that is farthest from the reference point $\bar{x}$. One crisp cluster, C_t, is iteratively constructed, defined as the set of points that are closer to the seed point than to the reference point. After, the cluster seed of C_t is substituted by the cluster's gravity center and the procedure is reiterated until it converges. The procedure is reiterated over the residual data set taken as $Y^{t+1} = Y^t - C_t$ until any of the following stop conditions is reached: (C1) the residual data set, Y^{t+1}, is empty which means that all the entities had been clustered; (C2) the contribution of the t-th cluster to the data scatter (Eq. (8)) is too small, that is, less than a pre-specified threshold τ; (C3) the number of clusters, t, reached a pre-specified value K^*; (C4) the cardinality of t-th cluster is too small.

Taking Y as the standardized data set, the total scatter of all data points (row-vectors in the $N \times D$ matrix Y) is defined as $T(Y) = \sum_{i=1}^{N} \sum_{h=1}^{D} y_{ih}^2$. In [9] it is derived how the total data scatter $T(Y)$ can be decomposed into an explained part due to the cluster structure retrieved from data Y and the unexplained part which corresponds to the K-means clustering criterion to be minimized. From that, it is defined the relative contribution of each individual cluster, $(\mathcal{C}, \mathbf{v})$, to the data scatter, such as:

$$W((\mathcal{C}, \mathbf{v})) = \frac{n \sum_{h=1}^{D} v_h^2}{T(Y)} = \frac{n \sum_{h=1}^{D} v_h^2}{\sum_{i=1}^{N} \sum_{h=1}^{D} y_{ih}^2}, \tag{8}$$

with n the cardinality of cluster $\mathcal{C}$.

The prototypes $\{\mathbf{v}_k\}_{k=1}^{K^*}$ returned by AP can be set as the initial seeds to deterministically initialize the AA and FCPM algorithms, which will be referred to as AP-AA, AP-FCPM0 and AP-FCPM2, respectively.

4   Experimental Study

The two procedures, Anomalous Pattern (AP) vs. FurthestSum (FS), are compared as initialization methods for AA and FCPM algorithms. The versions of the algorithms are designated as AP- AA/FCPM0/FCPM2 and FS-AA/FCPM0/FCPM2. The algorithms will be analysed respecting the quality of found partitions exploring a collection of seven information-theoretic validity indices for fuzzy clustering from [7] as well as the soft version of the Adjusted Rand Index (ARI_s) as a baseline validity index. The Adjusted Mutual Information index [17] is also considered for analysis.

The study is conducted using both synthetic data generated from the FCPM-DG [13] and real world data from the UCI ML repository [16]. A collection of 99 data sets from the FCPM-DG had been taken. For a fixed pair, D and K_0, a group of 15 data sets (approximately) were generated with different numbers of entities and different original prototypes. The experiments comprised 7 such groups with D ranging from 5 to 180 and K_0 from 3 to 6. According to a previous study by one of the authors [12] the data sets are categorized in three cluster dimensionalities wrt the ratio $\frac{D}{K_0}$: low ($\frac{D}{K_0} \leq 5$); intermediate ($5 < \frac{D}{K_0} < 25$); and high ($\frac{D}{K_0} \geq 25$). From the 99 data sets 26 are of low, 58 of intermediate, and 15 of high cluster dimensionality. We assigned ground truth partitions to each data concordant with the cluster generation.

We consider the following 15 data sets from UCI repository, with reference to number of entities, number of features, and number of classes in the form (N, D, L): Banknote (1372, 4, 2), (Wisconsin) Breast Cancer (683, 9, 2), (Wisconsin) Diagnostic Breast Cancer (569, 30, 2), Ecoli (336, 7, 8), Glass (214, 9, 6), Glass2 (214, 9, 2), Heart (270, 13, 2), Indian Liver Patient (579, 10, 2), Iris (150, 4, 2), Pima Indians Diabetes (768, 8, 2), Seeds Kernel (210, 7, 3), Vehicle silhouettes

(793, 18, 4), VertebralCol (310, 6, 3), Wine Recognition (178, 13, 3), and Zoo (101, 16, 7). Two other data sets from the field of psychiatry are taken from [13]: Mental Disorders (44, 17, 4) and Augmented Mental Disorders (80, 17, 4). The attribute class was removed from all the data sets and for the data sets Indian Liver Patient and Wisconsin Breast Cancer the data cases with missing values were removed (4 and 16, respectively). The ground truth partitions of these data sets have been created according to their labeled subsets. For the Glass data set we had considered two instances: one ground truth with six labels and another with two labels. All data sets had been standardized by subtracting the features mean and divided by features range.

4.1   Performance of Information-Theoretic Validity Indices

The AA, FCPM0 and FCPM2 algorithms had been initialized with FS and run five times each starting from the same initial seeds. The 99 synthetic data were given to the former algorithms, looking for the number of clusters from $K = 2$ to $K = 3L$. For each data set, the fuzzy K-partition is selected according to the optimum of the corresponding validation index. After, it is counted as success if the best $K^* == L$. This way, the overall success rate of an index is the total number of successes across the 99 data sets divided by the total number of partitions.

The left graphic of Fig. 1 shows the results of the eight CVIs for the three algorithms. The NMI_{SM} index fits the best for FS-FCPM2 and FS-FCPM0 with success rates of 100% and 64.7% while the AMI index shows to be more adequate for FS-AA with 100% success rate. The evaluation by ARI_S decreases the values to 98.8% (FS-FCPM2), 41.2% (FS-AA) and 52.9% (FS-FCPM0). The CVIs MI_S and NMI_{Sm} have the lowest success rates.

Looking at the success rate of these CVIs over the collection of real data (right graphic of Fig. 1), we observe that NMI_{SM} better fits FCPM2 and FCPM0 algorithms with success rates of 61.1% and 55.6%, respectively. The AMI index evaluates 66.7% of FS-AA partitions as success. The superior performance of NMI_{SM} and ARI_S indices against the other seven soft CVIs is concordant with the study in [7].

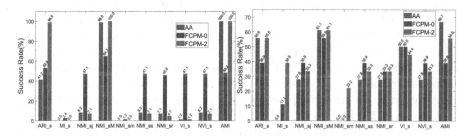

Fig. 1. Overall success rate of the CVIs for FS- AA/FCPM0/FCPM2 on: (left) 99 FCPM-DG data sets; (right) 18 UCI data sets.

We conducted a similar study applying the five internal fuzzy validity indices: PE, PC, NPC, FSI, and Xie-Beni explored in [8]. In the present experiment we enlarged the range of analysed partitions from $K = 2, \cdots, L + 1$ to $K = 2, \cdots, 3L$. In this scenario the Partition Entropy (PE) is the more adequate indice for the three algorithms with success rates of 44.4% for AA and FCPM2 and 38.9% for FCPM0. So, the information-theoretic validity indices better fit these algorithms.

4.2   Modeling Anomalous Pattern Stop Condition

Exploring the AP algorithm to model the number of clusters to be retrieved from data requires a careful analysis of its stop condition. For that, the algorithm was run over the collection of the 99 synthetic data sets till no unclustered entities remained (stop condition C1).

Figure 2(a) shows, for the clusters sequentially extracted by AP, the average clusters contribution (Eq. (8)) for the collection of FCPM-DG data sets generated from $K_0 = 3, 4, 5, 6$ prototypes. The clusters' contributions fall off to zero (approximately) after recovering the correct number of clusters (cases $K_0 = 3, 4, 6$), holding stop condition (C2) or the total number of extracted clusters coincides with the number of generated clusters (case $K_0 = 5$) holding stop condition (C1).

The situation on real data is not that straightforward. For these data, the stop condition of AP had been experimentally modeled by the relative cumulative cluster contribution, defined as the ratio of the sum of the clusters contributions by the total cluster contribution. As can be seen in the graphic on Fig. 2(b) this condition fits well to determine the number of clusters from which data had been generated.

4.3   Comparing Anomalous Pattern with FurthestSum Initialization

The performance of AP initializing the algorithms AA, FCPM2 and FCPM0 is compared against the FS initialization. The analysis considers: (i) improvement of the quality of found partitions by the algorithms measured by the three best CVIs: NMI_{sM}, AMI and ARI_s; and (ii) the convergence improvement measured by the number of iterations taken by the clustering algorithms.

For the FCPM-DG 99 data sets the AP-AA slightly outperforms AP-FCPM2, mainly in high dimensional data. However, this finding is meaningless since the difference of values of the three validity indices never exceeds 0.095. For real world data sets AP outperforms FS initialization for all algorithms. Indeed, the numbers of data sets (out of 18) at which the AP initialization improves the quality of found partition compared with FS initialization measured by NMI_{sM} index are: AA- 17/18, FCPM2- 16/18, FCPM0- 13/18.

The mean number of iterations taken by AP/FS- AA, FCPM0, FCPM2 algorithms running over FCPM-DG data are: **125**/184, **1900**/3520, 238/**176**, while

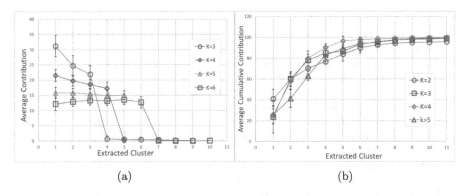

(a)   (b)

Fig. 2. (a) Anomalous Pattern (AP) mean values of clusters' contributions for FCPM-DG data sets; (b) mean values of relative cumulative clusters contribution for the 18 benchmark data sets.

running over real data are: **205**/211, **67**/70, 28/**24**. Therefore, the AP initialization outperforms FS for AA and FCPM0. The FCPM2 slightly improves its convergence when initialized by FS.

5   Conclusion

In most real world problems the granularity to look at groups in data depends on the ultimate goal of clustering and/or the user's "perspective". This work shows how the Anomalous Pattern initialization offers a good modeling strategy to determine the number of clusters with success on initializing AA and FCPM algorithms.

A normalized version of the soft mutual information (NMI_{sM}) and the Adjusted Mutual Information (AMI) indices are shown to be the best overall choices to assess FCPM and AA partitions, respectively. This result is concordant with other studies involving FCM and EM clustering algorithms.

We will advance the study of Anomalous Pattern composed stop condition to automate the process to obtain the number of cluster from data.

Acknowledgments. S.N. acknowledges the support by FCT/MCTES, NOVA LINCS (UID/CEC/04516/2019). The authors are thankful to the anonymous reviewers for their insightful and constructive comments that allowed to improve the paper.

References

1. Bezdek, J.: A Primer on Cluster Analysis: 4 Basic Methods That (Usually) Work, 1st edn. Design Publishing (2017)
2. Chen, Y., Mairal, J., Harchaoui, Z.: Fast and robust archetypal analysis for representation learning. In: 2014 IEEE Conference on Computer Vision and Pattern Recognition, pp. 1478–1485 (2014)

3. Cutler, A., Breiman, L.: Archetypal analysis. Technometrics **36**(4), 338–347 (1994)
4. Eugster, M., Leisch, F.: From Spider-man to hero - archetypal analysis. R. J. Stat. Softw. **30**(8), 1–23 (2009)
5. Eugster, M.: Performance profiles based on archetypal athletes. Int. J. Perform. Anal. Sport. **12**(1), 166–187 (2012)
6. Hannachi, A., Trendafilov, N.: Archetypal analysis: mining weather and climate extremes. J. Clim. **30**(17), 6927–6944 (2017)
7. Lei, Y., Bezdek, J., Chan, J., Vinh, N.X., Romano, S., Bailey, J.: Extending information-theoretic validity indices for fuzzy clustering. IEEE Trans. Fuzzy Syst. **25**, 1013–1018 (2017)
8. Mendes, G.S., Nascimento, S.: A study of fuzzy clustering to archetypal analysis. In: Yin, H., Camacho, D., Novais, P., Tallón-Ballesteros, A.J. (eds.) IDEAL 2018. LNCS, vol. 11315, pp. 250–261. Springer, Cham (2018). https://doi.org/10.1007/978-3-030-03496-2_28
9. Mirkin, B.: Clustering for Data Mining: A Data Recovery Approach. Chapman & Hall /CRC Press, Boca Raton (2005)
10. Mirkin, B.: Core Data Analysis: Summarization, Correlation, and Visualization. Undergraduate Topics in Computer Science, 524 pp. Springer, Cham (2019). https://doi.org/10.1007/978-3-030-00271-8
11. Mørup, M., Hansen, L.: Archetypal analysis for machine learning and data mining. Neurocomputing **80**, 54–63 (2012)
12. Nascimento, S., Mirkin, B., Moura-Pires, F.: Modeling proportional membership in fuzzy clustering. IEEE Trans. Fuzzy Syst. **11**(2), 173–186 (2003)
13. Nascimento, S.: Fuzzy Clustering Via Proportional Membership Model, 178 pp. IOS Press, Amsterdam (2005)
14. Nascimento, S., Franco, P.: Segmentation of upwelling regions in sea surface temperature images via unsupervised fuzzy clustering. In: Corchado, E., Yin, H. (eds.) IDEAL 2009. LNCS, vol. 5788, pp. 543–553. Springer, Heidelberg (2009). https://doi.org/10.1007/978-3-642-04394-9_66
15. Porzio, C., Ragozini, G., Vistocco, D.: On the use of archetypes as benchmarks. Appl. Stoch. Model. Bus. Ind. **54**(5), 419–437 (2008)
16. UCI Machine Learning Repository. http://archive.ics.uci.edu/ml. Accessed 27 July 2018
17. Vinh, N.X., Epps, J., Bailey, J.: Information theoretic measures for clusterings comparison: variants, properties, normalization and correction for chance. J. Mach. Learn. Res. **11**, 2837–2854 (2010)
18. Wynen, D., Schmid, C., Mairal, J.: Unsupervised learning of artistic styles with archetypal style analysis. In: Proceedings of the 32nd Conference on Neural Information Processing Systems (NIPS 2018), pp. 6584–6593 (2018)

Special Session on Machine Learning Towards Smarter Multimodal Systems

Multimodal Web Based Video Annotator with Real-Time Human Pose Estimation

Rui Rodrigues[1,2(✉)], Rui Neves Madeira[1,2], Nuno Correia[2],
Carla Fernandes[3], and Sara Ribeiro[3]

[1] Sustain.RD Center, ESTSetúbal, Polytechnic Institute, Setúbal, Portugal
{rui.rodrigues,rui.madeira}@estsetubal.ips.pt
[2] NOVA LINCS, DI, FCT, NOVA University of Lisbon, Lisbon, Portugal
nmc@fct.unl.pt
[3] ICNOVA, FCSH, NOVA University of Lisbon, Lisbon, Portugal
{carla.fernandes,sribeiro}@fcsh.unl.pt

Abstract. This paper presents a multi-platform Web-based video annotator to support multimodal annotation that can be applied to several working areas, such as dance rehearsals, among others. The CultureMoves' "Motion-Notes" Annotator was designed to assist the creative and exploratory processes of both professional and amateur users, working with a digital device for personal annotations. This prototype is being developed for any device capable of running in a modern Web browser. It is a real-time multimodal video annotator based on keyboard, touch and voice inputs. Five different ways of adding annotations have been already implemented: voice, draw, text, web URL, and mark annotations. Pose estimation functionality uses machine learning techniques to identify a person skeleton in the video frames, which gives the user another resource to identify possible annotations.

Keywords: Multimodal video annotations · Real-time human pose estimation · Machine learning for creativity

1 Introduction

Taking notes is a very old human activity, which has been done on multiple platforms, including clay tablets, pieces of wood or slate and digital notebooks in the last decades. Creative processes in the performative arts, or other similar activities, could involve periods of several practice iterations where video-based annotation tools can give significant advantages in the process over traditional methods.

The CultureMoves is a user-oriented project that aims to develop a series of digital tools that will enable new forms of touristic engagement and access to educational resources by leveraging the re-use of Europeana content [14]. In the context of the CultureMoves project, we have developed "Motion-Notes", which enables to record, replay and add new information to the video footage itself; this is done by adding multiple annotations.

"Motion-Notes" allows users to use multiple devices over different platforms, giving the opportunity to have new user experiences that are uniquely well suited on

© Springer Nature Switzerland AG 2019
H. Yin et al. (Eds.): IDEAL 2019, LNCS 11872, pp. 23–30, 2019.
https://doi.org/10.1007/978-3-030-33617-2_3

how people naturally work as a team, with the opportunity to share their annotations. Motion and pose are essential elements in graphical video footage, especially in the performing arts, and so annotations could be improved by motion tracking methods. Therefore, this work explores text graphics video annotations across multiple platforms with the integration of machine learning algorithms for pose estimation. Human Pose Estimation is an exciting topic that allows several new applications in areas like Human Computer Interaction, Multimodal Systems and Computer Vision among others. Microsoft was revolutionary with Kinect in the late 2000s, which is used widely for several pose estimation studies. One of the disadvantages is the dependency that exists in terms of hardware and software, others are related with the system price, current availability and sometimes with difficulties to configure the right settings.

Our proposal is to perform real-time pose predictions in the client environment, by using one pre-trained neural network. This approach has many advantages, like being accessible in any device with Web browser and camera installed. The system will be evaluating as much frames as it will be possible and match them to a correct set of reference poses with the propose of returning matching scores.

2   Related Work

Video capture is essential to analyse and study human body movements. In order to add and share information, video annotators are one important tool in the performance development [3]. It is known that annotations quality can increase by giving some extra technology to annotator users. Related work is here introduced exploring the idea of integrating annotations and motion tracking, by using machine learning techniques.

2.1   Annotation and Motion Tracking

Motion tracking has been a popular research area for many years and several different types of solutions were studied [4]. One of the most frequent issues with object tracking is the amount of computation processing time that this task requires. Therefore, this could be a major issue if our context has real-time user interaction, despite the proposed solution and used technology.

Microsoft Kinect is a popular technology for pose tracking, which is currently used in several research studies [5]. The Kinect device features an RGB camera, a depth sensor and a microphone. From a single depth image, it can detect a small set of 3D position candidates for each skeletal joint, giving a good accuracy in a real-time environment. Nevertheless, Kinect was not designed for mobile devices, due to its size and absence of a battery.

Recent developments on machine learning using neural networks are allowing accurate real-time pose estimation in RGB images [6]. These motion estimation methods require massive amounts of classified training data, which will be used to train the algorithm (neural network). It is also possible to use models that are already trained and ready to predict new cases as the PoseNet Initiative [7].

Also, the recently launched OpenPose [8] is another approach that provides developers with human skeleton identification from video frames using a Convolutional Neural Network. It was proposed by the Carnegie Mellon University in 2017, being developed in C++ and it uses the OpenCV library. It is possible to get position data of human body easily from OpenPose on several computer operating systems and deep learning frameworks merely with an ordinary camera. Therefore, this paper proposes taking advantage of these pretrained models, by integrating PoseNet in the "Motion-Notes" tool.

2.2 Tools for Creative Processes

Video capture is essential to analyse and study human body movements. In order to add and share information, video annotators are one important tool in the performing arts. Several types of video annotators, with various features, were developed over the last years. Different types of annotations add real information to video content, improving processes and work methodologies, which are in part related with the usage of digital technology.

The MRAS [9] was one of the first solutions allowing collaborative audio and text annotations over the Internet. On the other hand, AntV [11] explored multimedia annotations, in the form of text, images or videos, over the main video window. In the work of Goldman et al. [12], graphical video annotations are combined with an object tracking algorithm, but this work does not consider the case of pen-based annotations and the tracking feature needs a long time of video pre-processing.

Cabral et al. presented the concept of pen-based video annotations using frame differences in order to track motion in video features [1]. This object tracking method can be fast enough for real-time annotations, even on modest machines, but there are several limitations to its actual usefulness. Among other issues, frame differences cannot distinguish two or more moving objects, which is crucial in order to maintain the correct association between an annotation and the annotated video feature. The older MediaDiver project [13] allows users to access information on tracked hockey players and to follow them around a hockey rink, across multiple camera views. By using several instances of a feature's position over time, users can also train the system to follow that feature. However, it does not allow them to associate visual notes with the players on the video window. Its tracking capabilities also need a pre-calibrated, multi-camera setup in order to work. In contrast, Silva et al. presents a work [2] that combines real-time people and object tracking with pen-based video annotations, by applying two object trackers: one based on a depth camera and another using a machine learning technique to video annotation, exploring two methods for real-time video annotation.

Finally, ELAN is an annotation tool that is popular among linguistic and gesture researchers [10], developed under The European Distributed Corpora Project (EUDICO). ELAN has a tier-based data model that supports multi-level, multi-participant annotation of time-based media, which can be archived in a friendly XML-based data format.

2.3   Machine Learning Tools

Machine learning is becoming a significant tool in multiple activities and, in several cases, it is used to improve the results of existing software. TensorFlow is a very popular machine learning library which allows very hard matrix calculation using both CPU and GPU. Although it was originally written in C++, it was also made compatible with Phyton.

Moreover, recently, an initiative was started to rewrite all TensorFlow code in JavaScript. This new library is called TensorFlow.js and can run on client-side, via Web browser with WebGL combability. The JavaScript developer's community is increasing way faster than any other programming language community [15]. Most of that popularity was achieved in last years, given the fact that now it is possible to produce code both from client side and server side.

3   Motion-Notes and Pose Estimation Integration

"Motion-Notes" supports the capture and multimodal annotation of video content in real-time. The annotations associated to video content can be text, ink strokes, URLs, audio or user-configured marks. The prototype was developed for any device capable of running a modern Web browser, exploring multiple input modes like keyboard and touch interaction. The interface is responsive to allow users with different screen sizes to have an adequate human-computer interaction.

The main challenge of this preliminary work is to study the possibility of providing some assistance to the annotator users, by helping them increasing the annotation quality. For this purpose, we are suggesting the inclusion of machine learning techniques in "Motion-Notes", starting by identifying human movement included in the videos. Our proposal is to perform real-time pose predictions in the client environment, by using one pre-trained neural network. The system will be evaluating as much frames as it will be possible and match them to a correct set of reference poses with the propose of returning matching scores.

3.1   Tool Overview

Users can use the system to analyse and improve their videos, by recording and annotating them, for a later review or for sharing notes with other people.

The proposed tool is a single page application (SPA). This tool is being developed in a client-server architecture, which is a mandatory requirement, in order to allow users to share their videos and annotations. Web browser is working as the application client because it is widely available and is compatible with almost every user device with Internet access. This software is implemented using HTML5, CSS3 and Java-Script (ES6) for client side. Regarding the server side, Node.js and the Express.js framework were used for development (see Fig. 1).

The "Motion-Notes" user interface is composed of a video display area, showing the current video recording or a previous recorded video (replay mode), in which it is possible to add new or update current annotations (Fig. 2).

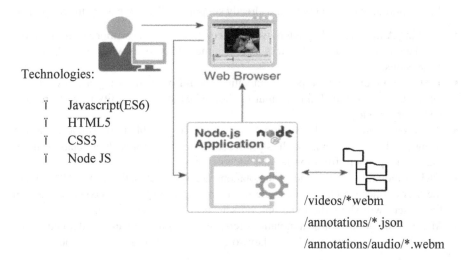

Technologies:

ï   Javascript(ES6)
ï   HTML5
ï   CSS3
ï   Node JS

/videos/*webm

/annotations/*.json

/annotations/audio/*.webm

Fig. 1. "Motion-Notes" client-server architecture.

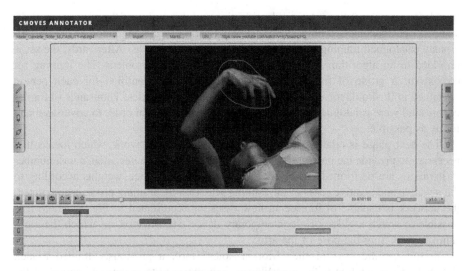

Fig. 2. Motion-Notes User Interface's large version.

In order to maximize the user experience, the representation of the annotations was placed right below the video area and was built making use of the full screen width. This annotation tracks area (Fig. 2) shows all annotations associated to the video and it is composed of several tracks, each one associated to an annotation type. The duration of any given annotation is displayed in its annotation track, allowing browsing the video content and annotations. Annotation properties - such as duration, start time, colour, line width and style, text size - are updatable by interacting with the annotation representation rectangles (move, stretch/shrink, select). Just after the creation of a new annotation, its data is sent to the Web server in JSON format and stored in the filesystem.

Five different annotation types are already supported. These types are the following:

- Drawing Annotation - To sketch over a video, in order to allow more freedom in the annotation process. Basically, it is a set of timed ink strokes, with spatial/temporal dimensions.
- Text Annotation - This type of annotations could be made using a physical or a virtual keyboard. This kind of input method gives the opportunity to write regular text over the video.
- Audio Annotation - With "Motion-Notes" it is also possible to do sound recording using the device microphone. It is produced a sound file for each annotation made, totally independent from the regular video ambient sound.
- URL Annotation - This kind of annotations are Websites defined by a URL. First, the user defines which resource s/he wants to display in the video content and then the system will show the hyperlink on screen in the selected time.
- Marks Annotation - These annotations represent meaning and are pre-defined by the user. Marks are represented by a keyword and an icon of a chosen colour.

3.2 Pose Estimation Integration

In order to improve the annotation identification and the quality of the annotation itself, a supervised machine learning technique was applied to the annotator. It uses a previously catalogued training set with positive and negative cases. Afterwards, the set is provided to an algorithm in order to train it to solve this problem. The training set consists of a group of frames where twelve points are identified for each person contained in it: shoulders, elbows, wrists, hips, knees and ankles. Thousands of frames of this kind were catalogued with several different positions, in order to cover as many forms as possible.

The next phase is related to the training of a neuronal network, which means it is necessary to provide the previously classified training set. As a next step, a vast number of iterations are performed on the network nodes to adjust their weights according to the negative and positive cases of the training set. After the training phase is completed, the network is capable of foreseeing new cases, as represented in Fig. 3, where it is possible to watch a video of a contemporary dance performance, on which the algorithm is predicting the pose of the artist.

In the specific case of the "Motion-Notes" tool, a previously trained model, PoseNet, in conjunction with tensorFlow.js is being used to make predictions on the client's own machine side. Predicted points are drawn on an HTML canvas object located in the same position as the video but at a higher layer position. Finally, straight lines are calculated between the respective points, as we can see in Fig. 3, which gives us a skeletal approximation.

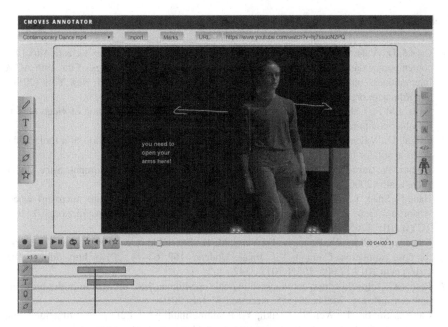

Fig. 3. "Motion-Notes" with machine learning pose estimation.

4 Conclusions and Future Work

"Motion-Notes" is a real-time Web-based multimodal video annotator, which was presented in this paper. The software explores a client-server architecture and is being developed for any device capable of running a modern Web browser. The tool allows to use five different types of multimodal inputs: text, drawings, audio, URL and marks annotation, improving the tool's flexibility. In addition, it is provided with a visual and temporal structure to organize all annotations in tracks, which gives the possibility of navigation between annotations.

We carried out the first steps towards the goal of helping annotator users by integrating a smarter module, based on machine learning techniques, for adding pose estimation to the annotation tool. One of the possible advantages is to offer a better understanding of human movements; therefore, enabling better quality annotations. At this stage, this functionality is still at a prototype stage and needs further developments, but we consider this feature and others also based on machine learning techniques will leverage human creativity activities. We will conduct a thorough evaluation plan in order to optimise the machine learning algorithm and assess the new feature with the participation of end-users framed in the CultureMoves project.

Acknowledgements. This work was supported by the project CultureMoves, Grant Agreement Number: INEA/CEF/ICT/A2017/1568369, Action No: 2017-EU-tA-0171.

References

1. Cabral, D., Valente, J., Silva, J., Aragão, U., Fernandes, C., Correia, N.: A creation-tool for contemporary dance using multimodal video annotation. In: Proceedings of the 19th ACM International Conference on Multimedia, MM 2011, pp. 905–908. ACM, New York (2011). http://doi.acm.org/10.1145/2072298.2071899
2. Silva, J.M.F., Cabral, D., Fernandes, C., Correia, N.: Real-time annotation of video objects on tablet computers. In: MUM 2012, p. 19 (2012)
3. Cabral, D., Valente, J., Aragão, U., Fernandes, C., Correia, N.: Evaluation of a multimodal video annotator for contemporary dance. In: AVI 2012
4. Yilmaz, A., Javed, O., Shah, M.: Object tracking: a survey. ACM Comput. Surv. **38**(4), 13:1–13:45 (2006)
5. Han, J., Shao, L., Xu, D., Shotton, J.: Enhanced computer vision with microsoft kinect sensor: a review. IEEE Trans. Cybernet. **43**(5), 1318–1334 (2013). https://doi.org/10.1109/TCYB.2013.2265378
6. Kawana, Y., Ukita, N., Huang, J.-B., Yang, M.-H.: Ensemble convolutional neural networks for pose estimation. Comput. Vis. Image Underst. **169**, 62–74 (2018). https://doi.org/10.1016/j.cviu.2017.12.005. ISSN 1077-3142
7. PoseNet. https://medium.com/tensorflow/real-time-human-pose-estimation-in-the-browser-with-tensorflow-js-7dd0bc881cd5. Accessed 31 July 2019
8. Cao, Z., Simon, T., Wei, S., Sheikh, Y.: Realtime multi-person 2D pose estimation using part affinity fields. In: 2017 IEEE Conference on Computer Vision and Pattern Recognition (CVPR), Honolulu, HI (2017). https://doi.org/10.1109/cvpr.2017.143
9. Bargeron, D., Gupta, A., Grudin, J., Sanocki, E.: Annotations for streaming video on the Web: system design and usage studies. Comput. Netw. **31**(11–16), 1139–1153 (1999). ISSN 1389-1286
10. Lausberg, H., Sloetjes, H.: Behav. Res. Methods **41**, 841 (2009). https://doi.org/10.3758/BRM.41.3.841
11. Correia, N., Chambel, T.: Active video watching using annotation. In: Proceedings of the Seventh ACM International Conference on Multimedia (Part 2) (MULTIMEDIA 1999), pp. 151–154. ACM, New York (1999)
12. Goldman, D.B., Gonterman, C., Curless, B., Salesin, D., Seitz, S.M.: Video object annotation, navigation, and composition. In: Proceedings of the 21st Annual ACM Symposium on User Interface Software and Technology, UIST 2008, New York, USA (2008)
13. Marshall, C.C.: Toward an ecology of hypertext annotation. In: Proceedings of the Ninth ACM Conference on Hypertext and Hypermedia, HYPERTEXT 1998. ACM, New York (1998)
14. Europeana. https://www.europeana.eu/portal/pt. Accessed 31 July 2019
15. Stackoverflow. https://insights.stackoverflow.com/survey/2019#most-popular-technologies. Accessed 31 July 2019

New Interfaces for Classifying Performance Gestures in Music

Chris Rhodes[1](✉) [iD], Richard Allmendinger[2] [iD],
and Ricardo Climent[1] [iD]

[1] NOVARS Research Centre, University of Manchester, Manchester, UK
{Chris.rhodes,Ricardo.climent}@manchester.ac.uk
[2] Alliance Manchester Business School (AMBS),
University of Manchester, Manchester, UK
Richard.allmendinger@manchester.ac.uk

Abstract. Interactive machine learning (ML) allows a music performer to digitally represent musical actions (via gestural interfaces) and affect their musical output in real-time. Processing musical actions (termed performance gestures) with ML is useful because it predicts and maps often-complex biometric data. ML models can therefore be used to create novel interactions with musical systems, game-engines, and networked analogue devices. Wekinator is a free open-source software for ML (based on the Waikato Environment for Knowledge Analysis – WEKA - framework) which has been widely used, since 2009, to build supervised predictive models when developing real-time interactive systems. This is because it is accessible in its format (i.e. a graphical user interface – GUI) and simplified approach to ML. Significantly, it allows model training via gestural interfaces through demonstration. However, Wekinator offers the user several models to build predictive systems with. This paper explores which ML models (in Wekinator) are the most useful for predicting an output in the context of interactive music composition. We use two performance gestures for piano, with opposing datasets, to train available ML models, investigate compositional outcomes and frame the investigation. Our results show ML model choice is important for mapping performance gestures because of disparate mapping accuracies and behaviours found between all Wekinator ML models.

Keywords: Interactive machine learning · Wekinator · Myo · HCI · Performance gestures · Interactive music · Gestural interfaces

1 Motivation

Interactive music is an artform that uses gestural interfaces to build new instruments [1] – or augment existing ones [2] – to generate a musical output. A gestural interface is a device which captures physical gestures to control digital systems. Typically, they detect biometric and inertial measurement unit (IMU) data. Through their use, novel forms of human-computer interaction (HCI) can be explored in music composition and performance.

© Springer Nature Switzerland AG 2019
H. Yin et al. (Eds.): IDEAL 2019, LNCS 11872, pp. 31–42, 2019.
https://doi.org/10.1007/978-3-030-33617-2_4

In the concise history of interactive music, technological developments made with gestural interfaces have always evoked a creative response. In 1917, the Theremin permitted people to create music with novel actions [3]. In the 1970s, the analogue synthesiser allowed users to affect parameters of sound through interacting with modules [4]. In 1987, Jon Rose [5] developed a violin bow (K-bow), which used the Musical Instrument Digital Interface (MIDI) protocol to digitise violin performance. In 1993, Atau Tanaka et al. [6] created Sensorband; an ensemble investigating how wearable sensors could affect the digital signal processing (DSP) of real-time sound generation, showing music composition could be embodied. In recent years, electroencephalography has been used as a brain-computer interface to further investigate the embodiment of music and drive music composition [7]. However, as the interface uses several electrodes attached to the scalp, it is not ideal for interactive music composition. This is because of the limited mobility of the interface. However, in 2015, the Myo armband [8] offered a mobile interface for improved gestural control when investigating interactive music. This allowed researchers to access two forms of raw biometric data: IMUs and Electromyographic (EMG). Although access to IMU data offered by the Myo is not novel, EMG data is useful because it measures electrical activity of skeletal muscles, allowing the user to map muscular data to digital systems.

Previous literature using the Myo interface has seldom explored the application of ML to process data, but rather conditional statements [9]. Although conditional statements are a feasible way of using interface data for music composition, they are not efficient because performance gestures can be misclassified if an ML approach to gesture classification is not adopted [9]. An ML approach would therefore be more accurate when interacting with music systems and evoking a sonic output. It is possible to realise this via an interactive ML software called Wekinator [10].

Wekinator is an ML environment built on WEKA. It is useful because it allows for real-time data input from gestural interfaces and has a simplified GUI for configuring ML models, providing a better accessible tool for HCI applications. Previous research has used Wekinator for the classification of instrument articulations [11], mapping sound from colour within virtual reality [12] and duetting with an LED gestural interface [13]. However, there is a lack of study regarding the classification of performance gestures in Wekinator with an informed-model approach; only a demonstrative (trial-and-error) model training method [11]. There is also a lack of literature on model evaluation when using EMG data to train ML models in Wekinator.

In this work we provide a solution to the problem that there are different ways to process and create ML models from performance gestures using the Myo interface within Wekinator. We also address that EMG data can be used to create bespoke physical gestures to interact with music systems, instead of research limitations surrounding the use of pre-defined gestures for music research via the Myo Software Development Kit (SDK) [9, 14]. It seeks to investigate which models are the most useful for predicting a sonic output for use in interactive music performance. If solved, this will improve knowledge Regarding the use of the Myo interface (plus similar interfaces) with Wekinator when choosing an ML model.

Our results establish that different ML models are more effective when chosen based on performance gesture characteristics and data type for model training. Our results also show model choice has an impact on controlling a musical output. The next section of this paper will detail technical aspects of the gestural interface we are using (the Myo armband), followed by ML models offered by Wekinator, the study methodology, results and conclusions derived.

2 Gestural Interfaces and Machine Learning

Gestural Interfaces allow us to access raw data, representing performance gestures, and build ML models. The gestural interface we use for this investigation is the Myo armband.

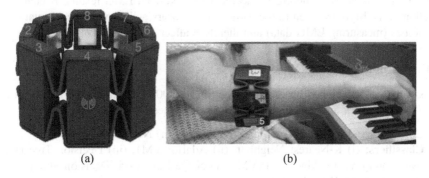

(a)   (b)

Fig. 1. (a) The Myo armband showing 8 channels of EMG electrodes. (b) A photo of the standardized placement of the Myo armband when collecting raw IMU and EMG data.

2.1 Gestural Interface: The Myo Armband

Technical Specification. The Myo armband is a gestural interface which allows us to access raw IMUs and 8-channels of 8-bit EMG data. It communicates raw data via Bluetooth and streams IMU data at 50 Hz and EMG data at 200 Hz [9]. It communicates all data to a Bluetooth dongle with a distance of <15 m [15]. Data is taken from the Myo library within the SDK created by Thalmic Labs [14]. The EMG data provided by the Myo is unique to the interface. This is because a gestural interface offering access to raw EMG data is hard to acquire on the consumer market.

2.2 Raw Transmitted Data

The Myo allows access to two forms of biometric information: IMUs and EMG data. IMU data from the Myo SDK has three main parameters for measuring orientation: acceleration, gyroscope and quaternions. Acceleration and gyroscope data parameters use 3 axes (X, Y, Z) to measure orientation. However, quaternions use 4 axes to do so (X, Y, Z, W). The Myo has 8 individual electrodes for measuring EMG activity around the arm, as seen in Fig. 1a.

2.3   Myo Usage Issues

Data retrieved from the Myo is subject to issues affecting data validity. These are: The placement of the Myo armband and user calibration.

Calibration. Because muscle activity in a Myo user's arm is as unique as a fingerprint [16], the Myo must be calibrated by the user so that data is reliable. If the Myo is not calibrated before use, the data collected will not be accurate as measurement is skewed. An example of inaccuracy is not providing a point of origin or data parameter limits for IMU data.

Placement. Careful placement of the Myo must be observed and standardised. See Fig. 1b. This is because incorrect placement will skew all EMG results. A change in rotation of the Myo, between future participants or users, will change the orientation of electrodes (measuring EMG data) and therefore affect data or system validity.

2.4   Machine Learning: Wekinator

Wekinator was developed by Rebecca Fiebrink in 2009 [10]. It is a GUI built on the WEKA framework and contains three ML model types for input data:

1. **Continuous:** (i) Linear/Polynomial Regression, (ii) Neural Network.
2. **Classifiers:** (i) K-Nearest Neighbor, (ii) AdaBoost.M1, (iii) Decision Tree (J48), (iv) Support Vector Machine (SVM), (v) Naive Bayes, (vi) Decision Stump.
3. **Dynamic Time Warping (DTW)**

3   Methodology

The methodology behind evaluating the efficacy of Wekinator for predicting sonic output via actioning performance gestures is as follows.

3.1   Planning Performance Gestures for Model Testing

Performance gestures chosen for predicting a music output were selected based on their ability to augment piano practice. Using an instrument would provide an initial focus for the application of gesture prediction and interactive music composition. Both gestures can be considered an extended technique; meaning a non-orthodox method of playing the instrument. For ease of understanding the results, we use a single Myo worn on the right arm. To train different ML models, we use two different performance gestures with opposing datasets. This is because continuous models, classifiers and DTW algorithms process gestural data in a very different way. For example, a performance gesture that travels between two spatial positions on the y-axis – states y_1 and y_2 – can be considered either a continuous or static gesture. This is because we may wish to measure the metric of the two states in different ways. From a data perspective, this would be represented as either a series of floating-point values (continuous position) or a single integer value (classifier position).

However, both states combined (to a single unifying value) can be considered a necessary input for a DTW model. This is because a DTW algorithm is a continuous model that predicts a performance gesture - per iteration - and not as a stream of floating-point values. Given the problem of deciding the metric, it is therefore clear that the performance context (for the application of model output) would be the principal factor when deciding a suitable model for training data. This culminated in the creation of two performance gestures for investigation:

1. **Right arm positioned above head (gesture 1).** This gesture involves extending the right arm out straight above the head. This gesture is measured using single parameter acceleration data (IMU) on the y-axis; it is also linear, meaning data variation will be low. Low variation in the data will make model prediction more accurate. As a result, model output (using this data) will closely align to model input during mapping. Therefore, this gesture is thus most apt for a continuous/DTW model. This is because continuous data analysis is integral to how each model makes an accurate prediction.
2. **Spread fingers and resting arm (gesture 2).** This gesture involves: (a) Fully extending fingers on the right-hand outwards (b) returning the arm to a limp resting position. Gesture position (b) was created as a tool to test the efficacy of the model predicting position (a) when switching between both states. As this gesture operates between two static (non-continuous) states it became clear a classifier model was most apt for this gesture. This gesture uses two EMG data parameters targeting electrodes 3 and 4 on the Myo interface (see Fig. 1b).

3.2   Data Acquisition and Structuring

Raw data was structured from the Myo, worn on the right arm, via software built by the first author in the Max 8 development environment[1]. The software built routed the Myo C++ API information, from the Myo SDK, to Max 8. Two sets of software were built in Max 8: A program to acquire raw data from the Myo and a program to send filtered Myo data to Wekinator and then apply DSP. After receiving the raw data within Max, data was pre-processed, structured and then exported as a csv file. The first author performed all gestures for acquisition and testing.

Acquisition. Raw data was recorded from the Myo SDK at 10 ms per data entry. This was deemed accurate enough to see meaningful patterns in the data. Each gesture was recorded with 4 iterations over 16 s. A digital metronome was synchronised with the data acquisition software to keep timing when performing gestures.

Pre-processing and Structuring. Data was pre-processed during acquisition by unpacking and structuring (labelling) each data type (IMU and EMG) and respectful parameters to an array (see Sect. 2.2 for a summary). After acquisition, the data was exported from the array to a csv file.

---

[1] Max 8 is a GUI programming environment primarily used by artists and musicians. More information regarding Max 8 can be found here: https://cycling74.com/.

3.3 Data Post-processing, Visualisation and Filtering

Data Post-Processing. EMG data needed to be processed via taking an average of each EMG parameter and then applying an absolute function. Using this function would make EMG data much clearer to use and visualise [17]. As IMU relies on vector movement for each axis (to show orientation), IMU data was not post-processed.

Data Visualisation and Analysis. The data visualisation process involved plotting the data within suitable graphs to spot any significant trends when performing gestures. Applying a preliminary analysis would show key data and begin to manually filter the data.

Data Filtering. Data was manually filtered after collection to train ML models in Wekinator. This is because each physical gesture is unique. Therefore, different data parameters will best represent each gesture performance. For example, training a model to predict a music output via data detailing the placement of the arm above the head will not require EMG data but data regarding the orientation of the arm.

3.4 Training ML Models in Wekinator

Table 1. Model types and parameters used when training all available models in Wekinator.

Model type	Model	Available model parameters in Wekinator and their settings	Data type/range
Continuous (soft limits)[a]	Neural Network	1 Hidden layer. 1 node per hidden layer	
	Linear Regression	Linear inputs. No feature selection used. Colinear inputs removed	[0,1] (float)
	Polynomial Regression	Polynomial exponent = 2. No feature selection used. Colinear inputs not removed	
Classifier	K-Nearest Neighbor	Number of neighbors (k) = 1	{1,2} (integer)
	AdaBoost. M1	Training rounds = 100. Base classifier = Decision tree	
	Decision Tree (J48)	Model not customisable	
	SVM	Kernel: Linear. Complexity constant: 1	
	Naïve Bayes	Model not customisable	
	Decision Stump	Model not customisable	
DTW	DTW	Threshold for prediction set in GUI to be sensitive enough to predict output once per gesture iteration	Single-fire

[a]Meaning the maximum value set for the model range [0,1] can be exceeded.

After filtering, training data was sent from a real-time Myo signal in Max 8 to each model input in Wekinator via Open Sound Control (OSC) and User Datagram Protocol (UDP). UDP was then used to send the OSC data packet to a specific port that Wekinator was listening to. All models were measured against performance gesture 1 and 2 (detailed in Sect. 3.1). Each gesture used c.8000 data entry examples to train ML models at a frequency of 10 ms per example. See Table 1 for a detailed list of parameters used when training each ML model with both gestures. A model evaluation was used (provided by the Wekinator GUI) for each model type over both gestures. The evaluation method used was cross-validation, with 10 folds, and model training accuracy was also measured.

4 Results

4.1 Model Evaluations

Continuous Models. Results from training the NN, linear regression and polynomial regression models in Wekinator elucidated interesting results across both gestures.

The NN showed to be the most accurate model when mapping model prediction to input data for both gestures (see Table 2 for full list of model accuracies). An example of mapping accuracy with the NN can be seen in Fig. 2. Results for continuous models also showed that NN and polynomial regression models apply curvature to the mapping of gesture input data, whereas the linear regression model applies a linear mapping. When looking at Fig. 3, it is clear that the linear regression model maps best the movement of gesture 1. Observing this difference between the NN/polynomial model and regression model is significant for interactive music practice. This is because the rate at which data increases will create a disparate musical output when affecting the DSP of an audio signal.

A further observation also shows that data type used to train continuous models is more accurate than other data types. Table 2 clearly shows that IMU data (gesture 1) is more effective than using EMG data (gesture 2) to train continuous models. This is further elucidated when comparing model mapping in Figs. 2 and 3. All continuous models averaged 8029 examples of input training data for both performance gestures.

Table 2. Accuracy of continuous models after training, measured in root mean square (where 0 is optimal).

Gesture no.	Model type	Training accuracy	Cross-validation (10 folds)
1	NN	**0.00**	**0.00**
	Linear Regression	0.04	0.04
	Polynomial Regression	0.04	0.04
2	NN	**0.12**	**0.12**
	Linear Regression	0.32	0.32
	Polynomial Regression	0.32	0.32

Classifier Models. Results for the classifier models showed a varying level of accuracy when predicting gesture because of the data type used to train models. When looking at Table 3, we can see that gesture 1 is more accurate in model prediction than gesture 2. However, we can also observe that models trained with EMG data (gesture 2) show disparate levels of accuracy. In particular, the SVM and decision stump models perform the poorest, in comparison to all available classifier models. This is clear when looking at their erratic misclassification of gesture 2 in Fig. 4. Results taken from performing gesture 1 reported a striking accuracy with all classifier models. They also showed that individual models display an earlier prediction than others. The decision tree and adaboost. M1 models reported (in unison) a prediction of 70 ms (first instance) and 60 ms (second instance) ahead of all other classifier models. All classifier models averaged 8022 examples of input training data across both performance gestures.

Table 3. Accuracy of classifier models after training, measured in percentage (where 100% is optimal).

Gesture no.	Model type	Training accuracy	Cross-validation (10 folds)
1	K-Nearest Neighbor	100%	100%
	AdaBoost.M1	100%	100%
	Decision Tree (J48)	100%	100%
	Naive Bayes	100%	100%
	SVM	100%	100%
	Decision Stump	100%	100%
2	K-Nearest Neighbor	99.36%	99.36%
	AdaBoost.M1	99.36%	99.36%
	Decision Tree (J48)	99.36%	99.36%
	Naive Bayes	98.86%	98.86%
	SVM	97.75%	97.57%
	Decision Stump	94.78%	94.78%

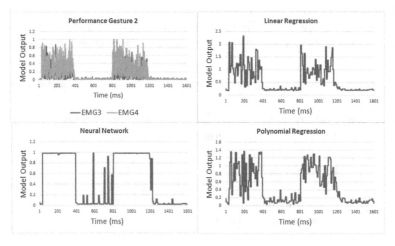

Fig. 2. A graph showing all continuous model outputs and their output activity when performing gesture 2 over a period of 16 s.

Dynamic Time Warping (DTW). Observations from using the DTW model indicate different measures of accuracy through different datasets. Observation of Fig. 5 shows inaccuracy in predicting the onset of a performance gesture. It also shows inaccuracy when making several predictions of the same gesture iteration. This is because EMG data is a more complex dataset than IMU data. However, data taken suggests the DTW model is very effective in predicting gesture onsets with IMU data.

Weaknesses of using the DTW model are concerned with the GUI. When recording examples for training each model, the user is unable to know how long each training sample is. This is because the metric is not shown (unlike classifier and continuous models). The 'threshold' available in this mode, via the GUI, is also arbitrary. It is designed for the user to adjust model prediction sensitivity. However, the metric is inaccessible. The DTW model trained to predict gesture 1 used 67 examples of input data, whereas the model trained to predict gesture 2 used 336 examples.

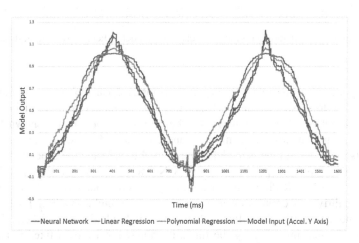

Fig. 3. A graph showing a comparison between all continuous models and their outputs vs model input over a period of 16 s when performing gesture 1.

4.2   Model Output: Interactive Music Composition

Model results applied to interactive music composition can show: (a) Continuous models alter the data scaling (mapping) of gestures during model output (compare all model outputs in Fig. 2), (b) classifier models run specific DSP algorithms by creating layers within gestures (where each layer is defined via an arbitrary number of integers), and (c) DTW models run a single DSP algorithm when predicting a gesture performed over time.

Continuous models are useful for the real-time control of music parameters. Albeit, each model is shown in this study to have differences in model output mapping.[2]

---

[2] *Mapping* is an integral part of interactive music composition and computer music. It means to use a transformed parameter to control or affect another parameter [18]. Here, a model output to affect a DSP process (e.g. delay time, reverb, etc) of an audio signal.

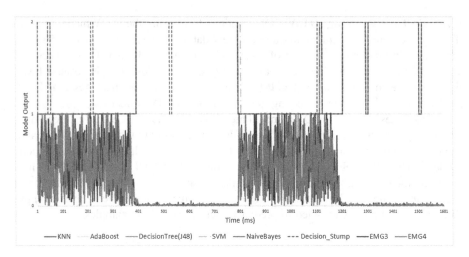

Fig. 4. A graph showing model output of all classifier models when performing gesture 2 over a period of 16 s.

This is important because interactive music is composed via applying multi-parametric control of DSP to music material [19]. The mapping (contour) from each model output will therefore affect sonic output. This is because data contour has shown to affect parametric control in two ways, smoothly and erratically. For example, creating a 1:1 mapping between a continuous model output (NN, LR, PR) and the generation of a sine wave will affect sonic output, disparately. A NN would smoothen the cycle between frequencies and the LR/PR would make the cycle behaviour erratic. However, other DSP events thrive off erratic movement in data (i.e. granular synthesis) and others a smoother movement (i.e. delay lines), due to their individual sonic aesthetic. This can be evidenced when investigating the timbral difference between a fixed music event (i.e. piano chord) and individual continuous model output.

Classifier models can be used to distinguish between stages of gesture performance. This is useful because a gesture can be used to trigger several pre-defined algorithms, at different stages of performing a gesture. This is useful for interactive music composition because it triggers a pre-defined music output, allowing the composer greater control over the music output. However, the output can be processed by using other datasets in real-time, once the algorithm (for each integer class) has been triggered. This can therefore augment music parameters of the running algorithm.

Alike classifier models, DTW models make use of a pre-defined DSP algorithm, albeit without being able to control parameters within the algorithm after triggering the event. As a single-fire event, with no parametric control over the DSP process, this model is only ideal for music composition within a system containing a rigid DSP architecture. Interactive music composition therefore benefits marginally from DTW.

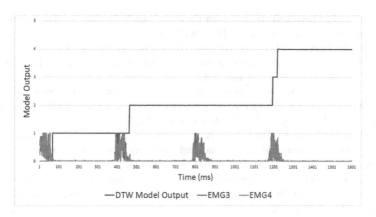

Fig. 5. Graph showing the accuracy of the DTW model when performing gesture 2 (4 times) in regular intervals over a period of 16 s.

5 Conclusions

It is apparent that model performance is unique to the data used to train each model. This is because data variation dictates prediction accuracy for ML models. This is evident when observing IMU data (low variation) and all model performances in this study. However, it is also clear that there are noticeable differences within ML model type; Variances in data mapping (continuous models) and model prediction strengths (classifier models). Music practitioners investigating interactive ML should therefore observe data variation when training ML models, model prediction strengths (see Tables 2 and 3) and pair model choice to desired music outcome (see Sect. 4.2).

Acknowledgements. This work was supported by the Engineering and Physical Sciences Research Council [2063473].

References

1. Tanaka, A.: Sensor-based musical instruments and interactive music. In: The Oxford Handbook of Computer Music, pp. 233–257. Oxford University Press, USA (2011)
2. Miranda, E., Wanderley, M.: New Digital Musical Instruments: Control and Interaction Beyond the Keyboard, 1st edn. A-R Editions, USA (2006)
3. Hayward, P.: Danger! Retro-Affectivity! The Cultural Career of the Theremin. Convergence **3**(4), 28–53 (1997)
4. Finamore, E.: A Tribute to the Synth: How Synthesisers Revolutionised Modern Music. https://www.bbc.co.uk/programmes/articles/3ryZCdlXtpkNG3yRl3Y7pnh/a-tribute-to-the-synth-how-synthesisers-revolutionised-modern-music. Accessed 28 July 2019
5. Rose, J.: K-bow: The Palimpolin Project. http://www.jonroseweb.com/e_vworld_k-bow.html. Accessed 28 July 2019
6. Bongers, B.: An interview with sensorband. Comput. Music J. **22**(1), 13–24 (1998)

7. Wu, D., et al.: Music composition from the brain signal: representing the mental state by music. Comput. Intell. Neurosci. **2010**, 1–6 (2010)
8. Jackson, B.: Thalmic Labs Myo Armband Hits Consumer Release, for Sale on Amazon. https://www.itbusiness.ca/news/thalmic-labs-myo-armband-hits-consumer-release-for-sale-on-amazon/54056. Accessed 26 July 2019
9. Nymoen, K., et al.: MuMyo – evaluating and exploring the MYO armband for musical exploration. In: Proceedings of the International Conference on New Interfaces for Musical Expression, vol. 2015, pp. 215–218. NIME, USA (2015)
10. Fiebrink, R., Trueman, D., Cook, P.R.: A meta-instrument for interactive, on-the-fly machine learning. In: Proceedings of the International Conference on New Interfaces for Musical Expression, vol. 2009, pp. 280–285. NIME, USA (2009)
11. Fiebrink, R., Schedel, M.: A demonstration of bow articulation recognition with Wekinator and K-Bow. In: Proceedings of the International Computer Music Conference (ICMC), vol. 2011, pp. 272–275. ICMC, UK (2011)
12. Santini, G.: Synesthesizer: physical modelling and machine learning for a color-based synthesizer in virtual reality. In: Montiel, M., Gomez-Martin, F., Agustín-Aquino, Octavio A. (eds.) MCM 2019. LNCS (LNAI), vol. 11502, pp. 229–235. Springer, Cham (2019). https://doi.org/10.1007/978-3-030-21392-3_18
13. Hantrakul, L., Kondak, Z.: GestureRNN: a neural gesture system for the roli lightpad block. In: Proceedings of the International Conference on New Interfaces for Musical Expression, vol. 2018, pp. 132–137. NIME, USA (2018)
14. North. Myo Connect, SDK and Firmware Downloads. https://support.getmyo.com/hc/en-us/articles/360018409792-Myo-Connect-SDK-and-firmware-downloads. Accessed 23 July 2019
15. North. What is the Wireless Range of the Myo?. https://support.getmyo.com/hc/en-us/articles/202668603-What-is-the-wireless-range-of-Myo. Accessed 23 July 2019
16. North, Creating a Custom Calibration Profile for Your Myo Armband. https://support.getmyo.com/hc/en-us/articles/203829315-Creating-a-custom-calibration-profile-for-your-Myo-armband. Accessed 26 July 2019
17. Soares, S.B., et al.: The use of cross correlation function in onset detection of electromyographic signals. In: ISSNIP Biosignals and Biorobotics Conference, vol 2013, pp. 1–5. IEEE, Brazil (2013)
18. Hunt, A., Wanderley, M.M.: Mapping performer parameters to synthesis engines. Organised Sound 7(2), 97–108 (2002)
19. Winkler, T.: Composing Interactive Music: Techniques and Ideas Using Max, 1st edn. MIT Press, Cambridge (1998)

Special Session on Data Selection in Machine Learning

Classifying Ransomware Using Machine Learning Algorithms

Samuel Egunjobi[✉], Simon Parkinson[✉] [iD], and Andrew Crampton[✉] [iD]

Department of Computer Science, School of Computing and Engineering, University
of Huddersfield, Queensgate, Huddersfield HD1 3DH, UK
{samuel.egunjobi,simon.parkinson,andrew.crampton}@hud.ac.uk

Abstract. Ransomware is a continuing threat and has resulted in the
battle between the development and detection of new techniques. Detection and mitigation systems have been developed and are in wide-scale
use; however, their reactive nature has resulted in a continuing evolution
and updating process. This is largely because detection mechanisms can
often be circumvented by introducing changes in the malicious code and
its behaviour. In this paper, we demonstrate a classification technique
of integrating both static and dynamic features to increase the accuracy of detection and classification of ransomware. We train supervised
machine learning algorithms using a test set and use a confusion matrix
to observe accuracy, enabling a systematic comparison of each algorithm.
In this work, supervised algorithms such as the Naïve Bayes algorithm
resulted in an accuracy of 96% with the test set result, SVM 99.5%, random forest 99.5%, and 96%. We also use Youden's index to determine
sensitivity and specificity.

Keywords: Ransomware · Malware · Machine Learning

1 Introduction

This wide-scale use of computers has resulted in computing systems being used
abusively for illegal activities [17]. Ransomware is a type of malicious software
(malware), which when deployed on the computer encrypts or locks a computer
or files, requesting that a ransom to be paid to the author of the ransomware for
the successful decryption and release of the users data and system. It is intended
to compromise the availability, confidentiality and integrity of the victim's data,
demanding payment from the user in order to have their files unencrypted and
accessible [18].

There are two main types of ransomware this paper will be focusing on:
the first one, Locker Ransomware, is designed to deny access to the victim's
computer, to prevent them from using the system as a whole or services on
the computer. The second is the Crypto Locker Ransomware which encrypts
personal files to make them inaccessible to its victims [19]. In the case of Lock-
ers Ransomware, the victim loses the ability to use the whole computer or a

© Springer Nature Switzerland AG 2019
H. Yin et al. (Eds.): IDEAL 2019, LNCS 11872, pp. 45–52, 2019.
https://doi.org/10.1007/978-3-030-33617-2_5

particular piece software on the computer. Crypto Locker Ransomware selects document, pictures or target files that have a favourable type. This causes a lot inconvenience and panic to victims as threats (e.g., blackmail, extortion) are made to facilitate quick response and payment by victims [18].

Typical techniques for detecting and classifying malwares often involve the use of static or dynamic features for analysis and detection [9]. One challenge with the methods of classification is the obfuscation of features and is explained in a recent paper [19]. Furthermore, this process is reactive, meaning that once a paper discussing malware detection is published, malware developers upgrade their form of attack so as to reduce or nullify the detection rates when run against anti-malware software and successfully carry out the intended action [9]. The proposed solution to this problem is increasing the alertness of anti-malware software to features of the malware software that increased reliable and detection accuracy. This includes utilising both static and dynamic features. Static features are those observed without executing the software, e.g file size, whereas dynamic features are those related to have the malware interacts with the system, e.g requests to operating system functionality or file interaction [6].

The second problem is the detection of false positives [9]. The introduction of malware that functions in a different way to those used during training of the machine learning [11] often results in mis-classification. These reduce the efficiency of the machine learning algorithm and its ability to accurately distinguish ransomware. In recent research, the authors propose a solution to increase the number of samples for both normal (goodwares) and malicious (malwares) software to cover a wide range of possible feature being taught to the machine [12].

In this paper, we explore the possibility of increasing the accuracy of detection and classification by integrating both static and dynamic features of ransomware, train machine learning algorithms and introduce test set with the aim to achieve a higher percentage of classification. A confusion matrix is used to observe the best algorithm with the least margin of error. The paper is structured as follows: First we discuss related work, and in the next section we explain the proposed technique. Following on, we present the research methodology for acquiring empirical understanding. We then give details about the samples used and how the classification was performed, followed on by the presentation of the results. Finally, a conclusion and recommendations as to future areas of research are provided.

2   Related Works

In this section, we review current research literature related to malware and ransomware analysis, with a particular focus on their research and analysis on the identification and classification of ransomwares.

2.1   Malware Analysis

An extensive review of the static feature analysis is provided in [11]. Static feature analysis is done using static features such as 'Application Programming Interface (API) calling, binary code, control graph, etc. to find suspicious behavioural patterns of the malware. There are three different types of features used for the extraction process as mentioned in [9]: Byte sequence n-grams, string features, and portable executable. The byte sequence n-grams technique extracts n bytes sequences from an executable file; The string features refer to encoded text from the program file. The portable executable approach extracts from the dynamic link library located within the Win32 portable executable binaries [4]. Furthermore, some antivirus and anti-malware tools are detecting malware based on their signature. However, their accuracy is reduced due to obfuscation techniques which are easily implemented to prevent matching [16].

Despite the use of the static feature extraction to distinguish between malicious and benign software, there are some downsides to this technique. It is largely restricted due to the obfuscation method being carried out by adversaries [11,12]. Furthermore, the entire process is reactive as it requires extensive manual human effort, time and expertise to formally describe unwanted behaviour. This can slow down the whole process of analysis especially when dealing with high volumes of malware variants to analyse [15].

An alternative approach to static analysis is dynamic feature analysis, which is performed while the malware is being executed. This focuses on system call sequences to provide information about the program at run-time, which is hard to obfuscate and uses various techniques like dynamic binary instrumentation, virtual machine inspection, information flow tracking and function call monitoring [16]. It involves the use of behavioural pattern during the period of the program execution to distinguish the malware and it is most often utilized to provide fast analysis on the malware's behavioural pattern [19].

Although it has been concluded that the dynamic feature extraction technique is harder to obfuscate, research has demonstrated that it does have some drawbacks [10]. Firstly, when analysing a program using the dynamic analysis, the behavioural characteristics vary on the condition which it was executed in therefore making a single execution of reveal a section of the program's behaviour [15]. Furthermore, while using the dynamic feature extraction technique, the program is identified as being malicious software based on sensitive APIs; however, many benign applications may also invoke this APIs which could lead to the issue of false positives in the result of the analysis [11].

Shijo et al. [19] proposed using both static and the dynamic feature analysis to produce an improved result accuracy compared to using each method individually. The paper states that by reason of the continuous increase in malware samples, security vendors depend on automated malware tools. It points out that the main advantage of static analysis is that the binary code contains useful information about the malware behaviour and the main advantage of the dynamic analysis is that it analyses the runtime behaviour of a malware which is hard to obfuscate and gives a clearer result [11]. Unlike other research work which used static feature or dynamic feature for their analysis [3,8,19], we explore the possibility of increasing the accuracy of detection and classification by integrating both static and dynamic features of ransomware, analyse these features using machine learning algorithms aiming to achieve a higher percentage of classification while also taking note of the elements in the confusion matrix (True positive, True Negative, False positive and False negative) to observe the best algorithm with the least margin of error.

2.2   Machine Learning Techniques

Machine learning is a widely used mechanism for malware detection which is heavily reliant on the selection of features to makeup data for analysis [18]. In this research, we utilise the machine learning procedure to enable the machine to make further predictions on distinguishing malicious software from benign software [7]. In this work, the algorithms that will be used to carry out the machine learning procedure include: instance-based (IB1) [13], Random Forest (RF) [16], Naïve Bayes and Support vector machine (SVM) [1,2]. Based on our research [8,21], these algorithms help produce accurate detection.

In machine learning, there are metrics used to evaluate the effectiveness of each algorithm [14]. In this work, we use a confusion matrix [20] and Youden's index the weighted mean of these metrics are also taken into consideration since it finds the mean by assigning the weight of each element such that each element contributes to the final result based on how much importance it carries [5]. True positive (tp), false positive (fp), true negative (tb), and false negative values (fn) are used to calculate the following performance measures:

1. True Positive Rate/recall/sensitivity (*tpr*): the fraction of malware samples correctly identified as ransomware;
2. False Positive Rate (*fpr = 1 - tnr*): the fraction of goodware samples incorrectly identified as being malware;
3. True Negative Rate/specificity (*tnr*): the fraction of goodware samples correctly identified as goodware;
4. False Negative Rate (*fnr = 1 - tpr*): the fraction of ransomware samples incorrectly classified as goodware; and
5. *Accuracy* is reported as the fraction of all samples correctly identified. More specifically, $Accuracy = \frac{tpr+tnr}{tpr+tnr+fpr+fnr}$;
6. *Precision* is calculated as $precision = \frac{tp}{tp+fp}$; and
7. *Youden's index* is calculated as $Y = tpr + tnr - 1$

2.3   The Data Set

Virus Share (virusshare.com) was used to access a ransomware repository and goodware were acquired from Portable Apps (portableapps.com). Kali Linux was used for analysing of the malware samples.

In total, 400 executable files (200 malwares and 200 goodware) are used. First, the executables were put through an online intelligence platform (virustotal.com) as shown in Fig. 1. The use of virtual machine was a necessity to be able to see the features of the malware and goodware properly without the risk of damaging or condemning the host machine being used for the analysis. The virtual machine

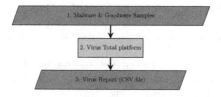

Fig. 1. Data extraction process overview

was running a default installation of Windows 10 and the networking was disabled [19]. The features extracted from the virus total platform which make up the dataset for the machine learning algorithms contained values such as detection rate, hash value, file size, dll calls, mutexes, pe info etc. which included both static and dynamic features.

Features were extracted using a script written by Didier Stevens[1] to automate the extraction of feature data from the virus total platform.

3 Experimental Analysis

The classification of the dataset was done using the WEKA which is a dedicated tool for machine learning [14]. Machine learning algorithms were selected involving the selection of appropriate classifier for the classification of the dataset into goodware and malwares, the classifiers used for this experiment are support vector machine (SVM) by sequential minimal optimization, instance-based (IB1), Naïve Bayes and Random Forest. The analysis carried out on the ransomware dataset classification was in two stages of training and testing, where each stage makes use of the above-mentioned classifiers. Normalization is a feature of WEKA which is applied to the attributes in the data pre-processed stage.

The training data set include 100 ransomware samples that are variants of Locker and Crypto Locker and also 100 goodware. The training of the machine in this stage involves using labelled data, whereas in the test stage, new and previously unseen dataset are provided for classification. The result is checked to see if it was classified accurately. The sections below show the result of the analysis in both training stage and test stage.

The training set was analysed using the WEKA classifiers which showed a perfect classification with confusion matrix and an accuracy of 100% which is shown in the Table 1. Furthermore, the table provides the fine-grained accuracy results necessary to understand the algorithm's capability. The confusion matrix of this analysis shows that 100 of the ransomware instances were correctly classified under the ransomware class and 100 of the goodware instances were correctly classified under the goodware class. This demonstrate the algorithms capability to correctly learn key characteristics that differentiate goodware and malware.

---

[1] Didier Steven's script available at: https://blog.didierstevens.com/programs/virustotal-tools/.

Table 1. Classifier performance on both training and test sets

Classifier	TP	TN	FP	FN	Youden's index	Accuracy		Precision		Recall		F-Measure	
						Training	Test	Training	Test	Training	Test	Training	Test
SVM	100	99	1	0	0.99	100	99.5	1.00	0.99	1.00	0.99	1.00	0.99
IB1	97	95	4	4	0.91	100	96	1.00	0.96	1.00	0.96	1.00	0.96
RF	100	99	1	0	0.99	100	99.5	1.00	0.99	1.00	0.99	1.00	0.99
NB	100	92	1	7	0.91	100	96	1.00	0.96	1.00	0.96	1.00	0.96

After analysing the training set using each algorithm, we go on to analyse the test set using each algorithm with which the training set was analysed in the sections below. It is important to highlight that in this section we use the previously unseen 100 malware and goodware samples. The detection and classification accuracy of the WEKA classifiers were empirically evaluated to choose the machine learning algorithm with the highest detection accuracy and lowest false positive rate and the result is shown in Table 1. This test analysis was carried out using the 10-fold cross validation for its test analysis due the reputation of the K-fold cross validation method, this method allows the dataset to be iteratively analysed which gives the algorithms a better chance to be less biased, the results demonstration that the Naïve Bayes algorithm has an accuracy of 96% with the test dataset. The IB1 produced an accuracy of 96%, demonstrating that it is not the best machine learning algorithm to use for classifying this type of dataset. This is due to 4 false positive instances and 4 false negative instances from the dataset. SVM and the random forest both gave an accuracy of 99.5% which means these algorithms performed well in comparison with Naïve Bayes and IB1. The training set demonstrates a 100% detection rate and accuracy, hence it can be deduced that a larger dataset produced a better result since a larger percentage of dataseta are used for the training while the remaining was used for the test set. Furthermore, the algorithms demonstrate that regardless of the available number of datasets, they result in a higher detection rate since it performs analysis regardless of the position of the data in the dataset.

Table 1 provides a comparison of the weighted average of the recall, precision and f-measure for the different algorithms when using the testing set. It also illustrates that RF and SVM with the highest recall and f-measure and a joint highest in accuracy. This research also considered the samples that were incorrectly classified in the WEKA result in relations to the other metrics which was why we calculated the informedness of the algorithms used. Informedness also refers to the Youden's index which takes into consideration the sensitivity and specificity, which is shown in Table 1. The table illustrates that SVM and RF both had the joint highest value of 0.99 and IB1 and the Naïve Bayes both had 0.91, which when compared to the accuracy clearly gives a better representation of how the algorithms performed. It was noted that considering the correctly classified class does not give a clear picture of the efficiency of detection and classification compared to a metric which factors all elements of the confusion matrix. Algorithms with high Youden's index are considered more efficient which is why we have selected SVM and RF to be better classifiers for this research.

A number of factors could be responsible for the error in classification ranging from use of wrong classifiers to error in dataset; however, this research has been considered to have minimal error in classification because the false positive and false negative rate is below 10% of the entire class. It could be that the malware and the goodware samples are of the same sample type which means that there is a likelihood of the classifier detecting a goodware as a malware or vice-versa due to the sample size. It could also be the case that the 100 goodware and malware samples for the training set was too restrictive, reducing the classifier's ability to learn the characteristics of the training set. There is every likelihood that the classifiers need more samples to cover enough characteristics before the test sample is introduced.

4 Conclusions

Due to the increasing development and use of ransomware, there is a strong requirement for improved detection rate. This paper sets about trying to contribute to achieve this goal. This research aimed to reduce the rate of false positive in detection, which is why samples of ransomware were collected and analysed using multiple machine learning algorithms, and finally tested in WEKA. The algorithms were used to train and test the dataset to be able to better detect the ransomware family of malwares.

The random forest learning algorithm gave a relatively high detection accuracy of 99.5%, but its analysis is not reliable because it chooses the modal class and if the modal class was of the wrong class it would still have given a relatively high detection accuracy. The support vector machine globally replaces every missing value and changes nominal characteristics into a binary representation meaning it is also not reliable. The aim of this research was achieved by the random forest and SVM algorithm and the analysis carried out has a very low false positive and false negative classification. We have however concluded that the number of instances used in the classification and detection is not enough to give an overall reliable result as there is need for the algorithms use to be tested on a larger scale so as to expose the training set to a wide variant of ransomware for analysis to be able to give an accurate and reliable analysis, also, there should be inclusion of older ransomware samples so as to improve the detection rate as the ransomware samples used are from a recent collection. Future works would look into using of more features from the ransomware samples like ssdeep hash, dlls, opcodes etc. with unsupervised machine learning algorithms for classification.

References

1. A. Kumar, K.S.K., Aghila, G.: A learning model to detect maliciousness of portable executable using integrated feature set. J. King Saud Univ. - Comput. Inf. Sci. (2017)

2. Mohaisen, A., Alrawi, O., Mohaisen, M.: Amal: high-fidelity, behavior-based auto-mated malware analysis and classification. Comput. Secur. **52**, 251–266 (2015)
3. Alazab, M.: Profiling and classifying the behavior of malicious codes. J. Syst. Softw. **100**, 91–102 (2015)
4. Shahzad, F., Shahzad, M., Farooq, M.: In-execution dynamic malware analysis and detection by mining information in process control blocks of linux OS. Inf. Sci. (Ny) **231**, 45–63 (2013)
5. Gatz, D.F., Smith, L.: The standard error of a weighted mean concentration-i. Bootstrapping vs other methods. Atmos. Environ. **29**(11), 1185–1193 (1995)
6. Grant, L., Parkinson, S.: Identifying file interaction patterns in ransomware behaviour. In: Parkinson, S., Crampton, A., Hill, R. (eds.) Guide to Vulnerability Analysis for Computer Networks and Systems. CCN, pp. 317–335. Springer, Cham (2018). https://doi.org/10.1007/978-3-319-92624-7_14
7. Lu, H., Wang, X., Zhao, B., Wang, F., Su, J.: Endmal: an anti-obfuscation and collaborative malware detection system using syscall sequences. Math. Comput. Model. **58**(5), 1140–1154 (2013)
8. Zhang, H., Xiao, X., Mercaldo, F., Ni, S., Martinelli, F., Sangaiah, A.K.: Classifi-cation of ransomware families with machine learning based on n-gram of opcodes. Futur. Gener. Comput. Syst. **90**, 211–221 (2019)
9. Islam, R., Tian, R., Batten, L.M., Versteeg, S.: Classification of malware based on integrated static and dynamic features. J. Network Comput. Appl. **36**(2), 646–656 (2013)
10. Deepa, K., Radhamani, G., Vinod, P.: Investigation of feature selection methods for android malware analysis. Procedia Comput. Sci. **46**, 841–848 (2015)
11. Sun, M., Li, X., Lui, J.C., Ma, R.T., Liang, Z.: Monet: a user-oriented behavior-based malware variants detection system for android. IEEE Trans. Inf. Forensics Secur. **12**(5), 1103–1112 (2017)
12. Milosevic, N., Dehghantanha, A., Choo, K.K.R.: Machine learning aided android malware classification. Comput. Electr. Eng. **61**, 266–274 (2017)
13. Burnap, P., French, R., Turner, F., Jones, K.: Malware classification using self organising feature maps and machine activity data. Comput. Secur. **73**, 399–410 (2018)
14. Patil, T.R., Sherekar, M.S.S.: Performance analysis of naive bayes and j48 classi-fication algorithm for data classification. Int. J. Comput. Sci. Appl. **6**(2), 256–261 (2013)
15. Provataki, A., Katos, V.: Differential malware forensics. Digit. Investig. **10**(4), 311–322 (2013)
16. Das, S., Liu, Y., Zhang, W., Chandramohan, M.: Semantics-based online malware detection: towards efficient real-time protection against malware. IEEE Trans. Inf. Forensics Secur. **11**(2), 289–302 (2016)
17. Schultz, M.G., Eskin, E., Zadok, F., Stolfo, S.J.: Data mining methods for detection of new malicious executables. In: Proceedings 2001 IEEE Symposium on Security and Privacy, S&P 2001, pp. 38–49. IEEE (2000)
18. Sharma, A., Sahay, S.K.: An effective approach for classification of advanced mal-ware with high accuracy. arXiv preprint arXiv:1606.06897 (2016)
19. Shijo, P.V., Salim, A.: Integrated static and dynamic analysis for malware detec-tion. Procedia Comput. Sci. **46**, 804–811 (2015)
20. Townsend, J.T.: Theoretical analysis of an alphabetic confusion matrix* (1971)
21. Zhang, H.: The optimality of naive bayes

Artificial Neural Networks in Mathematical Mini-Games for Automatic Students' Learning Styles Identification: A First Approach

Richard Torres-Molina[1](✉)[ID], Jorge Banda-Almeida[1][ID],
and Lorena Guachi-Guachi[1,2][ID]

[1] Yachay Tech University, Hacienda, San José, Urcuquí 100119, Ecuador
richard.torres@smartmathlabs.com
[2] SDAS Research Group, Urcuquí, Ecuador
www.sdas-group.com

Abstract. The lack of customized education results in low performance in different subjects as mathematics. Recognizing and knowing student learning styles will enable educators to create an appropriate learning environment. Questionnaires are traditional methods to identify the learning styles of the students. Nevertheless, they exhibit several limitations such as misunderstanding of the questions and boredom in children. Thus, this work proposes a first automatic approach to detect the learning styles (Activist, Reflector, Theorist, Pragmatist) based on Honey and Mumford theory through the use of Artificial Neural Networks in mathematical Mini-Games. Metrics from the mathematical Mini-Games as score and time were used as input data to then train the Artificial Neural Networks to predict the percentages of learning styles. The data gathered in this work was from a pilot study of Ecuadorian students with ages between 9 and 10 years old. The preliminary results show that the average overall difference between the two techniques (Artificial Neural Networks and CHAEA-Junior) is 4.13%. Finally, we conclude that video games can be fun and suitable tools for an accurate prediction of learning styles.

Keywords: Artificial Neural Networks · Learning Styles · Video games

1 Introduction

The main problem in the current educational system is the lack of personalized education according to the Learning Styles (LS) that identifies an individual [1]. Recognizing a student LS enables educators to create an appropriate learning environment. Furthermore, providing a personalized learning experience increases learning performance, enhance motivation and reduced learning time [2]. On this sense, LS are defined as the combination of cognitive and psychological factors that determine how a learner process, absorb, and retain new

© Springer Nature Switzerland AG 2019
H. Yin et al. (Eds.): IDEAL 2019, LNCS 11872, pp. 53–60, 2019.
https://doi.org/10.1007/978-3-030-33617-2_6

information [3]. Therefore, to identify LS, several ideas, theories, models and instruments have been developed. They are mainly static approaches characterized by their reliability and precision to discover LS such as questionnaires, tests, or self-reports based on psychologists cognitive styles theories [1]. However, they produce a lack of motivation from the students, and it is time-consuming in the answers collection. On the other hand, some Artificial Intelligence (AI) techniques have been used to support automatic methods to identify LS such as Artificial Neural Networks (ANN) [6], Expert Systems (ES) [7], and Genetic Algorithms (GA) [8].

Learning Styles Questionnaire (LSQ) [9] is one of the most used questionnaires to identify LS. It is focused on the Experiential Learning Theory (ELT) and based on the Learning Style Inventory (LSI) questionnaire. LSQ model introduced four LS: Activist (learn by doing), Reflector (learn by observing), Theorist (prefer models and concepts) and Pragmatist (concerned with knowledge practical applications). A Spanish version of LSQ for university students named "Cuestionario Honey-Alonso de Estilos de Aprendizaje" (CHAEA) was introduced in [10]. With the aim of discovering the predominant learning profile in students from early ages (nine to fourteen years old), CHAEA-Junior, an adapted version of CHAEA was presented in [11].

Consequently, as LS change under the influence of certain situations as age and learning environment, adaptive educational approaches are required to efficiently detect LS [12]. An alternative to overcome this issue is the use of video games to exploit the inclusion of rich multimedia and highly interactive features, which facilitate student learning [13]. Therefore, ADOPTA (ADaptive technOlogy-enhanced Platform for eduTAinment) family of playing styles based on LSQ and ELT is stated in [14]. The aim of ADOPTA approach is to recognize playing styles, and consequently, the student learning style during a game session based on the following correlation: Competitor ⇔ Activist, Dreamer ⇔ Reflector, Logician ⇔ Theorist, and Strategist ⇔ Pragmatist. In this context [14], identification of playing styles with high precision through key performance indicators (result, efficiency, and difficulty) was found in an action video game named "Rush for Gold".

Reached performance by automatic and dynamic approaches existing in literature depends on the data gathered from video games especially in mathematics [15]. Mainly, ANN has also been used combining different behavior attributes to achieve maximum accuracy. For instance, ANN with Felder and Silverman model in a Conversational Intelligent Tutoring System (CITS) is stated in [6]. Another ANN approach to detect and adjust student's LS based on Markov chains and GA using a probability distribution is introduced in [8]. Montes in [10] use the CHAEA questionnaire and ES based on semantic networks, to assign a recommended strategy to each student focused on the learning preference and their particular characteristics to acquire knowledge. A Learning style Identifier (LSID) using ANN to recognize the LS in learning management systems is shown in [16], where the back-propagation is used as a training algorithm.

In order to diminish the lack of automatic LS detection with educational games, this work proposes a first automatic approach to detect the student percentages of LS through the use of ANN in mathematical Mini-Games. This method is compared with the most precise static method known as CHAEA-Junior [10] due to its robustness in children LS detection. ANN is chosen by its strong accuracy, reliability, speed of execution, and use in LS detection. In several works, ANN are composed of one input layer, where the attributes are the interactions in static websites [4], or data from CITS [6]; one hidden layer compounded of a different number of neurons; and the output layer represented by learning styles based on Honey and Mumford theory. This work mainly differs from the relevant state-of-the-art in the input parameters given by metrics of Mathematical Mini-Games in LS automatic detection, while existing approaches often use metrics of static websites.

2 Methods

2.1 CHAEA-Junior Questionnaire (CHAEA-JQ)

It is based on the LS (Activist, Reflector, Theorist, and Pragmatist) of LSQ applied on primary level education students [11]. These LS are identified by a set of questions according to the student psychological characteristics. For this work purposes, the questions were implemented in an interactive web-system different from the traditional paper-based folio approach. In the web-system, the student answered a set of 44 questions. If the user clicks on the "YES" button, the percentage is increased in that learning style, meanwhile if the student clicks on the "NO" button, the percentage is not modified.

The percentage for learning style recognition in the CHAEA-JQ is calculated by the Eq. 3, composed by the Summatory of Each Learning Style ($SELS_L$), and the Summatory of all questions related to the LS (SLS) given by Eqs. (1) and (2):

$$SELS_L = \sum_{j=1}^{11} q_j^L \tag{1}$$

where $q_j^L = 1$, if the "YES" button is clicked; and $q_j^L = 0$, if the "NO" button is clicked.

$$SLS = \sum_{j=1}^{44} q_j^L \tag{2}$$

$$PLS_{JQ} = \frac{SELS_L}{SLS} \tag{3}$$

2.2 Mathematical Mini-Games

The mathematical Mini-Games belong to the video game known as "Smart Math" [18], which was developed by the social start-up "Smart Math Labs" using music, and 2D sprites applied to this research. The mathematical Mini-Games are designed following the four ADOPTA playing styles (Competitor, Dreamer, Logician, and Strategist) [19]. Each one is composed of basic mathematical operations (sums and subtractions) with binary operations in two difficulty levels as seen in Fig. 1. The goal of each Mini-Game is to select the correct answer from the mathematical operations by different interactions given by the user. If the student answers the mathematical operation correctly, its score is increased at one point, and zero otherwise.

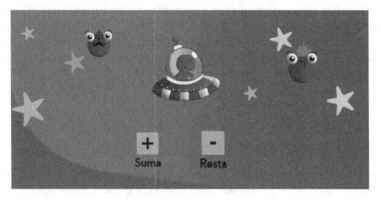

Fig. 1. Screen shot of "Smart Math" video game.

2.3 Artificial Neural Network (ANN)

The proposed network architecture is designed with three layers: input, hidden and output layer as shown in Fig. 2. The input layer takes the score and time from each mathematical Mini-Game, attributes selected as more relevant to student identification. The target corresponds to the percentages of each learning style PLS_{JQ}, calculated by the Eq. (3). After the model has been training with the students' group, the output layer represents the learning style percentages PLS_{ANN} of Activist, Reflector, Theorist and Pragmatist styles given by the Eq. (4). This work used the Back-Propagation (BP) algorithm [5] to find a local minimum of the error function. The weights were initialized randomly, where the error function gradient is applied to the initial weights correction.

The BP main equation is given by (4), where o_i is the neuron output belonging from the hidden neuron n_i; p represents the synaptic potential; w_{ij} are the synaptic weights between neuron i in the current layer and the neurons of the previous layer with activation $\widehat{o}_j$.

$$o_i = s(\sum_{j=1}^{n} w_{ij} \cdot \widehat{o}_j) = s(p) \tag{4}$$

The parameters $\eta = 0.05$, $\gamma=1/2$, and 1000 iterations have been used in the proposed ANN architecture to find the best model that fit the data. Several architectures have been tested and found their respective Mean Square Error (MSE), as a result, the higher performance was the hidden layer of 10 neurons.

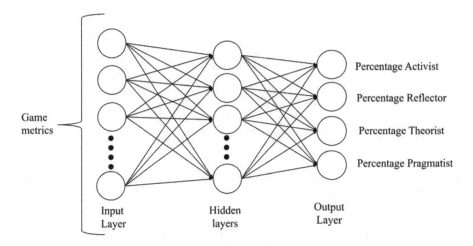

Fig. 2. Top-level architecture of the artificial neural network approach.

3 Experimental Setup

3.1 Dataset

It was collected data of 163 students between 9 and 10 years old of the schools "Teodoro Gómez de la Torre" and "San Francisco" (Imbabura-Ecuador). With the small available data, the BP model was trained with the normalized data set using a ten-fold cross-validation procedure to predict the LS percentages in order to avoid overfitting problems. Cross-validation allows to build a model to generalize an independent data set.

3.2 Experiment Design

At the beginning of the experiment, the students answered in the web-system, the 44 questions from the CHAEA-JQ. Then, they played mathematical Mini-Games based on ADOPTA playing styles. Finally, the recollected data was used in the ANN approach.

The average difference ($\in$) of detecting LS between ANN with CHAEA-JQ for the four LS (N = 4) was calculated by the Eq. (5).

$$\in= \frac{\sum_1^N \frac{abs(Pls_{ANN}-Pls_{JQ})}{Pls_{JQ}}}{N} \tag{5}$$

4 Results and Discussion

The proposed automatic approach based on ANN by using video games was compared with respect to CHAEA-JQ [11], a static and the most reliable traditional approach, to determine the precision reachable to recognize LS.

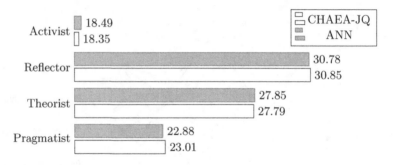

Fig. 3. Learning style percentage obtained by ANN and comparison with CHAEA-JQ from averaging the results on the ten-fold cross-validation testing datasets.

For this reason, the average of Pls_{JQ} and Pls_{ANN} values have been computed in each testing dataset from the ten-fold cross-validation procedure. The average overall difference between the two techniques is $\in= 4.13\%$. The average results furnished in Fig. 3 demonstrate that the majority of the students follow the order in LS as Reflector, Theorist, Pragmatist and Activist in both techniques. Particularly, for the Reflector learning style, CHAEA-JQ obtained 30.85, and ANN reached 30.78. Being the Reflector with the highest percentage due to the psychological evolution in children by following a traditional educational system.

The use of game metrics in the proposed ANN is a first approach to provide an enjoyable experience to LS detection. In addition, it can avoid the boredom, misunderstanding, and lack of interactivity in current questionnaires such as the CHAEA-JQ.

5   Conclusion

Latin America is one of the continents with the lowest ranking in mathematics performance in comparison with developed countries. Due to the lack of personalized learning to students in different areas of knowledge, and the absence of entertainment tools to avoid boredom in children. So, recognizing the LS at early ages is particularly significant as a potential solution to deal with learning issues such as difficult subjects learning.

In the comparison between a static (CHAEA-JQ) and an automatic (ANN) technique, the average overall difference was $\in= 4.13\%$ for percentages of Activist, Reflector, Theorist, and Pragmatist LS. Being video games a tool to an accurate prediction of learning and playing styles, and with less information needed to identify LS than traditional educational systems. As future work, after the student is detected with their learning style percentages, adaptability could be provided to a custom curriculum related to the student knowledge.

Acknowledgements. We would like to thank the directors Msc. Carlos Verdesoto and Arch Marco Lafuente from the schools "Teodoro Gómez de la Torre" and "San Francisco" (Ibarra-Ecuador) who have contributed in the data gathered in this work.

References

1. Coffield, F., Moseley, D., Hall, E.: Learning styles and pedagogy in post-16 learning: A systematic and critical review, 1st edn. Learning and Skills Research Centre, London (2004)
2. Kelly, D.: Adapting to intelligence profile in an adaptive educational system. Interact. Comput. **18**(3), 385–409 (2006)
3. Feldman, J.: Automatic detection of learning styles: state of the art. Artif. Intell. Rev. **44**(2), 157–186 (2015)
4. Jia, L., Pai, S.: Identification of learning styles online by observing learners' browsing behaviour through a neural network. Br. J. Educ. Technol. **36**(16), 43–55 (2005)
5. Torres, R., Ríofrio, A., Bustamante, C., Ortega, F.: Prediction of learning improvement in mathematics through a video game using neurocomputational models. In: Proceedings of the 11th International Conference on Agents and Artificial Intelligence, pp. 554–559. SciTePress, Prague (2019)
6. Latham, A., Crockett, K., Mclean, D.: Profiling student learning styles with multilayer perceptron neural networks. In: 2013 IEEE International Conference on Systems. Man, and Cybernetics, pp. 2510–2515. IEEE, Manchester (2013)
7. Olguín, J.A.M., García, F.J.C., de la Luz Carrillo González, M., Medina, A.M., Cortés, J.Z.G.: Expert system To *Engage* CHAEA learning styles, ACRA learning strategies and learning objects into an e-learning platform for higher education students. Advances on P2P, Parallel, Grid, Cloud and Internet Computing. LNDECT, vol. 1, pp. 913–922. Springer, Cham (2017). https://doi.org/10.1007/978-3-319-49109-7_89
8. Dorça, F.A.: Automatic student modeling in adaptive educational systems through probabilistic learning style combinations: a qualitative comparison between two innovative stochastic approaches. J. Braz. Comput. Soc. **19**(1), 43–58 (2013)
9. Mumford, A.: Questions and answers on learning styles questionnaire. Ind. Commer. Training **24**(7), 10–13 (1992)
10. Dziedzic, M., de Oliveira, F.B., Janissek, P.R., Dziedzic, R.M.: Comparing learning styles questionnaire. In: 2013 IEEE Frontiers in Education Conference (FIE), pp. 973–978. IEEE, Oklahoma (2013)
11. Sotillo, J.: CHAEA-Junior Survey or how to diagnose elementary and secondary students' learning styles. J. Learn. Styles **7**(1), 182–201 (2014)
12. Stash, N., De Bra, P.: Incorporating cognitive styles in AHA! (the adaptive hypermedia architecture). In: Proceedings of the Iasted International Conference Web-Based Education, pp. 378–383. Innsbruck (2004)
13. Chang, B.: The influences of an interactive group-based video game: cognitive styles vs. prior ability. Comput. Educ. **88**, 399–407 (2015)
14. Bontchev, B.: Playing styles based on experiential learning theory. Comput. Hum. Behav. **85**, 319–328 (2018)
15. Torres, R., Guachi, L., Guachi, R., Stefani, P., Ortega, F.: Learning style identification by CHAEA junior questionnaire and artificial neural network method: a case study. In: 1st International Conference on Advances in Emerging Trends and Technologies. Springer, Quito (2019)
16. Bernard, J.: Learning style identifier: improving the precision of learning style identification through computational intelligence algorithms. Expert Syst. Appl. **75**, 94–108 (2017)
17. Rivas, A.: Latin America after PISA: Lessons Learned about Education in Seven Countries, 1st edn. CIPPEC-Natura-Instituto Natura, Buenos Aires (2015)

18. Smart Math Labs. https://www.smartmathlabs.com. Accessed 4 July 2019
19. Aleksieva, A., Petrov, M.: ADOPTA model of learner and educational game struc-
    ture. In: 12th International Conference on Computer Systems and Technologies,
    pp. 636–640. ACM, Vienna (2011)

The Use of Unified Activity Records to Predict Requests Made by Applications for External Services

Maciej Grzenda[1]([⊠]) [ID], Robert Kunicki[1,2] [ID], and Jaroslaw Legierski[3] [ID]

[1] Faculty of Mathematics and Information Science, Warsaw University
of Technology, ul. Koszykowa 75, 00-662 Warszawa, Poland
{M.Grzenda,R.Kunicki}@mini.pw.edu.pl
[2] Digitalisation Department, The City of Warsaw, pl. Bankowy 2,
00-095 Warszawa, Poland
[3] IoT and Advanced Technologies, Orange Polska, ul. Wolumen 11,
01-912 Warszawa, Poland
jaroslaw.legierski@orange.com

Abstract. Many modern applications use services and data made available by provisioning platforms of third parties. The question arises if the use of individual services and data resources such as open data by novel applications can be predicted. In particular, whether initial software development efforts such as application development during hackathons can be monitored to provide data for the models predicting requests submitted to open data platforms and possibly other platforms is not clear.

In this work, we propose an iterative method of transforming request streams into activity records. By activity records, vectors containing aggregated representation of the requests for external services made by individual applications over growing periods of software development are meant. The approach we propose extends previous works on the development of network flows aggregating network traffic and makes it possible to predict future requests made to web services with high accuracy.

Keywords: Data aggregation · Prediction · Network flow · Web service

1 Introduction

Novel software applications and services frequently rely on third-party services and data. This includes the use of Application Programming Interfaces (APIs), which can be called by sending HTTP requests. Hence, the data on the use of network services can be collected and used for variety of purposes. Network flow records can be used to aggregate the data of multiple network packets. Standards such as NetFlow and IPFIX can be used to provide network flows representing varied volume of data transferred over the network [3]. Importantly, network flow vectors are composed of a fixed number of features i.e. are vectors of the

© Springer Nature Switzerland AG 2019
H. Yin et al. (Eds.): IDEAL 2019, LNCS 11872, pp. 61–69, 2019.
https://doi.org/10.1007/978-3-030-33617-2_7

same dimensionality. Hence, network flows were used as an input for machine learning methods. In particular, Kim et al. in [4] have shown that network flows can provide basis for network traffic classification i.e. distinguishing based on aggregated feature vectors major network applications, including WWW, FTP, and mail transmission. Similarly to other network flow studies, an important aspect was what aggregate numeric features to develop out of raw network traffic. Naive Bayes, decision trees and Support Vector Machines were among the methods tested in the study. Many network traffic aggregation methods are developed in cybersecurity domain. In [7], an overview of IP flow-based intrusion detection (ID) is provided. More recently, aggregate traffic features were used as an input for hybrid intrusion detection method for Internet of Things including the use of clustering and unsupervised approach [1]. Similarly, in [5], a hybrid ID solution for wireless infrastructures including data aggregation stage was proposed. Development of aggregate network features such as the number of bytes sent over TCP is also a vital step in preparing the data for botnet detection method involving the use of deep learning and proposed in [6].

The aforementioned studies are focused on the classification of network traffic and anomaly detection in complex network systems composed of multiple services. Typically, regular data transmission is assumed. In this study, we address the question whether aggregate data vectors developed based on raw request data streams can be used to predict the future number of requests made by individual client applications to network services. This question is related to the increase in the number of portals providing data in the Open Data model [8].

The key contribution of this study is the proposal of the way the sequences of requests made to different resources over different periods of time can be preprocessed to provide *activity records* of growing dimensionality, which can be used as an input for prediction methods. Some of the new challenges that are answered in this study are that (a) initial requests to APIs can be due to experimentation rather than ultimate functionality, (b) major gaps in development time have to be considered, and (c) development of different applications can start at different times or even days. Hence, non-standard network flow records have to be developed while addressing these aspects, which are not addressed when constant data flow to heavily loaded services such as mail servers is described with standard network flow records. The remainder of this paper is organised as follows. Iterative aggregation of request data streams into activity records is proposed in Sect. 2. This is followed by Sect. 3, which describes reference data and contains results of the tests performed with different scenarios of request data aggregation. Finally, the work is concluded in Sect. 4.

2   Iterative Aggregation of Request Data Streams

In this study, the input data for prediction purposes is a stream of requests made by different applications to different third-party services such as telco services and services providing open data. The requests can be requests made to web services such as RESTful services. The primary idea is to use requests collected during a period such as the period of initial software development to predict future application behaviour. The question arises how to pre-process the data of individual requests to use it to predict whether a service of interest will be used in the ultimate software product, and if so how frequently the provisioning portal will have to answer requests made to this service.

Let us denote requests registered in the logs of API portal or captured by collector module, by $T_1, T_2, ...$ Importantly, one request is made to a single API. Furthermore, let $t(T_i)$ denote the time of making T_i request to API portal counted in hours since the beginning of the monitored period. Moreover, let $a(T_i)$ denote the API key, uniquely identifying an application that triggered the request. Without loss of generality, we assume that every application is developed by a separate team. Thus, in the remainder of this study we use these two terms i.e. *team* and *application* interchangeably.

Before request aggregation is proposed, let us observe that every application may trigger a different number of requests. Moreover, individual teams can work at different periods of time and may suspend their work a number of times. In particular, they can finish their application development and testing at different times.

Input: S - the set of requests $S = \{S_1, S_2, ...\}$ made to individual resources, δ_I,
  P - duration of initiation elementary subperiod, and the number of
  these periods, respectively , Δ_E - duration of experimentation period
Data: $A(S) = \{a(S_i) : S_i \in S\}$ - the set of applications for which logs are
  available, $R(S) = \{r(S_i) : S_i \in S\}$ - the set of services that have ever
  been requested, $\Delta_I = P\delta_I$ - duration of initiation period
Result: $\Omega(\delta_I, P, \Delta_E) = \{(\mathbf{x}_i, y_i)\}$ - the set of application activity records, where
  $\mathbf{x}_i$ is a vector describing an application, a resource and including the
  features aggregating the requests to this resource by this application,
  whereas y_i is a discrete class label
begin
  for $a \in A(S)$ **do**
    for $r \in R(S)$ **do**
      $\mathbf{v}_j = [a, r, APIFeatures(r)]$;
      for $p = 1, ..., P$ **do**
        $\Gamma = \{S_i : \Delta_E + (p-1)\delta_I < t(S_i) \leq \Delta_E + p\delta_I \wedge a(S_i) = a \wedge r(S_i) = r\}$;
        $\mathbf{v}_j = [\mathbf{v}_j, DevelopFeatures(\Gamma)]$;
      end
      $\Lambda = \{S_i : \Delta_E + P\delta_I < t(S_i) \leq \tilde{t}_{\max}(a) \wedge a(S_i) = a \wedge r(S_i) = r\}$;
      $\Omega(\delta_I, P, \Delta_E) = \Omega(\delta_I, P, \Delta_E) \cup \{(\mathbf{x}_j, y_j) = (\mathbf{v}_j, DevelopLabel(\Lambda))\}$;
    end
  end
end

Algorithm 1. Iterative aggregation of request data

First of all, let us distinguish three stages, the first of which is *experimentation* i.e. the period starting when the first call to any resource is made by a team during which a team may try the use of different services not necessarily planned to be included in the ultimate application. Let us denote by $t_{\min}(a)$ the time of the first request made by team a to any of the services. The *experimentation stage* (ES) is followed by *initiation stage* (IS), which is followed by *development and deployment stage* (DDS). The DDS period is finished at the time $t_{\max}(a)$ of making last request during the monitored period by application a. The objective of the study is to develop prediction models capable of predicting the use of individual resources during development and deployment stage, based on the requests made during initiation period.

First, let us propose to eliminate gaps in team activity by creating the set of unified team requests composed of $S = \left\{ S_i : S_i = T_i, T_i \in T \wedge t(S_i) = \right.$ $\left. t(T_i) - t_{\min}(a(T_i)) + 1 - \sum_{j=t_{\min}(a(T_i))}^{t(T_i)} I(a,j) \right\}$, whereas $I(a,j) = 1$ when $\{T_i : t(T_i) = j \wedge a(T_i) = a\} = \emptyset$, and $I(a,j) = 0$ otherwise.

In other words, we map original times to unified times by eliminating all one hour periods for which no requests triggered by a team were observed. Furthermore, we set the time of the first request for a resource made by a team to the same value of 1. In this way the activities of all teams start at hour $t = 1$. Still, they may last for different periods of time, which are equal to the number of one hour clock periods during which some activity of a team was observed in the logs. Hence, $\tilde{t}_{\max}(a) = max(\{t(S_i) : a(S_i) = a\})$ will denote maximum unified time and will correspond to the number of one hour periods during which some activity of team a was observed.

Furthermore, let us propose to aggregate the data of requests in S to provide data for constructing the models predicting resource use by individual applications. Algorithm 1 defines the aggregation method we propose. Out of individual requests it develops *unified activity records* of fixed dimensionality describing access patterns of individual applications and referred to as activity records in the remainder of this work. The key proposition is to describe resource use during initiation period to provide the values of input features for a model and combine it with the ultimate aggregated use of this resource during evaluation stage, which is represented by a class label.

To evaluate the ability to predict which resources and how frequently will be used in evaluation period by individual applications, let us use cross-validation (CV). Let us observe that applications tend to use some of the services only. Hence, we propose to evaluate the prediction accuracy both for all combinations of applications and services ($A_T(D, P)$ accuracy) and only for activity records that are related to these combinations for which some requests were actually observed during initiation stage, which is denoted by $A_T^E(D, P)$.

3   Results

To evaluate iterative aggregation of request data streams, we designed experiments aimed to verify whether activity records can be used to predict ultimate

Input: $\Omega = \{(\mathbf{x}_i, y_i)\}$ - the set of application activity records, K - the number of CV folders

Data: D_i - a family of K sets: $D_i \cap D_j = \emptyset i \neq j$, D_T - testing data, D_L - training data, P_i, P_L, P_T - output features corresp. to D_i, D_L, D_T, respectively, U, U^E - the set of pairs $(y, \hat{y})$ of true and predicted label for all and only these instances for which requests in initiation stage were observed, $r(\mathbf{x}_k)$ - total number of requests for service observed in initiation stage

Result: $A_T(D, P)$, $A_T^E(D, P)$ - the average accuracy of predicted service use for all services and services used in initiation stage, respectively

begin

$\quad U = \emptyset; \ U^E = \emptyset;$

$\quad \{D_j, P_j\}_{j=1,\ldots,K} = \text{DivideSet}(D, P, K);$

$\quad$**for** $k = 1 \ldots K$ **do**

$\quad\quad D_T = D_k; \ D_L = \cup_{j \in \{1,\ldots,K\} - \{k\}} D_j;$

$\quad\quad M = train(D_L, P_L);$

$\quad\quad U = U \cup \{(\hat{y} = M(\mathbf{x}_k), y_k) : \mathbf{x}_k \in D_T \wedge y_k \in P_T\};$

$\quad\quad U^E = U^E \cup \{(\hat{y} = M(\mathbf{x}_k), y_k) : \mathbf{x}_k \in D_T \wedge y_k \in P_T \wedge r(\mathbf{x}_k) \neq 0\};$

$\quad$**end**

end

$$A_T(D, P) = \frac{|\{(\hat{y}, y) \in U : \hat{y} = y\}|}{|U|} \times 100\%; \quad A_T^E(D, P) == \frac{|\{(\hat{y}, y) \in U^E : \hat{y} = y\}|}{|U^E|} \times 100\% ;$$

Algorithm 2. The evaluation of the accuracy of predicted resource use

resource use by software applications. First, we convert real request data streams into activity records. Next, we construct classification models with the training data composed of activity records. This is to verify whether such models can predict how frequently individual services will be used based on initial requests for these services aggregated into activity records.

The request data used in the experiments come from 2015 and 2016 editions of long term BIHAPI software development competitions. During both editions, open data and telecom services were made available for developers. Further details on BIHAPI competitions can be found in [2]. All the data sets and services were made available using RESTful services, available over HTTP through a proxy server registering individual HTTP requests for both the data and services. Some of the data was exposed as most recent data e.g. bus locations rather than data set. Hence, we use the term service and resource interchangeably to refer to both telecom services and open data resources throughout the rest of this study. For every application that was developed and which was accessing services via a proxy server, a sequence of requests to individual services was recorded in server logs. The request data of 116 748 requests made by 61 applications to 67 services used at least by one of the applications provided basis for this study.

First BIHAPI dataset was converted into activity records, as defined in Algorithm 1. In the case of activity records, $\delta_I = 1\,h$ was used i.e. initiation stage was divided into one hour periods, each described with one feature of activity record. Alternatively, all requests were aggregated into one feature i.e. total request count $P = 1, \Delta_I = \delta_I$. We refer to the latter scenario as the scenario of not using activity records in the remainder of this study. In all cases, for every δ_I period, one feature i.e. request count during this elementary period was calculated based on raw Γ requests. Furthermore, we assigned the following service load class labels to individual activity records: 0 for no calls, 1 for 1–5 calls, 2 for 6–25 calls, and 3 for more than 25 calls during DDS. Next, the evaluation process defined in Algorithm 2 was used to develop prediction accuracy rates. Random forest (RF) composed of 100 trees was used to develop individual prediction models. We use RF due to its inherent ability to select relevant features and limited need to search for hyper-parameter settings, which makes it suitable for multiple experiments with varied feature sets. For comparison purposes, Naive Bayes (NB) classifier was also used. As individual teams used a subset of services rather than all services to develop applications, in this section by accuracy rates, the $A_T^E(D, P)$ rates developed in Algorithm 2 are meant. In this way, we can focus on non-trivial combinations of applications and services i.e. these combinations for which some requests were observed. The summary of trivial and non-trivial activity records is provided in Table 1. It shows that in 99.1% no use of a service in the IS was followed by the fact that this service was not used also in the DDS, which makes prediction trivial for such cases. Hence, in accuracy calculations, activity records that did not have calls in the IS were omitted.

Table 1. Class distribution of activity records developed for $\Delta_E = 0$ and $\Delta_I = 6$

Class	0	1	2	3
Activity records $\mathbf{x}_k : r(\mathbf{x}_k) = 0$	99.10%	0.64%	0.06%	0.19%
Activity records $\mathbf{x}_k : r(\mathbf{x}_k) > 0$	51.57%	13.84%	14.47%	20.13%

In order to analyse the accuracy of the predictions we have analysed various combinations of the two key parameters of Algorithm 2 i.e. the duration of ES and the duration of IS. First of all, the accuracy of predicting the use of individual services with RF during DDS when no proper activity records, but just one

Table 2. Prediction accuracy using the method without activity records ($P = 1$)

$\Delta_E[h]$	$\Delta_I = 2$			$\Delta_I = 12$			$\Delta_I = 24$		
	Correct	Over	Under	Correct	Over	Under	Correct	Over	Under
0	40.37%	0.00%	59.63%	75.00%	1.67%	23.33%	83.42%	1.04%	15.54%
4	42.25%	12.68%	45.07%	76.27%	0.85%	22.88%	76.74%	0.00%	23.26%
8	47.17%	7.55%	45.28%	68.18%	6.82%	25.00%	71.00%	0.00%	29.00%
12	42.22%	11.11%	46.67%	56.25%	1.56%	42.19%	60.00%	0.00%	40.00%
20	46.43%	10.71%	42.86%	31.82%	2.27%	65.91%	54.35%	0.00%	45.65%

Table 3. Prediction accuracy using the method with activity records ($P = \Delta_{\mathrm{I}}$)

$\Delta_{\mathrm{E}}[h]$	$\Delta_{\mathrm{I}} = 2$			$\Delta_{\mathrm{I}} = 12$			$\Delta_{\mathrm{I}} = 24$		
	Correct	Over	Under	Correct	Over	Under	Correct	Over	Under
0 (RF)	44.95%	9.17%	45.87%	79.44%	1.67%	18.89%	88.08%	4.66%	7.25%
0 (NB)	19.27%	56.88%	23.85%	9.44%	77.22%	13.33%	2.07%	84.46%	13.47%
8 (RF)	56.60%	15.09%	28.30%	76.14%	10.23%	13.64%	81.00%	11.00%	8.00%
8 (NB)	13.21%	49.06%	37.74%	9.09%	69.32%	21.59%	8.00%	75.00%	17.00%
20 (RF)	53.57%	17.86%	28.57%	65.91%	13.64%	20.45%	73.91%	6.52%	19.57%
20 (NB)	17.86%	21.43%	60.71%	22.73%	36.36%	40.91%	17.39%	69.57%	13.04%

overall request count for $P = 1$ was used as a key input feature are reported in Table 2. The table includes both the proportion of correctly predicted service load labels, over and underestimated. It follows from the table that the load label reflecting the number of requests made to individual services during DDS is frequently underestimated, especially for longer initiation stage periods e.g. $\Delta_{\mathrm{I}} = 24$ h. The best results are reported for ES set to $\Delta_{\mathrm{E}} = 0$ h.

The impact of Δ_{E} and Δ_{I} duration on accuracy rates for the case of not using vs. using activity records is illustrated in Fig. 1a and b, respectively. Table 3 shows the accuracy rates when activity records are used. It follows from this table and Fig. 1b that by developing activity records using the method proposed in this study with $\delta_{\mathrm{I}} = 1$h and RF, the use of services during DDS can be predicted with high accuracy. For long initiation periods it can reach the level of $A_{\mathrm{T}}^{\mathrm{E}} = 88.08\%$. Importantly, since every activity record contains a number of features describing the number of requests during one hour periods of initiation stage, the models avoid underestimating resource use during DDM period. A combination of unified time scale and one hour features makes it possible for the models to identify services used at the beginning of the initiation stage, but

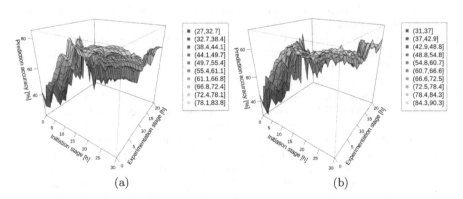

(a)   (b)

Fig. 1. Prediction accuracy when (a) activity records are not used, and (b) activity records are used. Prediction performed with RF.

no longer at the end of it, which may suggest that these services are not likely to be used in DDS period. In particular, the best accuracies are attained for $\Delta_E = 0\,h$. Hence, instead of explicitly skipping experimentation requests, we can use a model to let it identify whether first requests should be used to predict ultimate resource use or not and eliminate the search for Δ_E value. Accuracy rates observed for NB classifier are significantly lower, which shows that models incapable of skipping irrelevant features such as ES features may find it difficult to predict ultimate resource use.

4 Conclusions

Aggregated data on network traffic and requests it represents were used in the past to classify network flows. In this study, we proposed the way the stream of requests made to different services can be aggregated to provide data vectors needed for prediction purposes. First of all, we proposed the way the data describing development process composed of many disjoint periods can be mapped to a unified time scale. Moreover, we proposed how these data can be aggregated into activity records to enable the training of prediction models and the way these models can be evaluated.

In the future, the ability to predict normalised service use i.e. service use per time unit can be analysed. Secondly, prediction with other than RF techniques and at the level of service groups e.g. all map data services can be addressed.

Acknowledgment. This research has been partly supported by the European Union's Horizon 2020 research and innovation programme under grant agreement No. 688380 *VaVeL: Variety, Veracity, VaLue: Handling the Multiplicity of Urban Sensors.*

References

1. Bostani, H., Sheikhan, M.: Hybrid of anomaly-based and specification-based IDS for Internet of Things using unsupervised OPF based on MapReduce approach. Comput. Commun. **98**, 52–71 (2017)
2. Grabowski, S., Grzenda, M., Legierski, J.: The adoption of open data and open API telecommunication functions by software developers. In: Abramowicz, W. (ed.) BIS 2015. LNBIP, vol. 208, pp. 337–347. Springer, Cham (2015). https://doi.org/10.1007/978-3-319-19027-3_27
3. Hofstede, R.: Flow monitoring explained: From packet capture to data analysis with NetFlow and IPFIX. IEEE Commun. Surv. Tutor. **16**(4), 2037–2064 (2014)
4. Kim, H., Claffy, K., Fomenkov, M., Barman, D., Faloutsos, M., Lee, K.: Internet traffic classification demystified: myths, caveats, and the best practices. In: Proceedings of the 2008 ACM CoNEXT Conference, CoNEXT 2008, pp. 11:1–11:12. ACM, New York (2008). https://doi.org/10.1145/1544012.1544023
5. Otoum, S., Kantarci, B., Mouftah, H.: Adaptively supervised and intrusion-aware data aggregation for wireless sensor clusters in critical infrastructures. In: 2018 IEEE International Conference on Communications (ICC), pp. 1–6, May 2018
6. Pektaş, A., Acarman, T.: Deep learning to detect botnet via network flow summaries. Neural Comput. Appl. (2018). https://doi.org/10.1007/s00521-018-3595-x

7. Sperotto, A., Schaffrath, G., Sadre, R., Morariu, C., Pras, A., Stiller, B.: An overview of IP flow-based intrusion detection. IEEE Commun. Surv. Tutor. **12**(3), 343–356 (2010). https://doi.org/10.1109/SURV.2010.032210.00054
8. Thorsby, J., Stowers, G.N., Wolslegel, K., Tumbuan, E.: Understanding the content and features of open data portals in American cities. Gov. Inf. Q. **34**(1), 53–61 (2017)

Fuzzy Clustering Approach to Data Selection for Computer Usage in Headache Disorders

Svetlana Simić[1], Ljiljana Radmilo[2], Dragan Simić[3(✉)],
Svetislav D. Simić[3], and Antonio J. Tallón-Ballesteros[4]

[1] Faculty of Medicine, University of Novi Sad,
Hajduk Veljkova 1–9, 21000 Novi Sad, Serbia
svetlana.simic@mf.uns.ac.rs
[2] General Hospital "Dr Radivoj Simonović", 25000 Sombor, Serbia
ljiljanardml@gmail.com
[3] Faculty of Technical Sciences, University of Novi Sad,
Trg Dositeja Obradovića 6, 21000 Novi Sad, Serbia
{dsimic,simicsvetislav}@uns.ac.rs
[4] Department of Electronics, Computer System Engineering and Automation,
University of Huelva, Calle Dr. Cantero Cuadrado, 6, 21004 Huelva, Spain
antonio.tallon.diesia@zimbra.uhu.es

Abstract. This paper is focused on a new approach based on fuzzy clustering system for diagnosing headache disorders. The proposed fuzzy clustering system is based on two steps Gustafson-Kessel clustering system. Experimental data set consist of the frequency of the computer use and habits while using the computer and assessment of the adverse health effects due to computer use. The attribute selection for major features is done based on the experimental data set. The proposed fuzzy clustering system is tested on data set collected from patients in Clinical Centre of Vojvodina, in Novi Sad, Serbia.

Keywords: Attribute selection · Fuzzy clustering · Gustafson-Kessel algorithm

1 Introduction

Headache disorders are the most prevalent of all the neurological conditions and are among the most frequent medical complaints seen in general practice. More than 90% of the general population reports experiencing a headache during any given year, which is a lifetime history of head pain [1]. Daily activities related to work, education and leisure are changing due to the insertion of electronic devices into human life and society. This process is affecting all age groups because of the ever greater increase in their use of these devices [2].

Over the last decades artificial intelligence (AI) methods have gradually stepped into every area of life, but the most significant development has taken place in the field of healthcare. In recent times, data is the raw material from all information. This data and information in turn provide knowledge through modelling, data mining, analysis, interpretation, and visualization. It is not easy for a decision maker in decision-making processes to handle too much data, information and knowledge. It becomes increasingly

© Springer Nature Switzerland AG 2019
H. Yin et al. (Eds.): IDEAL 2019, LNCS 11872, pp. 70–77, 2019.
https://doi.org/10.1007/978-3-030-33617-2_8

necessary to extract useful knowledge and make scientific decisions for diagnosis and treatment of a disease. Data selection is usually divided into two phases. First, the experts – physicians select the variables from the database according to their experience, and in the second phase, the optimal number of variables is sought with the use of attribute selection method.

This paper presents the entire process of data selection for computer usage in headache disorder. They are defined with fourteen attributes. The neurologist, expert physician for diagnosing headache, classified the type of patient's headache in four groups: (1) *Migraine*; (2) *Tension-type headache* (TTH); (3) *Other primary headache disorders*; (4) *Secondary headache disorders*. By using fuzzy clustering approach based on Gustafson-Kessel clustering algorithm the proposed system creates four clusters (four groups) in which previously mentioned headache groups are classified. Also, this paper continues the authors' previous research in computer-assisted diagnosis and machine learning methods of headaches presented in papers [3–7].

The rest of the paper is organized in the following way: Sect. 2 provides related work, headache classification and attributes selection. Section 3 presents modelling the fuzzy clustering diagnosis model. Experimental results and their graphic representation are presented in Sect. 4. Section 5 provides conclusions and some points for future work.

2 Related Work, Headache Classification and Attribute Selection

Information and communication technology (ICT) has become an important part of the lives of the general population, the majority of whom regularly use computers for surfing the Internet, chatting, and playing games. At the same time, the prevalence of neck-shoulder and low back pain has increased among adolescents [8]. Studies among adolescents confirm a connection between musculoskeletal symptoms and the use of ICT, especially computers.

Headache, neck-shoulder pain and low back pain are more common among computer users than non-users. The risk of developing musculoskeletal pain increases with an increase in the amount of time spent on the computer [9]. Moreover, computer users agree that computer use causes these symptoms. The findings of several studies indicate that computer use induces pain and discomfort not only in the neck-shoulder and back area, but also in the hands, fingers, wrists, eyes, and head [10].

2.1 Headache Classification

The International Classification of Headache Disorders – The Third Edition (ICHD-3) established the uniform terminology and consistent operational diagnostic criteria for a wide range of the headache disorders [11]. The ICHD-3 provides a hierarchy of diagnoses with varying degrees of specificity, but short identification for just two important digit codes is presented in Table 1. All headache disorders are classified into two major groups: (**A**) *Primary headaches from ICHD-3 code 1. to 4.* and (**B**) *Secondary headaches ICHD-3 code from 5. to 12.*

Table 1. The international classification of headache disorders – the third edition

	ICHD-3 code	Diagnosis
A	1.	Migraine
	2.	Tension-type headache (TTH)
	3.	Trigeminal autonomic cephalalgias
	4.	Other primary headache disorders
B	5.	Secondary headache disorders
	⋮	
	12.	

When first meeting a patient, physicians who are more concerned with the detailed anamnesis and clinical examinations, apply ICHD-3 criteria and can easily establish the primary or secondary headache diagnosis.

2.2   Attribute Selection

In general, attribute selection – feature selection – feature reduction approaches are widely used in data mining and machine learning. On the other hand, feature selection could be divided on *Stochastic* and *no-Stochastic Feature Selection* methodology, a refinement of an initial stochastic feature selection task with a no-stochastic method to reduce a bit more the subset of features to be retained [12]. Therefore, feature selection plays a critical role in many domains for creating the model.

Four questionnaires were used in this study: (1) General questionnaire; (2) Scale for the assessment of the frequency of the computer use and habits while using the computer; (3) Headache questionnaire based on ICHD-3 diagnostic criteria; (4) Scale for the assessment of the adverse health effects due to computer use.

Table 2 presents two questionnaires: **(A)** Scale for the assessment of the frequency of the computer use and habits while using the computer; **(B)** Scale for the assessment of the adverse health effects due to computer use. Data selection is divided into two phases. First, the experts – physicians select the variables from the questionnaires according to their experience, and in the second phase, the optimal number of variables is sought with the use of attribute selection method and tested. Simple plus (+) denotes important selected attribute – feature by the considered method. **Bold** pluses (**+**) indicate a feature denoted as a very important (major) by the expert and a simple plus (+) indicates features that are used as an additional support in harder cases (minor). It could be concluded that the most important features are: **(1)**, **(4)**, **(5)**, **(8)**, **(9)**, **(10)**, **(11)**, **(12)**, and **(13)** which are marked with **bold** pluses (**+**).

The neurologist, expert physician for diagnosing headache, defines the type of patient's headache using the *Headache questionnaire based on ICHD-3 diagnostic criteria*. They are divided in four groups: (1) *Migraine*; (2) *Tension-type headache* (TTH); (3) *Other primary headache disorders*; (4) *Secondary headache disorders*.

Table 2. Comparison of a selection attribute for computer activity based on questionnaires: (**A**) Scale for the assessment of the frequency of the computer use and habits while using the computer; (**B**) Scale for the assessment of the adverse health effects due to computer use; (**a**) headache problems; (**b**) musculoskeletal pain symptoms; (**c**) vision disturbances

Questionnaire			Attributes	Important	Very important
1.	A		**1. I use computer since...**	+	+
2.			2. At work/In school/at university I use computer for... (how many hours a day)?	+	–
3.			3. At home, I use computer for... (how many hours a day)?	+	–
4.			**4. I use computer weekly for...** (how many hours a day)?	+	+
5.			**5. Do you take breaks after 1 h at the computer?**	+	+
6.			6. The break lasts for... (how long)?	–	–
7.			7. I use the break to ...	–	–
8.			**8. When I work at the computer my back is straight, and my hands are at the level of my elbows.**	+	+
9.	B	a	**9. Do you have a headache after work at the computer?**	+	+
10.		b	**10. Do you suffer from neck and shoulder stiffness after working on the computer?**	+	+
11.			**11. Do you have back pain after working on the computer?**	+	+
12.		c	**12. Do you suffer from vision disturbances?**	+	+
13.			**13. Do you use glasses or contact lenses?**	+	+
14.		a	14. Does your headache get worse after working more than 2 h on a computer?	+	–

3 Modelling the Fuzzy Clustering Diagnosis Model

The proposed fuzzy clustering diagnosis model for *Diagnosing Headache Disorder* implemented in this research is presented in Fig. 1. It has two phases, two steps in the Gustafson-Kessel clustering algorithm.

First phase includes: fuzzy clustering step where elements of input data set are divided in two groups. First group presents data set of *Primary headache* and the second group presents data set of *Secondary headache*. The second phase is realized also with *Gustafson-Kessel clustering algorithm*. The patients whose diagnosis undoubtedly confirms type of *Migraine* is marked in **blue bold**, *TTH* is marked in **pink bold**, *Other primary headache* is marked in **green bold**, and *Secondary headache* is marked in **black bold** on Fig. 1.

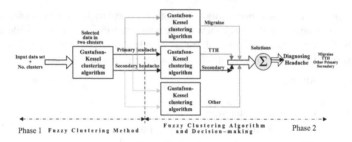

Fig. 1. A fuzzy clustering model for *Diagnosing Headache Disorder* (**Migraine** – Migraine; **TTH** – Tension Type Headache; **Other** – Other primary headaches; **Secondary** headaches) (Color figure online)

3.1 Gustafson-Kessel Clustering Algorithm

Gustafson-Kessel clustering algorithm extends the Fuzzy C-means algorithm by employing an adaptive distance norm, in order to detect clusters with different geometrical shapes in the data set. The Gustafson-Kessel (GK) algorithm associates each cluster with the cluster centre and its covariance. The main feature of Gustafson-Kessel clustering is the local adaptation of distance matrix in order to identify clusters. The objective function J_m of GK algorithm is defined in the following way [13]:

$$J_m = \sum_{j=1}^{n} \sum_{i=1}^{k} \mu_{ij}^m \, d_{ij}^2 \tag{1}$$

$$d_{ij}^2 = (x_j - z_i) A_i (x_j - z_i)^T \tag{2}$$

J_m defines the within-group sum of squared-errors i.e. it measures the total withingroup scatter or variance of X from k cluster centers $\{z_i, 1 \leq i \leq k\}$. J_m is used for finding the optimal pair (U, Z) where U denotes partitioning of the data, Z denotes the set of cluster centers of the partitions. A_i is a symmetric, positive definite matrix that induces for each cluster a norm of its own. In order to avoid singularity problem, A_i is constrained in such a way that $det (A_i) = \rho_i > 0$, ρ_i being fixed for each i permitting different sizes of cluster. The exponent $m \in (1, \infty)$ is a fuzzification parameter that controls the extent by which clusters may overlap. The objective function J_m is minimized using an alternating optimization (AO) technique. The AO optimization technique leads to the local optimum as it proceeds by fixing a set of parameters and optimizing the rest of parameters in an alternating manner. Iteratively updating in such a fashion yields the optimum value of J_m.

3.2 Data Set

The proposed fuzzy clustering system uses the data set collected from patients in Clinical Centre of Vojvodina, in Novi Sad, Serbia. A neurologist, expert physician for diagnosing headache, defined the type of patient's headache in four groups: (1) *Migraine* – 31 patients; (2) *Tension-type headache* (TTH) – 40 patients; (3) *Other primary headache disorders* – 28 patients; and (4) *Secondary headache disorders* – 37 patients. It can be concluded that 99 patients suffer of *Primary headaches* and other 37 patients suffer on *Secondary headaches*. That means that 136 patients were involved in this study and their answers are contained in two questionnaires: (A) Scale for the assessment of the frequency of the computer use and habits while using the computer; (B) Scale for the assessment of the adverse health effects due to computer use.

4   Experimental Results

The proposed fuzzy clustering system for diagnosing headache disorder, in our research, was tested on the data set collected from Clinical centre of Vojvodina, Novi Sad in Serbia. This data set is a part of large study [14]. As mentioned before, the most important features, concerning the frequency of the computer use and habits while using the computer; and the assessment of the adverse health effects due to computer use, are: (**1**), (**4**), (**5**), (**8**), (**9**), (**10**), (**11**), (**12**), and (**13**) presented in Table 2.

In the first phase of this experiment, by using Gustafson-Kessel clustering algorithm, the whole data set is divided in two groups, *Primary headaches* and *Secondary headaches*. The pairwise comparison classes (Primary – Secondary headaches) is done, and after 55 iterations, 72 instances are categorized as those who suffer from *Primary headaches* and 34 instances are categorized as those who suffer from *Secondary headaches*, which presents 78% of well classified instances. The optimal cluster centroids for *Primary headaches* is [2.01, 2.63] and for *Secondary headaches* is [1.94, 1.63] which is graphically presented in Fig. 2(a).

In the second phase the pairwise comparison classes (Migraine – Secondary headaches) is done, and after 37 iterations, 25 instances are categorized as those who suffer from *Migraine* and 30 instances are categorized as those who suffer from *Secondary headaches*, which presents 81% of well classified instances. The optimal cluster centroids for *Migraine* is [1.80, 2.73] and for *Secondary headaches* is [1.86, 1.68], which is graphically presented in Fig. 2(b). The pairwise comparison classes (TTH – Secondary headaches) is done, and after 32 iterations, 34 instances are categorized as those who suffer from *TTH* and 32 instances are categorized as those who suffer from *Secondary headaches*, which presents 84% of well classified instances. The optimal cluster centroids for *TTH* is [1.84, 2.73] and for *Secondary headaches* is [1.74, 1.58] which is presented in Fig. 2(c). The pairwise comparison classes (Other Primary headaches – Secondary headaches) is done, and after 45 iterations, 22 instances are categorized as those who suffer from *Other Primary headaches* and 31 instances are categorized as those who suffer from *Secondary headaches*, which presents 79% of well classified instances. The optimal cluster centroids for *Other Primary headaches* is [1.77, 2.87] and for *Secondary headaches* is [1.82, 1.90], which is graphically presented in Fig. 2(d).

The experimental data set which consists of the frequency, habits while using the computer and the assessment of the adverse health effects due to computer use, and implementation of two phase GK clustering system is very useful for correct categorization of every headaches type. The *overall accuracy* of this implemented system is 80%. Every experimental result can be improved; therefore, superinducing new AI techniques in existing system could improve the recent results.

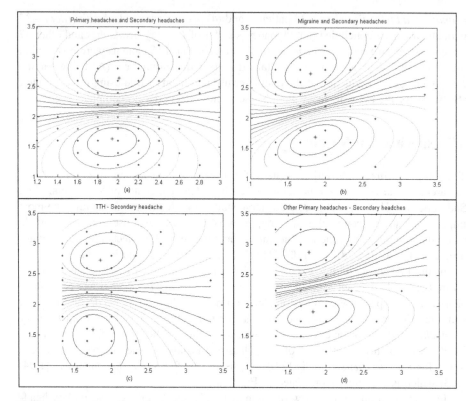

Fig. 2. Fuzzy clustering approach for *Diagnosing Headache Disorder* presented with two clusters - Pairwise comparison: **(a)** *Primary headaches* and *Secondary headaches*; **(b)** *Migraine* and *Secondary headaches*; **(c)** *TTH* and *Secondary headaches*; **(d)** *Other Primary headaches* and *Secondary headaches*

5 Conclusion and Future Work

The aim of this paper is to propose the new fuzzy clustering approach for diagnosing headache disorder. Experimental data set consist of the frequency of the computer use and habits while using the computer and the assessment of the adverse health effects due to computer use, and used for *Diagnosing Headache Disorder*. Data selection is done on the experimental data set and very important (major) features are used. The new proposed diagnosis system has of two phases, based on two step Gustafson-Kessel clustering algorithm. This system is tested on data set collected from patients in Clinical Centre of Vojvodina, in Novi Sad, Serbia. Preliminary experimental results encourage further research by the authors: because *overall accuracy* is about 80%.

Our future research will focus on creating new hybrid model to superinduce new AI evolutionary techniques which could improve the recent results. We expect the new model to be able to solve different well-known data sets and it will be tested on the real-world data sets from the Clinical centre of Vojvodina in Serbia.

References

1. Hagen, K., Zwart, J.-A., Vatten, L., Stovner, L.J., Bovin, G.: Prevalence of migraine and non-migrainous headache – head-HUNT, a large population-based study. Cephalalgia **20**(10), 900–906 (2000)
2. Costigan, A.S., Barnett, L., Plotnikoff, R.C., Lubans, D.R.: The health indicators associated with screen-based sedentary behavior among adolescent girls: a systematic review. J. Adolesc. Health **52**(4), 382–392 (2013)
3. Simić, S., Simić, D., Slankamenac, P., Simić-Ivkov, M.: Rule-based fuzzy logic system for diagnosing migraine. In: Darzentas, J., Vouros, G.A., Vosinakis, S., Arnellos, A. (eds.) SETN 2008. LNCS (LNAI), vol. 5138, pp. 383–388. Springer, Heidelberg (2008). https://doi.org/10.1007/978-3-540-87881-0_37
4. Krawczyk, B., Simić, D., Simić, S., Woźniak, M.: Automatic diagnosis of primary headaches by machine learning methods. Open Med. **8**(2), 157–165 (2013)
5. Simić, S., Banković, Z., Simić, D., Simić, S.D.: A hybrid clustering approach for diagnosing medical diseases. In: de Cos, J.F., et al. (eds.) HAIS 2018. LNCS, vol. 10870, pp. 741–752. Springer, Cham (2018)
6. Tallón-Ballesteros, A.J., Riquelme, J.C.: Tackling ant colony optimization meta-heuristic as search method in feature subset selection based on correlation or consistency measures. In: Corchado, E., Lozano, J.A., Quintián, H., Yin, H. (eds.) IDEAL 2014. LNCS, vol. 8669, pp. 386–393. Springer, Cham (2014). https://doi.org/10.1007/978-3-319-10840-7_47
7. Milutinović, D., Ćirić, E., Simić, D.: Nurses' attitudes toward computer use in healthcare and computer literacy. Pons **15**(1), 21–28 (2018)
8. Hakala, P., Rimpelä, A., Salminen, J.J., Virtanen, S.M., Rimpelä, M.: Back, neck, and shoulder pain in Finnish adolescents: national cross sectional surveys. Br. Med. J. **325**, 743–745 (2002). https://doi.org/10.1136/bmj.325.7367.743
9. Smith, L., Louw, Q., Crous, L., Grimmer-Somers, K.: Prevalence of neck pain and headaches: impact of computer use and other associative factors. Cephalalgia **29**(2), 250–257 (2008). https://doi.org/10.1111/j.1468-2982.2008.01714.x
10. Burke, A., Peper, E.: Cumulative trauma disorder risk for children using computer products: results of a pilot investigation with a student convenience sample. Public Health Rep. **117**(4), 350–357 (2002). https://doi.org/10.1016/S0033-3549(04)50171-1
11. The International Classification of Headache Disorders, 3rd edn. https://www.ichd-3.org/
12. Tallón-Ballesteros, A.J., Correia, L., Cho, S.-B.: Stochastic and non-stochastic feature selection. In: Yin, H., et al. (eds.) IDEAL 2017. LNCS, vol. 10585, pp. 592–598. Springer, Cham (2017). https://doi.org/10.1007/978-3-319-68935-7_64
13. Gustafson, D.E., Kessel, W.C.: Fuzzy clustering with a fuzzy covariance matrix. In: IEEE Conference on Decision and Control including the 17th Symposium on Adaptive Processes, pp. 761–766 (1978). https://doi.org/10.1109/cdc.1978.268028
14. Radmilo, Lj.: The effect of the computer usage on the occurrence of Primary headaches. (Uticaj upotrebe računara na pojavu primarnih glavobolja). Draft text Ph.D. theses, Faculty of Medicine, University of Novi Sad (2019)

Multitemporal Aerial Image Registration Using Semantic Features

Ananya Gupta[✉] , Yao Peng , Simon Watson , and Hujun Yin

The University of Manchester, Manchester, UK
{ananya.gupta,yao.peng,simon.watson,hujun.yin}@manchester.ac.uk

Abstract. A semantic feature extraction method for multitemporal high resolution aerial image registration is proposed in this paper. These features encode properties or information about temporally invariant objects such as roads and help deal with issues such as changing foliage in image registration, which classical handcrafted features are unable to address. These features are extracted from a semantic segmentation network and have shown good robustness and accuracy in registering aerial images across years and seasons in the experiments.

Keywords: Image registration · Semantic features · Convolutional neural networks

1 Introduction

Image registration is widely used for aligning images of the same scene. These images can be taken at different times, by different sensors and viewpoints and hence have appearance differences due to the varying imaging conditions [23]. Registration is particularly useful in remote sensing for aligning multitemporal and/or multispectral imagery for tasks such as multi-sensor data fusion and change detection [16]. It can also be used for Unmanned Aerial Vehicle (UAV) localisation by matching online UAV (query) images with the corresponding aerial (reference) images [3]. Image registration is also used in medical diagnosis for tumour monitoring or analysis of treatment effectiveness [23].

Image registration methods can be broadly divided into two categories: area-based and feature-based [23]. Area-based methods try to match corresponding patches in two images using similarity based metrics [21]. Feature-based methods, on the other hand, extract salient features from two images and then try to find corresponding features to estimate the transformation between the two images. Area-based methods are typically limited to images with different translations and/or minor rotations but cannot deal with scale variations. They also do not perform any structural analysis, hence multiple smooth areas can lead to

A. Gupta is funded by a Scholarship from the Department of Electrical and Electronic Engineering, The University of Manchester and the ACM SIGHPC/Intel Computational and Data Science Fellowship.

H. Yin et al. (Eds.): IDEAL 2019, LNCS 11872, pp. 78–86, 2019.
https://doi.org/10.1007/978-3-030-33617-2_9

incorrect correlation. Feature-based methods match more distinctive locations and are more robust. Hence, feature-based algorithms are more commonly used in registration, especially in remote sensing for matching areas with distinctive objects such as buildings and roads.

A number of popular feature-based methods are based on variants of the scale-invariant feature transform (SIFT) descriptors [8,13,17]. For instance, SAR-SIFT was developed to register synthetic aperture radar (SAR) images [4]. Fast Sample Consensus with SIFT was proposed as an improvement to random sample consensus to improve the number of correct correspondences for image registration [20]. A coarse to fine registration strategy based on SIFT and mutual information was proposed to achieve good outlier removal [9]. SIFT provides robust features in terms of translation and scale invariance. However, they can only take local appearance into account and lose global consistency.

Recently, deep learning is also being used to extract features from aerial imagery [22]. A method for combining SIFT features with Convolutional Neural Network (CNN) features has been developed for remote sensing image registration [19]. CNN features have also been used for registration of multitemporal images [18]. However, these methods were trained for image classification and do not encode fine-grained information about objects, hence they do not work well in registering high resolution images with fine details.

In this paper, we focus on registering multitemporal high resolution nadir aerial images that have large variations due to changing seasons, lighting conditions, etc. We propose to use semantic features extracted from a segmentation network for the purpose of aerial image registration. These segmentation-based semantic features (SegSF), as compared to handcrafted and classification-based CNN features, are more finely localised and more discriminative for multitemporal registration.

2 Methodology

The proposed methodology comprises of two main steps, SegSF extraction and class-specific feature matching, detailed as follows.

2.1 Segmentation-Based Semantic Feature Extraction

A LinkNet34 [2] network trained for road segmentation with aerial images as the input and binary road masks as the output is used for SegSF extraction. The network structure is shown in Fig. 1. The network is trained with a pixel-wise loss function defined by the binary cross entropy between predicted value and ground truth road mask. It provides a probability mask as output, which is converted to a binary road mask with a threshold of 0.5, so any pixels with a probability greater than 0.5 are regarded as road pixels and vice versa.

The features from the output of the 'Decoder3' block shown in Fig. 1 are extracted as descriptors. Additionally, keypoint locations on the image are given

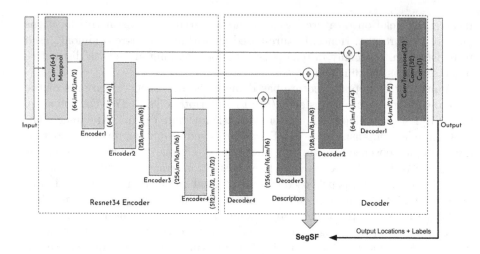

Fig. 1. LinkNet34 architecture for SegSF extraction. Each block in the encoder down-samples the image by a factor of 2 and each corresponding decoder block upsamples the image as shown by the block output sizes in the figure. The image width and height are each denoted by *im*.

by the centers of the effective receptive field of the descriptors and can be calculated using Eqs. 1–4.

$$j_l = j_{l-1} * s_l \tag{1}$$

$$rf_l = rf_{l-1} + (k_l - 1) * j_{l-1} \tag{2}$$

$$start_l = start_{l-1} + \left(\frac{k_l - 1}{2} - p_l\right) * j_{l-1} \tag{3}$$

$$outputLoc_l = featureLoc_l * rf_l + start_l \tag{4}$$

where subscripts l and $l-1$ are layer indices, s is the stride, p gives the padding size, k is the convolution kernel size and j is the "jump" or the effective stride for each layer as compared to the input. So the jump for the first layer is the same as its stride. The effective receptive field has size of rf and $start$ is the center coordinate of the receptive field of the first feature. Corresponding output locations ($outputLoc$) and hence labels can then be assigned to feature descriptors using Eq. 4, where $featureLoc_l$ indexes over all features in the feature map output from layer l.

In practice, since the encoder is a ResNet34 [6], the values of padding, kernel and stride are chosen so that the effective jump only changes between two encoder blocks and the $start$ value is always 0.5.

SegSF features are defined by three components: class label, descriptor and keypoint location. Each descriptor-keypoint pair is assigned a class label based on the location of the keypoint on the segmentation output, hence adding semantic knowledge to the feature descriptor.

2.2 Feature Matching

All SegSF descriptors are L2-normalised individually and their dimensionality is reduced to 100 using class-specific PCA, further followed by L2 normalisation to obtain the final descriptor.

Matching class descriptors between the query and reference images are found using nearest neighbour search in the per-class descriptor space where Euclidean distance is used as the distance metric. The correctness of the extracted correspondences is estimated using Lowe's ratio test [8], where the match is assumed to be correct if the distance ratio between the first neighbour and the second neighbour is less than 0.7.

The keypoints of the feature matches are then used to estimate the homography matrix between the two images using random sample consensus [5]. This feature matching process has been given as Algorithm 1.

Algorithm 1. SegSF matching

    Inputs: Query Image, Reference Image *query_image, ref_image*
    Output: Transformation Model *Transform_Model*
1: *all_ref_kp* = []
2: *all_query_kp* = []
3: **for** *class* in *num_classes* **do**
4:     *query_class_kp, query_class_des* = GetClassFeatures(*class, query_image*)
5:     *ref_class_kp, ref_class_des* = GetClassFeatures(*class, ref_image*)
6:     *query_class_des*.NormalisePCANormalise
7:     *ref_class_des*.NormalisePCANormalise
8:     *inlier_query_kp, inlier_ref_kp* = KNN(*query_des, ref_des*)
9:     *all_ref_kp*.append(*inlier_ref_kp*)
10:    *all_query_kp*.append(*inlier_query_kp*)
11: **end for**
12: *Transform_Model* = Affine TransformRansac(*all_ref_kp, all_query_kp*)

---

3 Experimental Setup

3.1 Datasets

The segmentation network was trained on images from different seasons and years to learn temporally invariant features. Aerial imagery datasets provided by the Australian Capital Territory Government [1] for the years 2015–2018 were used for this purpose. The images in this dataset are georeferenced, orthorectified and have a ground sampling distance of 10 cm with an expected error less than 20 cm. An area of 27.5 km^2 around Canberra was extracted for the experiments, with 90% of the images from 2015, 2016 and 2017 being used for training and validation of the segmentation network. The remaining 10% images from 2017 and all images from 2018 were set aside for testing.

The annotations for training were extracted from the OpenStreetMap (OSM) [10] where all polylines marked as one of motorways, primary, residential, secondary, service, tertiary and trunk and their respective links were assumed to be roads. The OSM roads were provided in vector format and were rasterised with a width of 2 m to obtain data suitable for training the segmentation network.

3.2   Training Details

The segmentation network was trained on image crops of 416×416 from the training dataset of 2600 images. It was trained for 200 epochs with the images being augmented by random horizontal and vertical flipping with a probability of 0.5. All image pixel values were normalised between 0 and 1. The Adam optimiser [7] with a learning rate of 0.0001 was used for optimisation. Pytorch [11] was used for creating and training the neural networks.

3.3   Testing Scheme and Metrics

The test images from 2017 were rotated around their center point by angles of $1°$, $2°$, $3°$, $4°$, $5°$, $10°$, $15°$, $20°$, $30°$ and $40°$. Corresponding images from the 2018 dataset were extracted and the methods were tested on their accuracy for registering these multitemporal images. Samples of the test pairs can be seen in Fig. 2. Note that the network was not trained on any images from the 2018 dataset or on any images from the testing region.

Since the image transformation parameters were known, a per-pixel metric has been reported for the experiments. The root mean squared error (RMSE) between the pixel positions using the predicted transformation and the actual transformation was used as the error metric. The mean values of the errors over all the images for the different transformations are reported in Sect. 4.

4   Results

We compared our results with that of CNN-Reg [18] which is based on extracting multi-scale CNN features for multitemporal remote sensing image registration. They utilise features from a VGG-16 [14] network trained on the ImageNet dataset [12] for classification. We have also reported the t-values and p-values from Welch's t-test [15] to obtain the significance of the results.

Table 1. Registration RMSE of CNN-Reg and SegSF on multi-temporal aerial imagery dataset over different rotation parameters

Method	1°	2°	3°	4°	5°	10°	15°	20°	30°	40°
SegSF	37.03	37.33	41.93	46.43	61.51	70.08	90.63	131.91	184.06	372.39
CNN-Reg	88.05	143.51	213.18	242.806	287.13	425.63	522.5	596.35	715.13	777.53
t-value	5.11	8.38	10.87	11.62	12.41	17.82	22.92	20.79	24.21	10.37
p-value	1.53e−6	3.07e−12	3.81e−18	5.55e−19	9.59e−22	8.58e−28	5.68e−37	8.53e−38	6.09e−38	1.46e−14

Reference Image	Query Image	CNN-Reg	SegSF

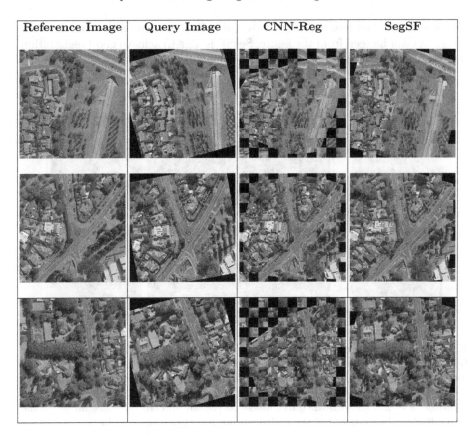

Fig. 2. Checkerboard results of registering multitemporal images using CNN-Reg and SegSF

As can be seen from the results in Table 1, SegSF performs much better than CNN-Reg for all the transformation parameters. The p-values for the tests show that the results are extremely significant.

We believe that the performance improvement can be explained by two factors. Firstly, since CNN-Reg is based on ImageNet, it extracts universal patterns for feature matching. It does not train specifically on any aerial images but finetuning on aerial datasets, as in the case of SegSF, can provide more relevant features. However, their method cannot be finetuned with our dataset since it trains for a classification-based loss and our dataset does not contain any classes. Secondly, the addition of the semantic labels makes SegSF less sensitive to image variations. Note that both methods struggle when the rotation is increased to 20° and beyond.

Checkerboards of some of the images registered using these two methods are shown in Fig. 2. The query images are from 2017 and the reference images are from 2018 and the images have varied foliage. The quality of the registration can be estimated by how well the roads and buildings align on the checkerboard image. As can be seen from the figure, SegSF is able to achieve good registration results even with foliage difference.

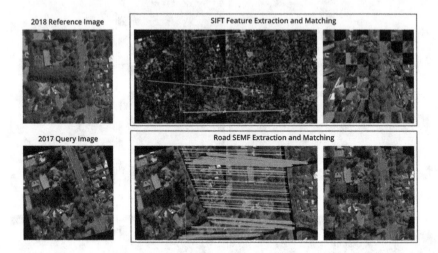

Fig. 3. Matching comparison between SegSF and SIFT features. *Left:* Reference image from 2018 and query image from 2017 rotated by 15° with colour and shadow variations due to different times of day and year. *Middle:* All keypoints extracted by the two methods are shown as black circles with the correspondences estimated by RANSAC shown with blue lines with the query image on the left and the reference image on the right. *Right:* Checkerboard of the registered images. (Color figure online)

We have also compared SegSF with SIFT features, as shown in Fig. 3. As can be seen, SIFT features do not deal well with seasonal variations. Conversely, SegSF features match correctly even with large seasonal variations. Only the road features have been used in this case for the SegSF to demonstrate the relevance of semantically meaningful features.

5   Conclusions

We have proposed a semantic feature extraction method for multitemporal aerial image registration to deal with variations such as changing tree foliage caused due to changing seasons, changing shadows due to different times of day, and pixel variations caused by different sensors. The features are extracted from a CNN trained for segmentation and are conditioned on the output class. We provide both quantitative and qualitative results from experiments and draw comparisons to previous work.

Our results show that the proposed features achieve better localisation precision due to the fine resolution allowed by the use of a segmentation network as compared to a classification CNN. The features are also capable of handling temporal variations that classical features, such as SIFT, struggle with.

This work is limited in that it is only applicable to images with visible roads. However, this can be improved by increasing the number of classes for feature extraction. Further improvements in the registration accuracy can be achieved by extracting features from later layers in the network which have smaller receptive fields. Multi-scale features can also be used for dealing with scale differences.

References

1. ACTmapi Aerial Imagery. www.ACTMapi.act.gov.au
2. Chaurasia, A., Culurciello, E.: LinkNet: exploiting encoder representations for efficient semantic segmentation. In: 2017 IEEE Visual Communications and Image Processing, VCIP 2017, vol. 2018, pp. 1–4, June 2018
3. Costea, D., Leordeanu, M.: Aerial image geolocalization from recognition and matching of roads and intersections, pp. 118.1–118.12, May 2017
4. Dellinger, F., Delon, J., Gousseau, Y., Michel, J., Tupin, F.: SAR-SIFT: a SIFT-like algorithm for SAR images. IEEE Trans. Geosci. Remote. Sens. **53**(1), 453–466 (2015)
5. Fischler, M.A., Bolles, R.C.: Random sample consensus: a paradigm for model fitting with applications to image analysis and automated cartography. Commun. ACM **24**(6), 381–395 (1981)
6. He, K., Zhang, X., Ren, S., Sun, J.: Deep residual learning for image recognition. arXiv.org **7**(3), 171–180 (2015)
7. Kingma, D.P., Ba, J.: Adam: a method for stochastic optimization. In: International Conference on Learning Representations (2015)
8. Lowe, D.G.: Distinctive image features from scale-invariant keypoints. Int. J. Comput. Vis. **60**(2), 91–110 (2004)
9. Gong, M., Zhao, S., Jiao, L., Tian, D., Wang, S.: A novel coarse-to-fine scheme for automatic image registration based on SIFT and mutual information. IEEE Trans. Geosci. Remote. Sens. **52**(7), 4328–4338 (2013)
10. OpenStreetMap Contributors: Planet dump (2017). https://planet.osm.org
11. Paszke, A., et al.: Automatic differentiation in PyTorch (2017)
12. Russakovsky, O., et al.: ImageNet large scale visual recognition challenge. Int. J. Comput. Vis. **115**(3), 211–252 (2015)
13. Sedaghat, A., Mokhtarzade, M., Ebadi, H.: Uniform robust scale-invariant feature matching for optical remote sensing images. IEEE Trans. Geosci. Remote Sens. **49**(11), 4516–4527 (2011)
14. Simonyan, K., Zisserman, A.: Very deep convolutional networks for large-scale image recognition. In: International Conference on Learning Representations, pp. 1–14 (2015)
15. Welch, B.L.: The generalization of 'Student's' problem when several different population variances are involved. Biometrika **34**(1–2), 28–35 (1947)
16. Dai, X., Khorram, S.: The effects of image misregistration on the accuracy of remotely sensed change detection. IEEE Trans. Geosci. Remote. Sens. **36**(5), 1566–1577 (1998)
17. Yang, K., Pan, A., Yang, Y., Zhang, S., Ong, S., Tang, H.: Remote sensing image registration using multiple image features. Remote Sens. **9**(6), 581 (2017)

18. Yang, Z., Dan, I., Yang, Y.: Multi-temporal remote sensing image registration using deep convolutional features. IEEE Access **6**, 38544–38555 (2018)
19. Ye, F., Su, Y., Xiao, H., Zhao, X., Min, W.: Remote sensing image registration using convolutional neural network features. IEEE Geosci. Remote Sens. Lett. **15**(2), 232–236 (2018)
20. Yue, W., Ma, W., Gong, M., Linzhi, S., Jiao, L.: A novel point-matching algorithm based on fast sample consensus for image registration. IEEE Geosci. Remote Sens. Lett. **12**(1), 43–47 (2015)
21. Zagoruyko, S., Komodakis, N.: Learning to compare image patches via convolutional neural networks. In: The IEEE Conference on Computer Vision and Pattern Recognition (CVPR) (2015)
22. Zhang, L., Zhang, L., Du, B.: Deep learning for remote sensing data: a technical tutorial on the state of the art. IEEE Geosc. Remote Sens. Mag. **4**(2), 22–40 (2016)
23. Zitová, B., Flusser, J.: Image registration methods: a survey. Image Vis. Comput. **21**(11), 977–1000 (2003)

Special Session on Machine Learning in Healthcare

Brain Tumor Classification Using Principal Component Analysis and Kernel Support Vector Machine

Richard Torres-Molina[1] , Carlos Bustamante-Orellana[1] ,
Andrés Riofrío-Valdivieso[1] , Francisco Quinga-Socasi[1] ,
Robinson Guachi[3,4] , and Lorena Guachi-Guachi[1,2(✉)]

[1] Yachay Tech University, Hacienda San José, Urcuquí 100119, Ecuador
lguachi@yachaytech.edu.ec
[2] SDAS Research Group, Urcuquí, Ecuador
[3] Department of Mechatronics, Universidad Internacional del Ecuador, Av. Simon Bolivar, 170411 Quito, Ecuador
[4] Department of Mechanical and Aerospace Engineering, Sapienza University of Rome, Via Eudossiana 18, 00184 Rome, RM, Italy
http://www.sdas-group.com

Abstract. Early diagnosis improves cancer outcomes by giving care at the most initial possible stage and is, therefore, an important health strategy in all settings. Gliomas, meningiomas, and pituitary tumors are among the most common brain tumors in adults. This paper classifies these three types of brain tumors from patients; using a Kernel Support Vector Machine (KSVM) classifier. The images are pre-processed, and its dimensionality is reduced before entering the classifier, and the difference in accuracy produced by using or not pre-processing techniques is compared, as well as, the use of three different kernels, namely linear, quadratic, and Gaussian Radial Basis (GRB) for the classifier. The experimental results showed that the proposed approach with pre-processed MRI images by using GRB kernel achieves better performance than quadratic and linear kernels in terms of accuracy, precision, and specificity.

Keywords: KSVM · Brain tumor classification · Image processing

1 Introduction

In medical image analysis, there are some important applications as automatic segmentation [1] and automatic detection of abnormal (malignant) tumors from MRI. The automatic brain cancer detection approaches typically work in three stages: pre-processing, feature extraction and classification, each one performing specific techniques to handle the critical situation as low quality, redundant and non-informative data, and low accuracy to categorize high-dimensional input data, respectively. The pre-processing stage usually involves reducing noise and

© Springer Nature Switzerland AG 2019
H. Yin et al. (Eds.): IDEAL 2019, LNCS 11872, pp. 89–96, 2019.
https://doi.org/10.1007/978-3-030-33617-2_10

enhancing images with traditional techniques as intensity normalization, histogram equalization, morphological operators, and image binarization. Features extraction gives a set of discriminative and informative data such as texture, auto-correlation, contrast, correlation, energy, dissimilarity, entropy, homogeneity, among others. Feature extraction methods usually are followed by dimensionality reduction methods to reduce the number of extracted features of an image at the minimum possible without losing too much information. This process is done to reduce computational time, storage memory, and provide the classifier only with the most important features about an image so that the classifier can offer a fast answer with reasonable accuracy. The classification step allows predicting the class/category/label of given data features from pixel images. In this sense, new approaches have emerged with the use of SVM [2], KSVM [3], among others.

Notably, among the wide range of machine learning techniques for classification tasks, Support Vector Machine (SVM) has proved to have higher accuracy and to need a lesser amount of data in biomedical image processing [2, 4]. In this sense, several algorithms by using SVM models have been adopted to build more accurately approaches through the use of pre-processing operations to enhance visual quality and perceptibility of the anomalies present in the brain magnetic resonance imaging (MRI). For instance, a SVM classification method to predict whether a patient has a brain tumor or not is presented in [2]. Besides, linear, quadratic, and polynomial kernels for SVM are evaluated, where the quadratic kernel function obtained the highest performance. In [4], SVM and Hybrid classifiers (SVM-KNN) were used for brain cancer classification; the MRI image has been converted into a grayscale and enhanced by using filters to remove the adjacent noise, then skull masking (a combination of dilation and erosion operations) was carried out to remove non-brain tissues from MRI image. This process was essential to achieve high accuracy levels in diagnosis and predictive procedures in clinical applications. An intelligent system was proposed in [5]. The pre-processing step is focused on noise removing and a segmentation technique, through a threshold, to find the tumor. Multi-SVM, with the optimization of its parameters through a genetic algorithm, is used as a classifier to improve accuracy and efficiency.

Although several existing works provide accuracy results obtained with a different set of techniques applied in each stage, there are two common issues: (i) Acquired MRI images often contain undesirable information that degrades image data; therefore, the accuracy of the approach depends mainly on the set of pre-processing techniques and (ii) There are some machine learning classifiers that should be adjusted to the application data in order to improve the accuracy achieved.

To deal with the problems mentioned above, the presented approach proposes a set of pre-processing techniques to enhance the quality of the acquired image for having dissimilarity between different types of brain conditions fundamentally. Then, SVM will be exploited to classify the most common brain tumors in adults: meningioma, glioma, and pituitary tumors from MRI images [6]. To determine the optimal classifier settings, SVM classifier is evaluated using three different kernels named linear, quadratic, and GRB. The kernel strategy is applied to SVMs due to their ability to construct a hyperplane in data classification at different sides of a hypersurface. The different kernels were chosen as the typical ones in the SVM algorithm. So, the input for the classifier are features extracted with Discrete Wavelet Transform (DWT) and reduced with PCA.

2 Materials and Methods

2.1 Dataset

The dataset was obtained from Nanfang Hospital at Guangzhou, China, and General Hospital at Tianjin Medical University, China [7]. The images have a resolution of 512 × 512 and a pixel size of 0.49 × 0.49 mm^2. For this work, the sample size is compounded by 105 images randomly selected from the dataset, 35 for each type of tumor (meningioma, glioma and pituitary). Then, from the selected images, 93 were used for training and 12 for validation purposes, with an equal number of images for each type of tumor.

2.2 Method

The computational flow of the proposed methodology is schematized in Fig. 1. The obtained images were enhanced before putting them into the classifier due to the low quality they had. Besides, a comparison between the results with processed and unprocessed images was performed to determine the improvement in accuracy when using processed images.

Image Pre-processing: Four procedures are applied to the images. Firstly, median filtering in a 3-by-3 neighborhood around the central pixel is performed for noise removal. Then, a sharpening filter is used to mitigate the smoothes and obtain an image free of impulsive noise. Second, histogram equalization is performed to increase the global contrast of the given image and distribute the intensities on the histogram to get a contrast-enhanced image. Third, the skull masking process is done to separate the brain tumor from the rest of the brain. It consists of two erosions and one dilation with a structure of 5 × 5. Finally, the images resulting from the skull masking are converted into binary to segment the brain tumor by using a threshold of 0.9.

Feature Extraction - 2D Discrete Wavelet Transform: It performs three levels of wavelet decomposition through 2D Discrete Wavelet Transform (DWT) by using Eq. (1), to reduce the image size and, therefore, reduce the time of computation.

2D DWT is the result of applying DWT [3] to all the rows and columns of the image separately and decomposing the result in four columns that are subband

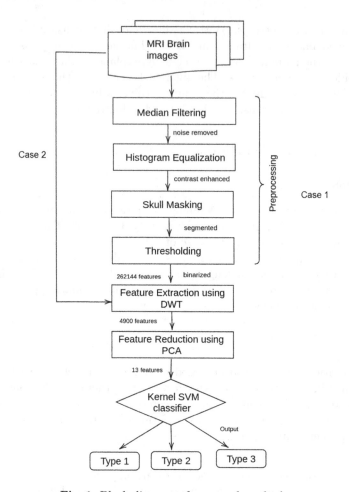

Fig. 1. Block diagram of proposed method.

images low-low (LL), low-high (LH), high-low (HL), and high-high (HH) which cover all the spectrum of the input image in the frequency domain. 2D DWT is used in processed (segmented binary images) and unprocessed (raw MRI images obtained from the data set) images.

$$\psi(t) = \begin{cases} 1 \text{ if } 0 \leq t < \frac{1}{2} \\ -1 \text{ if } \frac{1}{2} \leq t < 1 \\ 0 \text{ otherwise} \end{cases} \tag{1}$$

Feature Reduction - Principal Components Analysis (PCA): PCA is applied to the image obtained after applying 2D DWT to reduce the excessive number of features of the image that often slow down the algorithm's speed, retaining only the ones with the most information and variations. After applying

PCA from the resulting image, a matrix is formed with the 13 characteristics presented in Table 1.

Table 1. Characteristics used as input for the classifier.

Contrast	Correlation	Energy
$\sum_{i,j=0}^{G-1} \cdot (i-j)^2$	$\frac{\sum_{i,j=0}^{G-1}(i-\mu)\cdot(j-\mu)\cdot g(i,j)}{\sigma^2}$	$\sum_{i,j=0}^{G-1} g(i,j)^2$
Homogeneity	**Mean** (μ)	**Standard Deviation** (σ)
$\sum_{i,j=0}^{G-1} P(i,j)^2$	$\frac{1}{M_g N_g}\sum_{i,j=0}^{G-1} P(i,j)$	$\sqrt{\frac{1}{M_g N_g}\sum_{i,j=0}^{G-1}(P(i,j)-\mu)^2}$
Entropy	**RMS contrast**	**Variance**
$\sum_{i,j=0}^{G-1} P(i,j)\cdot log(P(i,j))$	$\sqrt{\frac{1}{MN}\sum_{i,j=0}^{M-1,N-1}((i,j)-\bar{I})^2}$	$\sum_{i,j=0}^{G-1}(i-\mu)^2 P(i,j)$
Smoothness	**Kurtosis**	**Skewness**
$1-(\frac{1}{1+\sum_{i,j=0}^{G-1} P(i,j)})$	$\frac{\sum_{i,j=0}^{G-1}(P(i,j)-\mu)^3}{M_g N_g \sigma^3}$	$\frac{\sum_{i,j=0}^{G-1}(P(i,j)-\mu)^4}{M_g N_g \sigma^4}$
	IDM	
	$\sum_{i,j=0}^{G-1}(i-\mu)^2 P(i,j)$	

where: M, N are the rows and column of the image. (i, j) are value of the ith and j^{th} position in the matrix of size MxN. M_g, N_g are the rows and column of the GLCM. $g(i,j)$ is the value of the ith and j^{th} position in the matrix GLCM of size $M_g x N_g$. $\sum_{i,j=0}^{G-1} = \sum_{i=0}^{M_g-1}\sum_{j=0}^{N_g-1}$. $P(i,j)$ is the probabilities calculated for values in GLCM and N is the size of GLCM. $\bar{I}$ is the average intensity of all pixel of the MxN matrix in the range $[0,1]$.

Kernel SVM Classifier (KSVM): KSVM classifier receives as input the matrix with 13 characteristics obtained in the previous stage and gives as result a classification for a type of tumor (glioma, meningiona or pituitary tumor). Binary SVMs learn decision function $f(.; w, b) : X \to \mathbb{R}$ defined by $f(x; w, b) =< w, \phi(x) > +b$. Let Y now be a set of $t > 2$ classes, which represents the tumor type classes. We now considered t output functions, one for each class, that quantify the confidence in the corresponding prediction. The output function for a class y ϵ Y is given by Eq. (2).

$$f_y(x; w, b) =< w, \phi(x, y) > +b_y \tag{2}$$

with $b = (b_1, ..., b_t)$. Thus, the predicted class $\hat{y}$ for a point x is chosen to maximize the confidence in the prediction, $\hat{y} = argmax_{y\epsilon Y} f_y(x; w, b)$. Training amounts to finding parameters w and b that lead, at least to large extent, to correct predictions, satisfying Eq. (3).

$$f_{yi}(x_i; w, b) > f_u(x_i; w, b) \tag{3}$$

for the training data points (x_i, y_i). Training can be implemented by the following optimization problem:

$$min_{w,b}\frac{1}{2}\parallel w \parallel^2 +C\sum_{i=1}^{n} max_{u\neq y_i}\{l(f_{y_i}(x_i; w, b) - f_u(x_i; w, b))\} \tag{4}$$

where n is the number of examples in the training set, C is the regularization parameter, and it is written $u \neq y_i$ short for $u \epsilon Y - \{y_i\}$.

The transform $\psi(x_i)$ is related to the kernel by the equation: $k(x_i, x_j) = \psi(x_i)\psi(x_j)$. w is in the transformed space, with $w = \sum_i \alpha_i \gamma_i \psi(x_i)$. The dot product can be computed by $w \cdot \psi(x) = \sum_i \alpha_i \gamma_i k(x_i, x)$. The three common kernels Linear [8], Quadratic [9], and GRB [10] applied to SVM are depicted in Table 2.

Table 2. Three common Kernels (Linear, Quadratic, and GRB) with their formula.

Name	Formula
Linear	$k(x_i, x_j) = (x_i \cdot x_j)$
Quadratic	$k(x_i, x_j) = (x_i \cdot x_j + 1)^2$
Gaussian Radial Basis (GRB)	$k(x_i, x_j) = exp(-\gamma \|x_i - x_j\|^2)$

2.3 Experimental Design

Two experimental tests are performed to evaluate the overall performance of the SVM classifier with three kernels (Linear, Quadratic, and GRB) in predicting brain cancer. In the first one (called Case 1), we use the output of the pre-processing step as the input for the feature extraction step. In the second experiment (called Case 2), we use raw images (not pre-processed images) from the dataset as the input for feature extraction step. So, in Case 1 the classifier is trained and tested with raw images, whereas in Case 2 the classifier is trained and tested with pre-processed images. Such procedure is done with the objective of comparing the difference in accuracy obtained when the images are processed before entering the classifier. The overall performance has been evaluated in terms of Precision (The proportion of predicted positive cases that were identified as correct), Specificity (The proportion of negative cases that are correctly identified), and Accuracy (The percentage of well-classified images cases over the total ones) by using Eqs. (5), (6) and (7), respectively. Where **TN** is the number of correct predictions that indicates the instance is negative, **FP** is the number of incorrect predictions that infers an instance as positive, **FN** is the number of incorrect of predictions that confirms the negative instance, and **TP** is the number of correct predictions that verifies the positive instance.

$$Precision = \frac{TP}{TP + FP} \tag{5}$$

$$Specificity = \frac{TN}{TN + FP} \tag{6}$$

$$Accuracy = \frac{TP + TN}{TP + TN + FP + FN} \tag{7}$$

3   Results and Discussion

For tests purposes, libraries such as pca, svmtrain, and svmclassify to learn and execute the portion of machine learning algorithms; and the image processing functions for the pre-processing stage provided by MATLAB are used in the implemented MATLAB software routines (Available in [11]). Results obtained from experimental tests on the validation set are depicted in Table 3. It can be observed that the proposed method obtains better accuracy for both study cases with GRB kernel. In Case 2, the highest accuracy of 61.11% is achieved by Linear and RGB kernels. Overall, Case 2 scored 57.41% of accuracy. This low accuracy is because the input validation data consists of raw MRI images. These images have low contrast which makes difficult to recognize the structures they contain. On the contrary, Case 1 outperformed Case 2. In this case, contrast and quality enhanced images were used. Case 1 holds an overall validation accuracy of 83.33%, precision of 81.85%, and Specificity of 87.50%. GRB kernel exceeds the performance of the other two kernels because the GRB kernel is based on exponential function, which can enlarge the distance between samples to the extent that Linear or Quadratic kernels cannot reach. Therefore, the idea of applying the RGB kernel to other industrial field is very attractive. It is clear that a pre-possessing stage (Case 1) improves the prediction capabilities significantly.

Table 3. Average accuracy, precision and specificity values.

Case study	Kernel	Accuracy	Precision	Specificity
Case 2 (Raw images)	Linear	61.11%	42.86%	70.83%
	Quadratic	50.00%	42.86%	62.50%
	GRB	61.11%	48.33%	70.83%
Average		57.41%	44.68%	68.05%
Case 1 (Pre-processed images)	Linear	83.33%	80.00%	87.50%
	Quadratic	77.78%	76.67%	83.33%
	GRB	88.89%	88.89%	91.67%
Average		83.33%	81.85%	87.50%

4   Conclusion

The use of tools like DWT and PCA has allowed to reduce the initial size of the used images significantly and extract only their most essential features. Besides, the use of DWT and PCA has proved to be more effective with the processed images, producing smaller enhanced images that preserve the form of the tumor and contain the most of the variance in the first six principal components.

This research work involves using KSVM to classify the input MRI image into meningioma, glioma, and pituitary tumor images without (Case 2) and with (Case 1) an image pre-processing stage. Overall accuracy, Case 2 = 57.41%, while Case 1 holds a validation accuracy of 83.33% shows the effectiveness of the two models. Besides, in all kernel types, we conclude that an image pre-processing stage before classification concerns increase the classification accuracy crucially.

As future work, the proposed model could be improved with pre-processing techniques of automatic parameterization. To improve the classification performance, new features could be added upon implementing more data and types of the common brain tumors in adults.

References

1. Guachi, L.: Automatic colorectal segmentation with convolutional neural network. Comput. Aided Des. Appl. **16**(5), 836–845 (2019)
2. Nandpuru, H.B., Salankar, S.S., Bora, V.R.: MRI brain cancer classification using support vector machine. In: 2014 IEEE Students' Conference on Electrical. Electronics and Computer Science, Bhopal, pp. 1–6. IEEE (2014)
3. Zhang, Y.D.: An MR brain images classifier via principal component analysis and kernel support vector machine. Prog. Electromagn. Res. **130**, 369–388 (2012)
4. Machhale, K., Nandpuru, H.B., Kapur, V., Kosta, L.: MRI brain cancer classification using hybrid classifier (SVM-KNN). In: 2015 International Conference on Industrial Instrumentation and Control (ICIC), Pune, pp. 60–65. IEEE (2015)
5. Keerthana, T.K., Xavier, S.: An intelligent system for early assessment and classification of brain tumor. In: 2018 Second International Conference on Inventive Communication and Computational Technologies (ICICCT), Coimbatore, pp. 1265–1268. IEEE (2018)
6. National Brain Tumor Society. https://braintumor.org/brain-tumor-information/understanding-brain-tumors/tumor-types/. Accessed 26 June 2019
7. Figshare. https://figshare.com/articles/brain_tumor_dataset/1512427/5. Accessed 26 June 2019
8. Dong, J., Suen, C.Y., Krzyzak, A.: Effective shrinkage of large multi-class linear SVM models for text categorization. In: 2008 19th International Conference on Pattern Recognition, Tampa, pp. 1–4. IEEE (2008)
9. Dagher, I.: Quadratic kernel-free non-linear support vector machine. J. Glob. Optim. **41**(1), 15–30 (2008)
10. Abd-Ellah, M.K., Awad, A.I., Khalaf, A.A.M., Hamed, H.F.A.: Classification of brain tumor MRIs Using a kernel support vector machine. In: Li, H., Nykänen, P., Suomi, R., Wickramasinghe, N., Widén, G., Zhan, M. (eds.) WIS 2016. CCIS, vol. 636, pp. 151–160. Springer, Cham (2016). https://doi.org/10.1007/978-3-319-44672-1_13
11. Github. https://github.com/RichardTorres2017/BrainTumorClassification. Accessed 16 July 2019

Modelling Survival by Machine Learning Methods in Liver Transplantation: Application to the UNOS Dataset

David Guijo-Rubio[1(✉)], Pedro J. Villalón-Vaquero[1], Pedro A. Gutiérrez[1], Maria Dolores Ayllón[2], Javier Briceño[2], and César Hervás-Martínez[1]

[1] Department of Computer Sciences, Universidad de Córdoba, Córdoba, Spain
{dguijo,i52vivap,pagutierrez,chervas}@uco.es
[2] Unit of Hepatobiliary Surgery and Liver Transplantation, Córdoba, Spain
lolesat83@hotmail.com, javibriceno@hotmail.com

Abstract. The aim of this study is to develop and validate a machine learning (ML) model for predicting survival after liver transplantation based on pre-transplant donor and recipient characteristics. For this purpose, we consider a database from the United Network for Organ Sharing (UNOS), containing 29 variables and 39,095 donor-recipient pairs, describing liver transplantations performed in the United States of America from November 2004 until June 2015. The dataset contains more than a 74% of censoring, being a challenging and difficult problem. Several methods including proportional-hazards regression models and ML methods such as Gradient Boosting were applied, using 10 donor characteristics, 15 recipient characteristics and 4 shared variables associated with the donor-recipient pair. In order to measure the performance of the seven state-of-the-art methodologies, three different evaluation metrics are used, being the concordance index (*ipcw*) the most suitable for this problem. The results achieved show that, for each measure, a different technique obtains the highest value, performing almost the same, but, if we focus on *ipcw*, Gradient Boosting outperforms the rest of the methods.

Keywords: United Network for Organ Sharing · Liver transplant · Survival analysis · Machine learning

1 Introduction

The Survival Analysis (SA) is a field traditionally tackled by statistical methods, aiming to model the data where the outcome is the time until the occurrence of an event of interest. One important characteristic of SA is that, for some of the instances, the outcomes are unobservable since these are no longer monitored or the study has finished previous to the occurrence of the event of interest; these instances are known as censored instances. Note that most of the SA books introduce the topic from a pure statistical point of view [1,2].

© Springer Nature Switzerland AG 2019
H. Yin et al. (Eds.): IDEAL 2019, LNCS 11872, pp. 97–104, 2019.
https://doi.org/10.1007/978-3-030-33617-2_11

In this paper, SA is applied to liver transplantation, which is an accepted treatment for patients with end-stage chronic liver disease. There are some previous works where SA is used for liver transplant. For example, Hong et al. [3] applied SA techniques to the Canadian Organ Replacement Registry analysing the 1-year-survival based on subsets of the database. Furthermore, Abolghasemi et al. [4] analysed SA methods applied to liver cirrhosis patients that were followed up for at least 5 years, concluding that bilirubin, INR, creatinine and white blood cell variables were effective for prediction. Martínez et al. [5] assessed the accuracy of the BAR score for liver transplant for predicting 3-months, 1-year and 5-years mortality. On the other hand, many machine learning (ML) algorithms have been recently adapted to handle survival data [6], using SVM [7] or Gradient boosting [8], among others. Up to our knowledge, ML methods have not been applied to model liver transplant survival.

In this way, the main objective of this study is to analyse the behaviour of SA techniques in large liver transplant databases, using ML methods. The dataset considered is the one provided by the United Network for Organ Sharing (UNOS) organisation. The opportunity to work with this huge database including thousands of donor-recipient pairs is crucial to ensure the applicability of ML classifiers in predicting liver transplantation survival.

The rest of the paper is organised as follows: Sect. 2 describes the UNOS database used for this study, then, in Sect. 3, all the methods are detailed, as well as the metrics used for measuring the performance. Section 4 exposed the results obtained and Sect. 5 closes the paper.

2   United Network for Organ Sharing Database

UNOS [9] is an organisation that administers the organ procurement and transplantation network in the United States of America (USA). This organisation manages the biggest database regarding transplants and organ sharing in the USA, with more than 9 kinds of transplants by organ type among which liver transplant is the second most popular, with 8250 cases in 2018, representing almost a 22% of the transplants made in the USA. The origin of the database dates back to January 1988, although the last update was made in June 2019. Specifically, we are considering a subset of the database from November 2004 until June 2015. This decision was made given that some variables changed in November 2004, and June 2015 is the date when we received the database.

This database contains more than 200,000 records and more than 380 different variables, including variables of the donor, the recipient and the transplant process. From this original database, a cleaning step was carried out because many of the pairs corresponded to patients in the waiting list and could not be used. Moreover, we have excluded from this study those transplants made with partial or split liver, those in which the donor was alive and those in which there were more than one organ transplanted. Furthermore, we removed all transplants with patients who were less than 18 years old. Regarding the variables, initially, all those variables with the same values information were deleted, as well as, those whose information was not updated. On the other hand, following expert knowledge, some variables such as the gender of both donor and recipient were unified (gender matching in this case) in order to keep valuable information and to reduce the huge amount of information.

After this pre-processing, the database contained 39,095 pairs donor-recipient and 29 variables. Finally, for recovering missing values, we used the mean for continuous variables, while, for categorical variables, the mode was used. It is considered that, for most of the transplants, the graft is in good conditions in the last follow-up. Thus, these instances have not suffered the event of interest and are censored patterns. This leads to, approximately, a 74% of censored instances. Furthermore, note that the mean survival time for uncensored patterns is 1.94 years, whereas, the mean censored time is 3.95 years. In conclusion, the difficulty of the problem is high, due to both, the high percentage of censored patterns and the high censored time.

3 Methodology

In this section, the basis of the SA are described as well as the metrics for measuring the performance of the methods. Finally, the different SA approaches used are briefly explained.

3.1 Survival Analysis

SA is a branch of statistics where the main objective is to model data where the output is the time until the occurrence of an event of interest (in our case, liver graft failure).

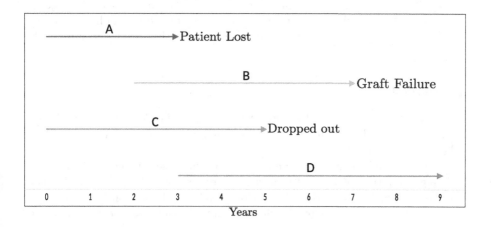

Fig. 1. Survival analysis samples for a 9-years-study.

Figure 1 shows a 9-years-duration study with 4 subjects where the event of interest is the graft failure: subjects A and C abandoned the study at the third and fifth years, respectively, without registering the occurrence; subject B ended with a graft failure at the seventh year (registering the event of interest), and, finally, subject D did not registered any occurrence until the end of the study. Thus, subjects A, C and D are censored instances, whereas, only for B, the time in which the event has occurred can be obtained.

Due to these censored instances, standard ML evaluation metrics are not suitable for SA. Furthermore, the outcomes are usually relative risks given by risk scores of arbitrary scale. Both facts make that specific metrics need to be used for measuring the quality of the models:

- The concordance index (c-index) [10] is defined as the proportion of all comparable pairs in which the predictions and outcomes are concordant.

$$\text{c-index} = \frac{1}{N} \sum_{i:\delta_1=1} \sum_{j:y_i<y_j} I[\hat{r}(\mathbf{X}_i) > \hat{r}(\mathbf{X}_j)], \tag{1}$$

where $\mathbf{X}_i$ the donor-recipient pair, $\hat{r}_i$ is the predicted risk, δ_i is a binary variable which is 1 if the instance had suffered the event of interest (0, otherwise), y_i is the observed time (survival or censored time, depending on δ_i), N is the total number of comparable pairs and $I[\cdot]$ is the indicator function.
- The concordance index ipcw [11] is an alternative to c-index, reducing the bias for datasets with a high amount of censoring. It is based on the inverse probability of censoring weights, that is, it depends of the distribution of censoring times.
- The cumulative/dynamic AUC [12] is based on the Receiver Operating Characteristic (ROC) curve and the Area Under the ROC curve (AUC) extended to survival data by defining sensitivity and specificity as time-dependent measures. The associated AUC quantifies how well the model can distinguish between subjects failing at a given time ($t_i \leq t$) and subjects failing after this time ($t_i > t$). In our case, we consider the mean AUC after evaluating 100 time points.

3.2   Methods

The SA models can be divided into three groups: (1) Cox's-regression-based models, (2) models based on Gradient Boosting and (3) adaptations of the SVM methodology to the SA.

Regarding Cox's-regression-based models, Coxnet uses an elastic net penalty [13] being a mixture of lasso (L1) and ridge regression (L2) penalties. On the other hand, CoxPH, uses a ridge regression penalty with Newton-Raphson optimiser and Breslow method to handle ties. Both models assume the following regression:

$$h(t, \mathbf{X}_i) = h_0(t) \exp(\mathbf{X}_i, \boldsymbol{\beta}), i \in \{1, 2 \ldots N\}, \tag{2}$$

where the base risk $h_0(t)$ is an arbitrary non-negative time function, $\mathbf{X}_i = (x_{i1}, x_{i2}, ..., x_{ik})$ are the variables considered, and $\boldsymbol{\beta}^T = (\beta_1, \beta_2, ..., \beta_k)$ is the vector of coefficients associated to the models. The inference of the model is made optimising the partial likelihood of Cox's proportional hazards model [14].

The accelerated failure time model [15] (IPCRidge) assumes a different regression form, and each sample is weighted by the inverse probability of censoring. It makes the assumption that censoring is independent of the features, i.e., censoring is random. It is defined as:

$$\log y = \beta_0 + \mathbf{X}\boldsymbol{\beta}^T + \epsilon, \tag{3}$$

where $L2$ penalty is applied to the coefficient vector $\boldsymbol{\beta}$ and ϵ is an error variable.

The second group is related to Gradient Boosting [16]. It is a ML technique combining *weak* models in order to build a *robust* model. In this paper, two based learners are considered:

- Component-wise least squares, which performs a linear least squares regression against the predictor variable which reduces more the residual sum of squares.
- Regression tree, which is a non-parametric supervised learning method used for classification and regression. The goal is to create a model that predicts the value of a target variable by learning simple decision rules inferred from the data features.

Given that the database has a high percent of censored instances, the loss function based in Cox's proportional hazards model is used, as it adapts better to the complexity of the problem. The partial likelihood of Cox's proportional hazards model [14] is defined:

$$L(\beta) = \prod_{i=1}^{N} \frac{\exp(\mathbf{X}_{j(i)}\boldsymbol{\beta})}{\sum_{j \in R_i} \exp(\mathbf{X}_j \boldsymbol{\beta})}, \tag{4}$$

where R_i is the set of instances for which $y_j \geq t_i$.

Finally, the third group consider survival support vector machines (SSVM) [17], which are closely related to RankSVM [18], combining regression and a ranking objective. It assumes some constraints for comparing censored instances, being only possible to compare them with non-censored instances with less survival time than the censoring time of the previous uncensored pattern. That is, a comparable pair is given by $P = [(i,j)|y_i > y_j \wedge \delta_j = 1]_{i,j=1}^n$. The following optimisation problem is solved during the training of SSVM:

$$\min_{\mathbf{w}} \frac{1}{2} \|\mathbf{w}\|_2^2 + \gamma \sum_{(i,j) \in P} \max(0, 1 - \mathbf{w}^T(\mathbf{X}_i - \mathbf{X}_j)), \tag{5}$$

where w are the coefficients of the model and $\gamma > 0$ is a user-defined parameter that weights the objective functions. We used two approaches, the first uses a linear SSVM (FastSurvivalSVM), whereas, the second one uses kernels (FastKernelSurvivalSVM).

4   Experiments and Results

This section shows the experimental settings and the results obtained using the dataset and methodology described in Sects. 2 and 3, respectively.

4.1   Experimental Settings

Regarding the experimental settings, a 5-fold cross-validation method is applied to the training set. To adjust the Cox's model with elastic net, the elastic net mixing parameter is set in the range $l1\_ratio \in \{0.05, 0.20, 0.50, 0.80, 1.00\}$, the number of alphas for the regularisation path is adjusted as $n\_alphas \in \{50, 112, 175, 237, 300\}$, and the tolerance takes values in $tol \in \{10^{-15}, 10^{-10}, 10^{-5}, 10^{-2}, 10^{0}\}$. Regarding the CoxPH model using L2 penalty, the regularisation parameter is set as $alpha \in \{0.00, 0.25, 0.50, 0.75, 1.00\}$, whereas the maximum number of iterations and the tolerance are set as $n\_iter \in \{50, 100, 150\}$ and $tol \in \{10^{-15}, 10^{-10}, 10^{-5}, 10^{-2}, 10^{0}\}$, respectively. For the IPCRidge model, the regularisation parameter takes values in the range $alpha \in \{-5.0, -1.0, -0.8, -0.5, -0.2, 0.0, 0.2, 0.5, 0.8, 1.0, 5.0\}$, as these values improve the conditioning of the problem and reduce the variance of the estimates. To adjust both gradient boosting models, Eq. (4) is used as loss function to be optimised, the number of boosting stages is set as $n\_estimators \in \{50, 87\}$, the learning rate and the dropout rate for each base learner are set based on $learning\_rate \in \{0.1, 0.5, 0.9\}$ and $dropout\_rate \in \{0.0, 0.1, 0.5, 0.9\}$, and the fraction of samples to be used for fitting the individual base learner takes values as $subsample \in \{0.01, 0.25, 0.50, 0.75, 1.00\}$. Finally, only for GradientBoosting model, the maximum depth of the individual estimators is set tas $max\_depth \in \{5, 10\}$. Regarding SSVM models, for the weight for penalising the squared hinge loss, $alpha \in \{0.001, 0.250, 0.500, 0.750, 1.000\}$, and, for the mixing parameter between regression and ranking objective, $rank\_ratio \in \{0.00, 0.25, 0.50, 0.75, 1.00\}$. Finally, RBF and polynomial kernels are considered for FastKernelSurvivalSVM models.

4.2   Results

The results obtained by the application of the SA techniques proposed in Sect. 3.2 are in Table 1, using the three performance evaluation metrics described in Sect. 3.1. In this Table, it can be seen that for each measure, a different technique is obtaining the highest value. For instance, in the case of c-index, Coxnet is reaching to the best performance (0.5906), and to the second one for ipcw and AUC, 0.5668 and 0.6078, respectively. On the other hand, the other model based on Cox's regression, CoxPH, achieves the best value in terms of AUC (0.6079) and the second best value in terms of c-index (0.5905). Finally, focusing on the ipcw, the best value is reached by Gradient Boosting, 0.5751.

These results show that the high degree of censoring of the database (almost 74% of the instances are censored) is a clear handicap for the problem. Most of the transplants made in the last 15 years survive more than 10 years, and, when the study finishes, the event of interest has not happened yet. However, GradientBoosting is the best method for this dataset, obtaining the more balanced results for ipcw and AUC.

Table 1. SA results obtained for the UNOS database.

Model	c-index	ipcw	AUC
Coxnet	$\mathbf{0.5906}_{\mathbf{0.0070}}$	$0.5668_{0.0193}$	$0.6078_{0.0078}$
CoxPH	$0.5905_{0.0069}$	$0.5665_{0.0194}$	$\mathbf{0.6079}_{\mathbf{0.0078}}$
IPCRidge	$0.4843_{0.0070}$	$0.5082_{0.0136}$	$0.4457_{0.0095}$
ComponentwiseGB	$0.5774_{0.0067}$	$0.5560_{0.0197}$	$0.5895_{0.0064}$
GradientBoosting	$0.5863_{0.0062}$	$\mathbf{0.5751}_{\mathbf{0.0151}}$	$0.5983_{0.0056}$
FastKernelSurvivalSVM	$0.4831_{0.0082}$	$0.5043_{0.0211}$	$0.5019_{0.0105}$
FastSurvivalSVM	$0.4793_{0.0062}$	$0.5008_{0.0155}$	$0.4616_{0.0099}$

5   Conclusions

This study considers the application of ML techniques in SA for a dataset containing more than $200,000$ records, provided by UNOS between the years 2004 and 2015. The data is collected by individual transplant centres, in such a way that a post-processing step is applied to the database, obtaining finally $39,095$ donor-recipient pairs, which can be considered a sufficiently large database from which worldwide applicability can be obtained. Furthermore, the database contains more than a 74% of censored patterns, which makes the problem very challenging. The results obtained are acceptable with respect to the three used measures, and Gradient Boosting stands out as the best method regarding the *ipcw* measure. On the other hand, Coxnet and CoxPH are able to obtain the more balanced results.

As future improvements, the application and adaptation of classical ML methods to the deep learning methodology would possibly bring a significant performance increase, as done in [19], where the Cox proportional hazards method is adapted to deep neural networks successfully.

Acknowledgement. This research has been partially supported by the Ministerio de Economía, Industria y Competitividad of Spain (Refs. TIN2017-90567-REDT and TIN2017-85887-C2-1-P). D. Guijo-Rubio's research has been supported by the FPU Predoctoral Program from Spanish Ministry of Education and Science (Grant Ref. FPU16/02128).

References

1. Kleinbaum, D.G., Klein, M.: Survival Analysis, vol. 3. Springer, New York (2010). https://doi.org/10.1007/978-1-4419-6646-9

2. Allison, P.D.: Survival analysis using SAS: a practical guide. SAS Institute (2010)
3. Hong, Z., et al.: Survival analysis of liver transplant patients in Canada 1997–2002. In: Transplantation Proceedings, vol. 38, no. 9, pp. 2951–2956. Elsevier, 2006 November
4. Abolghasemi, J., Toosi, M.N., Rasouli, M., Taslimi, H.: Survival analysis of liver cirrhosis patients after transplantation using accelerated failure time models. Biomed. Res. Ther. 5(11), 2789–2796 (2018)
5. Martínez, J.A., et al.: Accuracy of the BAR score in the prediction of survival after liver transplantation. Ann. Hepatol. 18(2), 386–392 (2019)
6. Wang, P., Li, Y., Reddy, C.K.: Machine learning for survival analysis: a survey. ACM Comput. Surv. (CSUR) 51(6), 110 (2019)
7. Kiaee, F., Sheikhzadeh, H., Mahabadi, S.E.: Relevance vector machine for survival analysis. IEEE Trans. Neural Netw. Learn. Syst. 27(3), 648–660 (2015)
8. Wang, Z., Wang, C.Y.: Buckley-James boosting for survival analysis with high-dimensional biomarker data. Stat. Appl. Genet. Mol. Biol. 9(1), 1–33 (2010)
9. Organ Procurement and Transplantation Network. United Network for Organ Sharing database (2019). https://unos.org
10. Harrell, F.E., Califf, R.M., Pryor, D.B., Lee, K.L., Rosati, R.A.: Multivariable prognostic models: issues in developing models, evaluating assumptions and adequacy, and measuring and reducing errors. Stat. Med. 15(4), 361–387 (1996)
11. Uno, H., Cai, T., Pencina, M.J., D'Agostino, R.B., Wei, L.J.: On the C-statistics for evaluating overall adequacy of risk prediction procedures with censored survival data. Stat. Med. 30(10), 1105–1117 (2011)
12. Lambert, J., Chevret, S.: Summary measure of discrimination in survival models based on cumulative/dynamic time-dependent ROC curves. Stat. Methods Med. Res. 25(5), 2088–2102 (2014)
13. Simon, N., Friedman, J., Hastie, T., Tibshirani, R.: Regularization paths for Cox's proportional hazards model via coordinate descent. J. Stat. Softw. 39(5), 1–13 (2011)
14. Cox, D.R.: Regression models and life tables (with discussion). J. R. Stat. Soc. Ser. B 34, 187–220 (1972)
15. Stute, W.: Consistent estimation under random censorship when covariables are present. J. Multivar. Anal. 45(1), 89–103 (1993)
16. Friedman, J.: Greedy function approximation: a gradient boosting machine. Ann. Stat. 29(5), 1189–1232 (2001)
17. Pölsterl, S., Navab, N., Katouzian, A.: Fast training of support vector machines for survival analysis. In: Appice, A., Rodrigues, P.P., Santos Costa, V., Gama, J., Jorge, A., Soares, C. (eds.) ECML PKDD 2015. LNCS (LNAI), vol. 9285, pp. 243–259. Springer, Cham (2015). https://doi.org/10.1007/978-3-319-23525-7_15
18. Chapelle, O., Keerthi, S.S.: Efficient algorithms for ranking with SVMs. Inf. Retr. 13(3), 201–215 (2010)
19. Katzman, J.L., Shaham, U., Cloninger, A., Bates, J., Jiang, T., Kluger, Y.: DeepSurv: personalized treatment recommender system using a Cox proportional hazards deep neural network. BMC Med. Res. Methodol. 18(1), 24 (2018)

Design and Development of an Automatic Blood Detection System for Capsule Endoscopy Images

Pedro Pons[1], Reinier Noorda[2(✉)], Andrea Nevárez[3], Adrián Colomer[1], Vicente Pons Beltrán[3], and Valery Naranjo[1]

[1] Instituto de Investigación e Innovación en Bioingeniería (I3B), Universitat Politècnica de València, Valencia, Spain
adcogra@i3b.upv.es

[2] iTEAM Research Institute, Universitat Politècnica de València, Valencia, Spain
reinoo@upv.es

[3] Unidad de Endoscopia Digestiva, Hospital Universitari i Politécnic La Fe, Digestive Endoscopy Research Group, IIS La FE, Valencia, Spain

Abstract. Wireless Capsule Endoscopy is a technique that allows for observation of the entire gastrointestinal tract in an easy and non-invasive way. However, its greatest limitation lies in the time required to analyze the large number of images generated in each examination for diagnosis, which is about 2 h. This causes not only a high cost, but also a high probability of a wrong diagnosis due to the physician's fatigue, while the variable appearance of abnormalities requires continuous concentration. In this work, we designed and developed a system capable of automatically detecting blood based on classification of extracted regions, following two different classification approaches. The first method consisted in extraction of hand-crafted features that were used to train machine learning algorithms, specifically Support Vector Machines and Random Forest, to create models for classifying images as healthy tissue or blood. The second method consisted in applying deep learning techniques, concretely convolutional neural networks, capable of extracting the relevant features of the image by themselves. The best results (95.7% sensitivity and 92.3% specificity) were obtained for a Random Forest model trained with features extracted from the histograms of the three HSV color space channels. For both methods we extracted square patches of several sizes using a sliding window, while for the first approach we also implemented the waterpixels technique in order to improve the classification results.

Keywords: Wireless capsule endoscopy · Blood detection · Machine learning · Hand-crafted features · Deep learning · Convolutional neural networks

© Springer Nature Switzerland AG 2019
H. Yin et al. (Eds.): IDEAL 2019, LNCS 11872, pp. 105–113, 2019.
https://doi.org/10.1007/978-3-030-33617-2_12

1   Introduction

1.1   Motivation

Currently, physicians have multiple techniques and instruments at their disposal to diagnose the many diseases that affect the human gastrointestinal tract (GI tract). Traditional endoscopy techniques enables access to some of the GI tract areas with both diagnostic and therapeutic purposes. However, they share the same limitation: they do not allow observation of the complete small intestine. This greater section of the GI tract is only accessible through different, more invasive techniques such as push enteroscopy or intraoperative endoscopy.

Wireless Capsule Endoscopy (WCE), first introduced in 2001, does allow for minimally invasively observation of the entire GI tract. The patient only has to swallow a pill-sized camera (Fig. 1), which goes through the GI tract driven by peristaltic movements. The camera, installed on one side of the capsule, captures 2 to 6 images per second, while a LED light source illuminates the scene.

One of the main problems when using WCE is the large number of video frames generated per exam (over 150.000). Physicians spend up to two hours reviewing these, resulting in not only a high cost, but also a high probability of a wrong diagnosis due to fatigue. This situation is aggravated by the variable appearance of abnormalities and that they sometimes only appear in a single or few frames, requiring high concentration.

This paper focuses on blood detection in WCE images recorded by the PillCam^{TM} SB 3 capsule, as presence of blood in the bowel is a symptom of many diseases such as polyps, tumors, ulcers or Crohn's disease. Therefore, blood detection is often a priority in analysis of WCE procedures. Even though RAPID^{TM} Reader, a software provided by the PillCam^{TM} manufacturer, contains an automatic blood detection tool, it cannot be used as a reliable tool for diagnosis, since several studies claim it has both a low specificity and a low recall [3,8].

1.2   Literature Review

Many researchers have taken up task of automatically detecting visible abnormalities in the GI tract, following different approaches. Regarding the type of features that are used, some authors try combining color features and texture

Fig. 1. Endoscopic capsule PillCam SB3.

features, although color features have been proven to be much more discriminative [5]. Most authors explore different color spaces aiming to find the features
that best differentiate between blood and healthy tissue. While many researchers
use the RGB color space because of its simplicity, it can be problematic as WCE
images usually have an uneven illumination, which drastically affects the three
RGB channels [1]. In contrast, in the HSV color space the inhomogeneous light-
ning only affects the V channel, which makes it a popular alternative.

There are also different approaches regarding the strategies for processing the
image. Pixel-level methods are the simplest and fastest ones, but they ignore the
existence of spatial information, usually returning poor and incoherent results.
Image-level methods benefit from spatial information, but fail to detect the
smallest anomalies, whose features are masked by those from the healthy tissue. Patch-level methods are an intermediate step between the previous ones.
While bigger patches are able to grasp more spatial information, smaller patches
tend to report better sensitivity values [7].

1.3 Objectives

In this work, we intend to develop an automatic blood detection system to reduce
the time needed to review WCE videos, following two different approaches. The
first consists in extracting "hand-crafted" features from the images and training machine learning algorithms capable of differentiating between blood and
healthy tissue. The second consists in deep learning, concretely in training convolutional neural networks (CNNs), capable of extracting the needed features by
themselves. Finally, we compared the results obtained with each procedure.

2 Materials and Methods

2.1 Data Collection

The available dataset consisted of 75 WCE images obtained with a
PillCam^{TM}SB3 capsule, all provided by the Digestive Endoscopy Center from
Hospital Universitari i Politécnic La Fe (Valencia). A total of 40 of these con-
tained blood, while the remaining 35 contained only healthy tissue.

2.2 Data Preparation

First, we manually created the ground truth of the images, i.e. the true target of
each pixel in the images (Fig. 2). Next we extracted square patches of different
sizes (32 × 32, 64 × 64 and 96 × 96 pixels) from each of the images in the data
set, using sliding windows with 50% overlap in both the horizontal and vertical
axis. The label of each patch was determined at the same time, labeling as blood
only those of which the corresponding area in the ground truth was at least 10%.

Both to reduce the dependency of the results on the way the data were split
and to improve the generalization of the models, we performed "nested cross-
validation" [10]. This technique consists in training several models on different

subsets of the data, through both an external and an internal cross validation loop. We used 5 folds in each loop. In each fold, a different partition of the set of all areas was used for testing, while the remaining areas were used for training. Results of the models were averaged to obtain a single value for each performance metric. In the external loop, we ensured that each whole image participated only in a single partition and never in the corresponding training set. In the internal loop, we again partitioned the training data into different validation sets, using one such partition as validation and the remaining areas for training as before. This loop we used for optimizing the hyperparameters of our models. Additionally, each training set in the external loop was balanced by down-sampling the majority class.

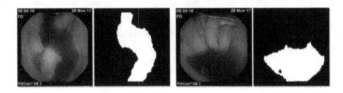

Fig. 2. Pixel-level ground truth of the images in the data set.

2.3   Hand-Crafted Feature Extraction

In our first approach, following the work done by other researchers, and after investigating whether either the LBP or the HOG histograms showed capacity to distinguish between the two classes, we chose to use only color features. Specifically, we used the RGB and HSV color spaces. We also tried to reduce the inhomogeneous illumination problem by converting the RGB images into the HSV space, applying a homomorphic filter to the V channel and converting the images back to RGB (Fig. 3). Concretely, the following features were tested to determine which of them behave better in our problem: histograms of all RGB channels, histograms of all HSV channels, histograms of the H and S channels of HSV and histograms of all RGB channels after our illumination correction.

After reviewing some of the most popular machine learning algorithms, we chose for Support Vector Machines (SVM) [4] and Random Forest [2] in this work because of their good performance in our early experiments, relative simplicity and resistance to overfitting. Regarding SVM models, we compared the performance of a simple lineal kernel and an RBF kernel, which usually outperforms the others and is relatively easy to implement. Regarding the Random Forest models, we estimated a convenient number of trees by observing that the out-of-bag error stayed practically constant from 80 trees onward.

The approach followed so far consisted in extracting features from square patches extracted from the images. However, the sliding window used for this purpose did not take the content of the image into account, thus sometimes capturing content of both classes in a single patch. In order to overcome this problem, we replaced this technique with a superpixel segmentation technique.

Superpixels are image regions with similar features, of which the borders tend to adjust to those of the objects in the image. Specifically, we used a variant based on the watershed technique, called waterpixels [6], which has been reported to obtain better results than other popular methods [2]. One of the most interesting aspects of waterpixels is the freedom of choosing a criterion for selecting the points from which the waterpixels start growing. In this work, we used the minimum gradient points. In case several minima existed in the same cell, the one with the highest surface extinction coefficient prevailed.

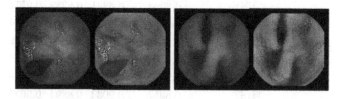

Fig. 3. WCE images after applying the homomorphic filtering.

2.4 Deep Learning

Our second approach was based on CNNs, capable of learning good features to extract, thus avoiding potential loss of relevant information due to hand-crafted, human feature selection. The CNN models were trained on exactly the same aforementioned subsets of our data as the previous algorithms: we thus trained five models for each patch size for their performance metrics to be averaged at the end. The CNNs were trained parting from an VGG19 [9] model that was pre-trained on the ImageNet data set, through a technique known as "fine-tuning". This consists in freezing weights of the first layers of a pre-trained model, while optimizing those in the last layers to adapt the model to the new problem.

3 Results

Performance test results for all of our trained models are reported here for each of the different region types (patches of different sizes or water pixels). Regarding the SVM models, only results obtained with an RBF kernel are shown (Table 1), since we obtained significantly superior results compared to the linear kernel. Results for Random Forest and CNN models are shown in Tables 2 and 3 respectively.

4 Discussion

Generally, we obtained worst results for histograms of the RGB channels directly extracted from the original image. However, if they were extracted after applying a homomorphic filter, the performance was significantly enhanced, with the results approaching those obtained for HSV. We also observed that the results

Table 1. Average test results for SVM models with an RBF kernel.

Processed Areas	Features	Accuracy	Recall	Spec.	AUC
32 × 32 pixels	RGB	0,9221	0,9137	0,9279	0,9685
	HSV	0,9051	0,9090	0,9030	0,9589
	HS	0,9175	0,9260	0,9128	0,9718
	RGB-filtered	0,9236	0,9181	0,9274	0,9709
64 × 64 pixels	RGB	0,9045	0,8587	0,9353	0,9566
	HSV	0,8968	0,8931	0,8982	0,9571
	HS	0,9132	0,9162	0,9104	0,9694
	RGB-filtered	0,9148	0,8880	0,9326	0,9689
96 × 96 pixels	RGB	0,8497	0,7954	0,8906	0,9303
	HSV	0,8817	0,9026	0,8658	0,9549
	HS	0,8957	0,9034	0,8894	0,9637
	RGB-filtered	0,8991	0,8524	0,9339	0,9640
Waterpixels	RGB	0,9257	0,9485	0,9132	0,9752
	HSV	0,9086	0,9526	0,8851	0,9734
	HS	0,9471	0,9119	0,8919	0,9688
	RGB-filtered	0,9325	0,9500	0,9258	0,9769

Table 2. Average test results obtained for random forest models.

Processed Areas	Features	Accuracy	Recall	Spec.	AUC
32 × 32 pixels	RGB	0,9266	0,9206	0,9313	0,9756
	HSV	0,9252	0,9363	0,9198	0,9806
	HS	0,9238	0,9372	0,9169	0,9802
	RGB-filtered	0,9292	0,9342	0,9275	0,9787
64 × 64 pixels	RGB	0,9276	0,9189	0,9343	0,9769
	HSV	0,9318	0,9483	0,9210	0,9828
	HS	0,9316	0,9443	0,9231	0,9822
	RGB-filtered	0,9365	0,9357	0,9386	0,9823
96 × 96 pixels	RGB	0,9273	0,9225	0,9317	0,9760
	HSV	0,9378	0,9568	0,9233	0,9844
	HS	0,9357	0,9549	0,9209	0,9841
	RGB-filtered	0,9389	0,9364	0,9424	0,9849
Waterpixels	RGB	0,9058	0,9448	0,8835	0,9701
	HSV	0,9062	0,9614	0,8750	0,9791
	HS	0,8966	0,9612	0,8621	0,9777
	RGB-filtered	0,9073	0,9548	0,8873	0,9770

Table 3. Average test results obtained for the CNN models.

Patches	Accuracy	Recall	Spec.	AUC
32 × 32 pixels	0,9496	0,9390	0,9560	0,9901
64 × 64 pixels	0,9494	0,9208	0,9688	0,9917
96 × 96 pixels	0,8943	0,7545	0,9838	0,8920

obtained for HS were very similar to, or sometimes even better than, those for all HSV channels. A possible reason for this is that the V channel, despite containing additional information, also brings in illumination problems.

In Fig. 4, where the best results of each classifier are compared, we can see that CNN models appear best regarding accuracy and AUC, but when considering recall, which is considered the most important metric in this study due to false negatives having worse consequences for diagnosis than false positives, Random Forest and SVM were significantly better. Using those recall values to compare the results obtained using different patch sizes, we observe that generally, greater patches resulted in lower recall. However, Random Forest models seem to be unaffected by changes in the patch size, since we can only observe differences in the order of hundreds and even appear slightly better for greater patches. For waterpixels we generally obtained higher recall than for patches, but while for SVM we also obtained similar or even slightly higher accuracy, both accuracy and specificity were significantly lower when using Random Forest.

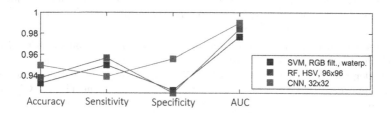

Fig. 4. Comparison of the best results from each type of model.

Figure 5 shows the results obtained on some of our test images. From each image we extracted patches (bottom row) or waterpixels (top row), which were classified and then combined to obtain a pixel-level classification of the image. Here we observe that the region labeled as blood appears to be more accurate pixel-wise when using waterpixels, as it does not produce block artifacts.

One of the greatest limitations found during this study was the lack data. The absence of a public database forces researchers to search for new images and get them manually labelled, which is a slow and tedious task. The pixel-wise segmentation obtained by our algorithm could be used as a labelling tool for easier creation of ground truth images in the future.

Waterpixels

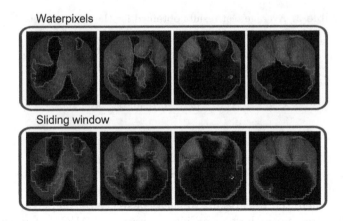

Sliding window

Fig. 5. Comparison of detected regions using waterpixels (top) and patches (bottom).

5   Conclusion

We trained different models capable of automatically detecting blood in WCE images, following both classical machine learning and deep learning approaches. The color features we found to be most useful in this work were the histograms of the H and S channels from the HSV color space and the three RGB channels after applying a homomorphic filter to correct illumination variances.

Of the patch-based models, we obtained greatest recall (95,68%) for a Random Forest model based on HSV histograms, with 92,33% specificity. Additionally, we found that using waterpixels detected areas had visually better borders.

Acknowledgments. This work was funded by the European Union's H2020: MSCA: ITN program for the "Wireless In-body Environment Communication - WiBEC" project under the grant agreement no. 675353. Additionally, we gratefully acknowledge the support of NVIDIA Corporation with the donation of the Titan V GPU used for this research.

References

1. Berens, J., Finlayson, G.D., Qiu, G.: Image indexing using compressed colour histograms. IEE Proc. Vis., Image Signal Process. **147**(4), 349–355 (2000). https://doi.org/10.1049/ip-vis:20000630
2. Breiman, L.: Random forests. Mach. Learn. **45**(1), 5–32 (2001). https://doi.org/10.1023/A:1010933404324
3. Buscaglia, J.M., et al.: Performance characteristics of the suspected blood indicator feature in capsule endoscopy according to indication for study. Clin. Gastroenterol. Hepatol. **6**(3), 298–301 (2008). https://doi.org/10.1016/j.cgh.2007.12.029
4. Cortes, C., Vapnik, V.: Support-vector networks. Mach. Learn. **20**(3), 273–297 (1995). https://doi.org/10.1007/BF00994018

5. Li, B., Meng, M.Q.H.: Computer-aided detection of bleeding regions for capsule endoscopy images. IEEE Trans. Biomed. Eng. **56**(4), 1032–1039 (2009). https://doi.org/10.1109/TBME.2008.2010526
6. Machairas, V., Faessel, M., Cárdenas-Peña, D., Chabardes, T., Walter, T., Decencière, E.: Waterpixels. IEEE Trans. Image Process. **24**(11), 3707–3716 (2015). https://doi.org/10.1109/TIP.2015.2451011
7. Novozámský, A., Flusser, J., Tachecí, I., Sulík, L., Bureš, J., Krejcar, O.: Automatic blood detection in capsule endoscopy video. J. Biomed. Opt. **21**(12), 126007 (2016). https://doi.org/10.1117/1.JBO.21.12.126007
8. Signorelli, C., Villa, F., Rondonotti, E., Abbiati, C., Beccari, G., de Franchis, R.: Sensitivity and specificity of the suspected blood identification system in video capsule enteroscopy. Endoscopy **37**(12), 1170–1173 (2005). https://doi.org/10.1055/s-2005-870410
9. Simonyan, K., Zisserman, A.: Very deep convolutional networks for large-scale image recognition. arXiv preprint arXiv:1409.1556 (2014)
10. Varma, S., Simon, R.: Bias in error estimation when using cross-validation for model selection. BMC Bioinform. **7**(1), 91 (2006). https://doi.org/10.1186/1471-2105-7-91

Comparative Analysis for Computer-Based Decision Support: Case Study of Knee Osteoarthritis

Philippa Grace McCabe[1]([✉]) (ID), Ivan Olier[1] (ID),
Sandra Ortega-Martorell[1] (ID), Ian Jarman[1] (ID),
Vasilios Baltzopoulos[2] (ID), and Paulo Lisboa[1] (ID)

[1] Department of Applied Mathematics, Liverpool John Moores University,
Byrom Street, Liverpool L3 3AF, UK
P.McCabe@2014.ljmu.ac.uk, {I.A.OlierCaparroso,
S.OrtegaMartorell,I.H.Jarman,P.J.Lisboa}@ljmu.ac.uk
[2] Research Institute for Sport and Exercise Sciences, Liverpool John Moores
University, Byrom Street, Liverpool L3 3AF, UK
V.Baltzopoulos@ljmu.ac.uk

Abstract. This case study benchmarks a range of statistical and machine learning methods relevant to computer-based decision support in clinical medicine, focusing on the diagnosis of knee osteoarthritis at first presentation. The methods, comprising logistic regression, Multilayer Perceptron (MLP), Chi-square Automatic Interaction Detector (CHAID) and Classification and Regression Trees (CART), are applied to a public domain database, the Osteoarthritis Initiative (OAI), a 10 year longitudinal study starting in 2002 (n = 4,796). In this real-world application, it is shown that logistic regression is comparable with the neural networks and decision trees for discrimination of positive diagnosis on this data set. This is likely because of weak non-linearities among high levels of noise. After comparing the explanations provided by the different methods, it is concluded that the interpretability of the risk score index provided by logistic regression is expressed in a form that most naturally integrates with clinical reasoning. The reason for this is that it gives a statistical assessment of the weight of evidence for making the diagnosis, so providing a direction for future research to improve explanation of generic non-linear models.

Keywords: Comparative analysis · Neural networks · Logistic regression · Decision trees · Osteoarthritis · Healthcare

1 Introduction

The use of machine learning (ML) models by application domain experts requires a method to interpret the operation and inference made by these complex methods, in a language that people can understand. It is critical for successful translation and to ensure safety of real-world applications, such as healthcare domains, that mathematical algorithms are capable of being integrated into human reasoning models [1].

© Springer Nature Switzerland AG 2019
H. Yin et al. (Eds.): IDEAL 2019, LNCS 11872, pp. 114–122, 2019.
https://doi.org/10.1007/978-3-030-33617-2_13

Machine Learning methods may be interpretable by design, typically in the form of rules in induction trees such as Chi-squared Automatic Interaction Detection (CHAID) and Classification and Regression Trees (CART) [2]. In contrast, more parameterized and sometimes less interpretable models require model calibration in order to be used effectively.

This paper reports a case study to compare and contrast the different approaches with focus on accuracy of discrimination and explanation of model inferences, for the purpose of diagnosing knee osteoarthritis (OA) at first clinical presentation.

Osteoarthritis is a degenerative bone disease that affects joints as a whole. OA is one of the most common diseases affecting people in old age. The prevalence in people 65 years and older ranges from 12% to 30% [3]. The disease is also the most common form of arthritis to cause pain and mobility limitations. OA most commonly affects the knee, and around 10% of people over 55 years old have knee OA. This statistic is not surprising as weight-bearing joints are where disease occurs most [4]. Weight is just one of the factors that can play a part in developing OA. Some of the other factors are genetics, past injury, loading and overuse of a specific joint [4]. Despite many people thinking that OA is a disease that only affects the elderly, everyone is susceptible, with younger people more likely to develop the disease following trauma. Most factors that cause OA are not things the patient can modify, however, there are a few modifiable factors that, if dealt with, can slow the progression of OA. One such factor that can be changed is weight [4].

There are five stages of OA according to the Kellgren-Lawrence (KL) scale [5]. These are differentiated between with the use of x-rays to determine the severity of the OA. Stage 0 is classed as no OA and Stage 4 is severe OA present in the joint. A clinician usually analyses and classifies images for diagnosis. By using both humans and ML there is the potential for more reliable diagnoses [6].

There are three research questions with particular clinical relevance: (i) initial diagnosis of OA at first presentation; (ii) risk of developing OA among the cohort initially diagnosed as disease free; (iii) characterizing progression through different stages among the population diagnosed with OA. This case study is focused on the first of these questions. The aim of this paper is to benchmark a range of statistical and ML methods relevant to computer-based decision support in clinical medicine, focusing on the first research question.

2   Description of the Data

The data used in this study is from the Osteoarthritis Initiative (OAI) [7–9]. Only including the subjects who received an initial KL grade resulted in a useable data set of 4,226 subjects.

For this study the data used was extracted from multiple tables. Of the initial 4,796 subjects, 289 were excluded as they did not have an initial KL grade recorded, and a further 281 subjects were removed due to missing values in risk factor measurements. The remaining 4,226 subjects were used as the study cohort as they had a full description of symptoms with a definite outcome. This cohort is made up of people who have either radiographic or symptomatic OA, or no recorded OA at the initial baseline clinical assessment.

The variables selected for use were from tables of the OAI dataset based on reviewing the literature for symptoms of knee OA. A total of 17 variables were identified using the extracted OAI data. Amongst others, the variables include the age, gender and BMI of the individual, along with information of family history, previous injuries and diagnoses of OA in other joints and general arthritis in the body. Continuous variables were banded before the analysis, thus all of them were categoric.

The outcome variable looks at the presence of OA in the knee (KL0/1 – No OA vs KL2-4 -early OA to severe OA).

3   Methods

This case study reports the benchmarking of statistical and probabilistic ML models, from the standpoint of binary classification performance and interpretation of the inferences made by the models. The different models are summarized in this section.

3.1   Logistic Regression

This is the most commonly used statistical model in medical decision support [10]. Although it is linear-in-the-parameters, careful discretization of continuous variables creates a piecewise linear model with the capability to model highly non-linear data, which are typical in clinical medicine. As a result, logistic regression models are often very competitive in discrimination accuracy compared with neural networks and other ML methods, except when interactions between variables have a significant role in decision making, in which case rule induction may be preferred.

Importantly, the odds of class membership can be expressed in a form that is a natural description of human Bayesian decision making. This illustrated by the following expression where for binary covariates $\{x_i\}$ the exponentials show explicitly the size of the effect of the variable on the odds-ratio:

$$\frac{P(class|X)}{1 - P(class|X)} = \prod_{i=0}^{n} e^{\beta_i x_i} = e^{\beta_1 x_1}.e^{\beta_2 x_2}\ldots.e^{\beta_n x_n}.e^{\beta_0 x_0} \tag{1}$$

This provides a direct interpretation of the weight on the diagnostic inference arising from each individual variable, measured by the terms $\beta_i x_i$.

3.2   Multi-Layer Perceptron (MLP)

The MLP must be carefully regularized to ensure optimal generalization. The MLP is a type of feedforward artificial neural network. Typically, the MLP is constructed of at least three layers: a layer of nodes dealing with the input, a hidden layer and a layer responsible for dealing with the output. MLP exploits the supervised learning approach for training called backpropagation [11].

3.3   Rule Induction

Two methods for rule induction are utilized:

- Classification and Regression Trees (CART) is a ML method that also determines univariate cut points. The CART used in this analysis is in Fig. 1.
- Chi-squared Automatic Interaction Detection (CHAID) which uses standard p-values from contingency tables to carry out univariate partitions of the data. A visualization is shown in Fig. 2.

3.4   ROC Framework

The ROC framework is used throughout to compare model performance between the used ML methods. It has a close parallel to the Bayesian framework for decision making defined by LogR, in that the Positive Predictive Value (PPV) has an odds ratio that separates out the chance effect, defined by the prior odds using the prevalence, from the Likelihood Ratio (LR) which measures how much better than chance the classifier performs [12].

4   Experimental Results

The original sample is split into a training set (n = 2,145) and a test set (n = 2,081) to measure generalization to new data. The prevalence of Knee OA is 44.2% and 44.6%, consistent for both cases. In all the results the probabilistic cut-off for binary classification was taken to be 0.5.

The baseline method is logistic regression (LogR), since this is arguably the most commonly used multivariate statistical model for medical research. The data available to fit the model easily meets the standard requirement of 10 Events Per Variable (EPV), where for the binary variables, the relevant number of degrees of freedom is the total number of states added across all of the variables in the initial pool, to which stepwise forward conditional model selection was applied.

All statistical algorithms used default stopping criteria, typically using p-values of 0.05 and 0.10 for variable inclusion and exclusion, respectively. This was followed in all cases by further pruning based on classification performance. Summary statistics for all models are listed in Table 1.

Table 1. Performance summary for binary classification

Test data	AUROC	SENSITIVITY	SPECIFICITY	PPV-PRECISION
LogR	74.2%	52.6%	83.1%	75.7%
MLP	67.8%	56.5%	71.5%	66.4%
CHAID	73.6%	46.9%	86.5%	73.3%
CART	72.8%	62.2%	74.7%	71.1%

The variables selected by LogR are AGE_bins, BMI_bins, B.Line_SYMP, KOOS. SYMP and GEN.A, as seen in Table 2.

Table 2. Coefficients of logistic regression

	β	Standard Error	Exp(β)
Intercept	−1.251	0.295	0.286
AGE_bins (1)	0.168	0.185	1.184
AGE_bins (2)	0.580	0.187	1.786
AGE_bins (3)	0.756	0.191	2.131
AGE_bins (4)	0.926	0.167	2.523
BMI_bins (1)	0.771	0.117	2.162
B.LINE_SYMP (1)	1.417	0.120	4.124
GEN.A (1)	0.439	0.099	1.551
KOOS.SYMP (1)	−0.513	0.250	0.599
KOOS.SYMP (2)	−0.915	0.245	0.400

Among the induction trees derived directly from the data, CHAID and CART both split the data first on B.Line_SYMP, then followed by AGE_bins, BMI_bins. The other CART variables are KOOS.Symp and EV.INJ. The remaining variables from the CHAID tress are KOOS.Pain, Gen.A, GENDER and KNEE.SURGERY.

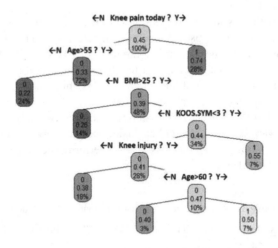

Fig. 1. This is the CART diagram generated on the test data showing the variable splits and the percentage of cases explained by a given rule set.

The Exp (β), as described in Table 2, are against the reference category for each variable. These are B.Line_SYMP (no knee pain exhibited on the day of the baseline assessment), AGE_bins (subjects aged 65 years and over), BMI_bins (any subject whose

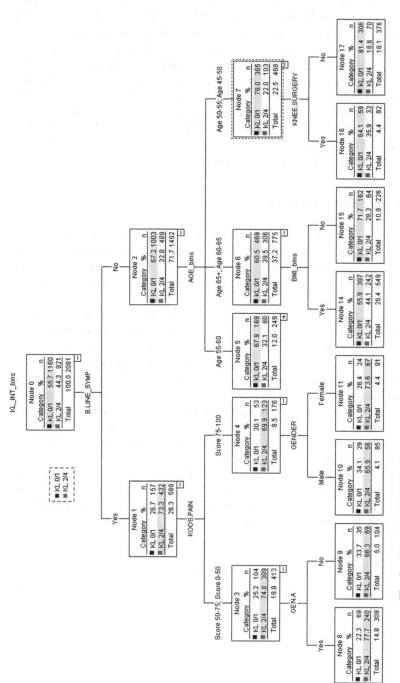

Fig. 2. This is the CHAID diagram generated on the test data showing the variable splits and the prevalence of OA in each case moving across through the decisions.

BMI was 25 and under), KOOS.SYMP (subjects with a score of 75–100 points) and GEN.A (subjects who had not previously been diagnosed with arthritis in any other joint).

It is possible to adapt the LogR model to a form that has clinical use – a nomogram. This visual display of the model would help a clinician demonstrate the strength of the effects to a subject seeking advice on their medical condition [13].

All of the interpretable models are consistent in model performance. However, the neural network is less transparent and has less predictive performance in this situation. In order to perform a comparison of the different models in order to determine which model is best, the McNemar test needs to be carried out. The results from this analysis show that the only pair of models that has no difference are the MLP and CART models. However, there is differences in the rest of the pairing. Due to the sample sizes more investigations would be required to establish what models perform the best.

5   Discussion and Conclusion

This broad range of commonly used models from conventional statistics and probabilistic ML shows broadly consistent performance in binary classification for diagnosis of Knee OA. They exploit the predictive power of a small set of covariates listed in the tables, which are therefore the concluding set of predictive factors for diagnosis at first presentation.

It remains to address the issue of model explanation. The rule sets are transparent by design. While the rules provide a filter for assigning patients to diagnostic categories, they do not provide a clear indication of the weight that each covariate has for the diagnostic inference made by the model for each individual patient.

This is provided only by logistic regression and can be conveniently expressed in the form of a nomogram, making the model easy to use and to interpret by clinicians. One such example in Fig. 3. is generated using the LogR model. Moreover, the explicit weighting of covariates also provides a tool to 'diagnose the model', by correlating these weights against prior clinical expertise about the expected influence of each covariate on the diagnostic outcome.

This provides a clear direction for future work. There is an urgent need to develop transparency for flexible probabilistic models such as neural networks. For all of the reasons explained above, it would be or particular benefit to be able to express their operation in terms of nomograms or equivalent.

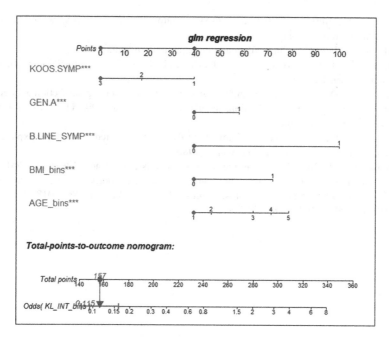

Fig. 3. The nomogram generated with the logistic regression model. This gives a point score to each level of the variable to show how it affects the output.

Acknowledgement. This work was funded under EU Grant OActive from the European Community's H2020 Programme. The OActive project looks to use advanced multi-scale computer models to better understand the risk factors associated with OA in order to prevent and delay the onset and progression of OA. Grant number 777159.

References

1. Vellido, A.: Neural Comput. Appl. (2019). https://doi.org/10.1007/s00521-019-04051-w
2. Song, Y., Lu, Y.: Decision tree methods: applications for classification and prediction. Shanghai Arch. Psychiatry. **27**, 130–135 (2015). https://doi.org/10.11919/j.issn.1002-0829. 215044
3. Peat, G., Mccarney, R., Croft, P.: Knee pain and osteoarthritis in older adults: a review of community burden and current use of primary health care. Ann. Rheum. Dis. **60**, 91–97 (2001). https://doi.org/10.1136/ard.60.2.91
4. Kohn, M.D., Sassoon, A.A., Fernando, N.D.: Classifications in brief: kellgren-lawrence classification of osteoarthritis. Clin. Orthop. Relat. Res. **474**, 1886–1893 (2016). https://doi. org/10.1007/s11999-016-4732-4
5. Jones, L., Golan, D., Hanna, S., Ramachandran, M.: Artificial intelligence, machine learning and the evolution of healthcare. Bone Joint Res. **7**, 223–225 (2017). https://doi.org/10.1302/ 2046-3758.73.BJR
6. Woolf, A.D., Pfleger, B.: Burden of major musculoskeletal conditions. Bull. World Health Organ. **81**, 646–656 (2003). https://doi.org/10.1590/S0042-96862003000900007

7. NIA: Osteoarthritis Initiative (OAI). https://www.nia.nih.gov/research/resource/osteoarthritis-initiative-oai

8. NIH: OAI - About OAI. https://data-archive.nimh.nih.gov/oai/about-oai

9. Hampton, T.: Osteoarthritis initiative. JAMA **291**, 1951 (2004). https://doi.org/10.1001/jama.291.16.1951-a

10. Shipe, M.E., Deppen, S.A., Farjah, F., Grogan, E.L.: Developing prediction models for clinical use using logistic regression: an overview. J. Thorac. Dis. **11**, 574–584 (2019). https://doi.org/10.21037/jtd.2019.01.25

11. Rumelhart, D.E., Hinton, G.E., Williams, R.J.: Parallel Distributed Processing: Explorations in the Microstructure of Cognition, vol. 1 (1986)

12. Marzban, C.: The ROC Curve and the Area under It as Performance Measures (2004)

13. Ing, E.B., Ing, R.: The use of a nomogram to visually interpret a logistic regression prediction model for giant cell arteritis. Neuro-Ophthalmol. **42**, 284–286 (2018). https://doi.org/10.1080/01658107.2018.1425728

A Clustering-Based Patient Grouper for Burn Care

Chimdimma Noelyn Onah[1(✉)], Richard Allmendinger[1], Julia Handl[1], Paraskevas Yiapanis[2], and Ken W. Dunn[3]

[1] University of Manchester, Manchester, UK
chimdimma.onah@postgrad.manchester.ac.uk
[2] Medical Data Solutions and Services, Manchester, UK
[3] University Hospital South Manchester, Wythenshawe, UK

Abstract. Patient casemix is a system of defining groups of patients. For reimbursement purposes, these groups should be clinically meaningful and share similar resource usage during their hospital stay. In the UK National Health Service (NHS) these groups are known as health resource groups (HRGs), and are predominantly derived based on expert advice and checked for homogeneity afterwards, typically using length of stay (LOS) to assess similarity in resource consumption. LOS does not fully capture the actual resource usage of patients, and assurances on the accuracy of HRG as a basis of payment rate derivation are therefore difficult to give. Also, with complex patient groups such as those encountered in burn care, expert advice will often reflect average patients only, therefore not capturing the complexity and severity of many patients' injury profile. The data-driven development of a grouper may support the identification of features and segments that more accurately account for patient complexity and resource use. In this paper, we describe the development of such a grouper using established techniques for dimensionality reduction and cluster analysis. We argue that a data-driven approach minimises bias in feature selection. Using a registry of patients from 23 burn services in England and Wales, we demonstrate a reduction of within cluster cost-variation in the identified groups, when compared to the original casemix.

Keywords: Patient casemix · Clustering · Data driven

1 Motivation

The NHS serves a wide population with varied demographic and medical histories, with the aim of providing health interventions to the population who need them. The provision and maintenance of these interventions is constrained by scarce resources and cost containment [1]. The pressure from binding budget constraints, and thus the need to control costs, has induced a shift in favor of prospective payments over retrospective payment systems.

The use of patient-level payment system transfers all cost burden to the payer, since the reimbursement is based on the real costs. In the context of such a system, even profit maximizing providers may be insufficiently motivated to decrease costs. In contrast,

© Springer Nature Switzerland AG 2019
H. Yin et al. (Eds.): IDEAL 2019, LNCS 11872, pp. 123–131, 2019.
https://doi.org/10.1007/978-3-030-33617-2_14

prospective payment systems (PPSs) determine the provider's payment rates ex ante without any link to the real costs of the individual provider [2]. This payment system is increasingly being adopted over retrospective systems, as it encourages cost containment and a shared burden with the providers. There is wide adoption of PPS globally, with approximately 70% of all OECD countries and more than 25 low-and middle-income countries having adopted some sort of casemix system for reimbursement purposes [3, 4].

Here, a casemix is a system of defining cohorts of related patients, which comprise cases that are homogenous by resource consumption pattern and at the same time, clinically similar. In the NHS, the National Casemix Office (NCO) is commissioned to develop and maintain a set of casemix groupings, called HRG (health resource group). This is a type of PPS where payment rate is determined as the average patient cost in each HRG. HRGs are generated using nationally mandated patient-level data, which primarily includes age, complications and comorbidities, diagnosis and procedures. Adopted in acute care, the groups are generated by transcribing expert advice into if-else rules, with the aim of capturing differing patient severity and length of stay (LOS).

Any reimbursement methodology based on generalizations across patient groups (i.e. determining payment rate as an average of cost in each HRG) will have weaknesses regarding its ability to fairly work across a variety of settings and HRGs are no exception to this. The use of LOS as an (imperfect) indicator of resource use contributes further to this weakness – it is known to be unreliable particularly for the case of surgical patients [5]. Finally, the identification of relevant factors based on expert advice alone carries the risk of ignoring other unknown (or less well established) factors that may account for the case complexity of certain patient sub-groups.

Our core hypothesis is that in-depth analysis of the available data should be used in conjunction with expert input to develop an evidence-based model that comprehensively captures the complexity of care provided by such services, and accurately classifies patients into homogeneous groups with respect to costs and patient characteristics. This dual approach was previously not possible due to a lack of availability of extensive patient-level cost data, and the resulting primary dependence on expert advice.

Our research aims to provide evidence for this hypothesis. First, we explore the accuracy of current HRGs in terms of actual resource usage. Second, we describe an analytical approach to the development of an alternative, data-driven grouper. Throughout our analysis, we use burn care as a base case. Burn services are selected as an example of a specialized service, which deals with rare and complex conditions and by necessity operates at high expenditure. Burn services are to be open regardless of the number of patients admitted, with a minimum number of staff, and they rely on the use of highly specialist equipment and interventions. We expect that the complex characteristics of this setting make them particularly sensitive to the impact of weaknesses in the current HRG classification.

The remainder of this paper is structured as follows. The next section introduces the data sets used to explore HRGs and generate the data-driven groups. We then introduce the analysis pipeline adopted, which includes data pre-processing, dimensionality reduction and the deployment of clustering approaches in two separate steps. In Sect. 3, we discuss the results, using visualizations and within cluster variation of costs to identify improvements. The final section includes a conclusion and discussion of future work.

2   Methodology

2.1   Data

This study uses comprehensive anonymized patient-level data that is nationally mandated for all burn units in England and Wales. The data covers a time period from 2003 to 2019 and captures 164 features for just over 100,000 patients. This includes features such as demographic characteristics (age, gender), burn characteristics (depth, total burn surface area, burn site, locality, type, source, category and injury group), pre-existing conditions (self-harm, alcohol usage, asthma, clotting disorder etc.), time from injury to admission, patient-level cost, LOS and index of multiple deprivation (IMD).

To highlight current variation in HRGs and as a benchmark for model performance, we use the 2017/18 average patient-level cost by HRG open data released by NHS Improvement. This is limited to one year as PLICS adoption was introduced just in 2017/18 data collection cycle. This data is at the burn service level and so represents average patient level cost in each service.

2.2   Analysis Pipeline

Step 1: Selecting Relevant Features and Cases. To ensure the use of quality features that reflect the clinical and cost differences of patients, the features selected for clustering were those identified as statistically significant in predicting patient-level cost and patient outcome. Cost prediction accuracy was improved with the removal of non-survivals, whose LOS and cost are lower compared to survivals with similar burn characteristics. Thus, is in line with the current grouper, the following analysis focuses on survival cases only. All cases with missing data were deleted, leaving just over 80,000 cases and 24 features after feature selection. Table 1 lists these features.

Table 1.   Selected features

Feature type (count)	Feature
Demographic (3)	Gender, Age, Index of Multiple Deprivation (IMD)
Burn characteristics (17)	Total burn surface area (TBSA); Presence of inhalation; Site of burn (leg; upper limb (UL); torso and thorax; face, hands, feet and perineum (FHPP); head and hand (HH); face, hands and feet (FHF)); Type of injury (contact, cold, flame, electrical, scald, chemical, friction, flash, radiation)
Comorbidity (2)	Number existing disorders, significance of existing disorder
Cost Features (2)	Adjusted LOS, Patient-level cost

We implement further dimensionality reduction to minimise noise, data complexity and reduce redundancy. Dimensionality reduction also helps reduce processing time and mitigates against the curse of dimensionality [6]. Linear discriminant analysis (LDA), a supervised approach to dimensionality reduction, is adopted. Here, this method is preferred over unsupervised dimension reduction models such as principal component analysis (PCA), as we wish to identify components that maximise cost separation rather than percent of variance alone.

Step 2: Deriving Target Feature for LDA. We derive a set of target classes for the LDA using a cluster analysis on multiple cost features, to reduce sensitivity to a single cost measure. This is achieved by using cost features: patient level cost and adjusted LOS as the target space. The target feature is then generated using k-means clustering algorithm (k = 38, same as number of HRGs) to partition the two-dimensional target space defined by adjusted LOS and patient-level cost.

Step 3: Segmentation by Age. The current grouper splits the data into young patients (< 16 years old) and older patients (>=16 years old). This reflects the burn care pathway, designed to treat pediatrics separately from adults as young age is identified as a significant complicator. The 2001 National Burn Care Review Report [7] highlights the unpredictable complication of seemingly simple burn injuries especially for pediatric patients. It argues and mandates the need for separate burn units for children and adults, due to the peculiar needs of children such as play specialist, teachers, family counselors and intensive psychosocial support. In line with the current grouper, we therefore further split the data by age group.

Step 4: Dimensionality Reduction using LDA. The comorbidity details, demographic and burn characteristics listed in Table 1 are used as the input features for the LDA. We retain the first two LDA components. Therefore, the output of this analysis is a projection from the original feature space into a two-dimensional manifold spanned by orthogonal components that maximise separation by the target feature constructed in Step 2. This is done on each segment derived in Step 3.

Step 5: Segmentation into Homogeneous Patient Groups. With these preprocessing and dimensionality reduction steps completed, an unsupervised clustering method is deployed to derive homogenous patient groups. This paper uses an unsupervised clustering method, as we assume that the true class of patients are unknown. The use of a supervised method, for example, using cost labels may create groups that are homogenous in terms of cost only. This therefore does not meet the clinical relevance criteria.

In particular, we deploy an agglomerative hierarchical clustering (HAC) algorithm using the LDA components generated on each age segments (< 16 years old and >=16 years old) as input data to generate 13 and 25 patient groups respectively. The group numbers reflect the number of segments generated by the current grouper, to facilitate comparison.

We use the Euclidean distance as the metric used to compute the distance between data points. Average link is used as the linkage criterion, i.e. the distance between clusters is calculated as the average of the pairwise distances between all observations in the two clusters. At each step in the hierarchy, the algorithm merges the pair of clusters that minimises this criterion. Average link HAC clustering and all preprocessing steps were implemented using the relevant scikit-learn module on python, with other parameters left as default [8]. Figure 1 depicts a summary of the analytical pipeline.

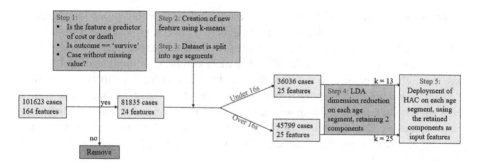

Fig. 1. Summary of analysis pipeline

3   Results and Analysis

We explore the patient-level cost by HRG, as generated by the National Casemix office. In Fig. 2, we visualize this data using boxplots that show the distribution of patient-level cost broken up by HRG. The wider the boxplot, the more variable are the costs within that group. There are 25 adult HRGs, with JB40A being the most complex and JB71A the least complex in terms of injury severity. Whereas, the child segment includes 13 HRGs, with JB50A as the most complex and JB71B the least complex in terms of injury severity. These HRGs shown in Fig. 2 are split into the two age segments and ordered in decreasing order of injury complexity.

Overall, we would expect the boxplots to slowly go down in terms of patient-level cost (for adults and children) as the injury complexity reduces. While this trend is generally apparent, there is scope to achieve a better match between injury complexity

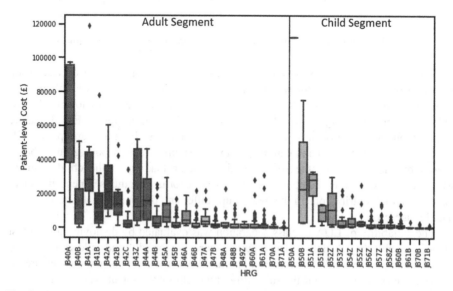

Fig. 2. 2017/18 HRG by patient-level cost. Ordered in decreasing order of injury complexity

and costs for both adult and children. The results show low within-group homogeneity and we also observe a low degree of heterogeneity between groups. For a model that accurately reflects resource usage and complexity, we would expect these two aspects to be more pronounced. For example, JB40B, a more complex group than JB41A should have the highest average cost amongst the two, but we observe that this is not the case.

Thus, this exploratory analysis appears to support the need for a grouper with a better performance at minimizing within-group cost variance and maximising between-group separation. Due to limitations of the dataset used, we do not have access to HRG grouping at a patient level and are therefore not able to link back costs to individual patient characteristics.

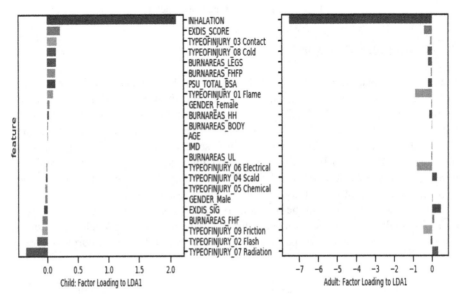

Fig. 3. Factor loadings on 1st linear discriminant: Child vs Adult segment

One of the key steps in the analysis pipeline was the split of data on age group reflecting the burn care pathway. The importance of this approach is highlighted in Table 2 and Fig. 3. Table 2 shows that child groups, although with lower adjusted LOS, have higher average costs relative to comparable adult groups. In Fig. 3, feature contribution to separation in data captured by LDA across both age groups are different. Where presence of inhalation was the most important in both groups, burns caused by radiation have the second highest factor loading in the child segment whereas injury caused by flame is second highest factor in the adult segment. It is clear that the feature contributions in the two groups differ and this supports the need to split data by age group.

The suitability of LDA in class separation was evaluated by deploying a train test model, to enable the calculation of an accuracy score. For the adult and child segments, we get a score of 73% and 72% respectively. This highlights the ability of non-cost features in capturing resource usage thus the suitability of the LDA components generated.

The resultant groups derived using our data-driven pipeline, are shown in Fig. 4. In particular, the model identifies well-defined high cost clusters in both adult (Adult0, Adult8, Adult13 and Adult17) and child (Child8, Child3) groups. Investigating further, we compare the within cluster variance (calculated using the coefficient of variation of patient-level cost in each group) between HAC generated groups and HRGs. Figure 5 highlights the improvement obtained from the data-driven model: as evident from the histogram, the HAC grouper tends to produce groups with a higher cost similarity. Specifically, the HAC groups (green) on average have a lower within cluster variation (0.89) compared to that of the HRGs at 1.19. The overall results therefore suggest an improved differentiation between groups.

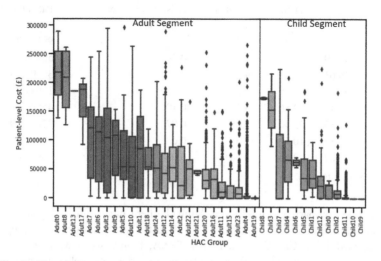

Fig. 4. Identified groups by patient-level cost ordered by decreasing average cost

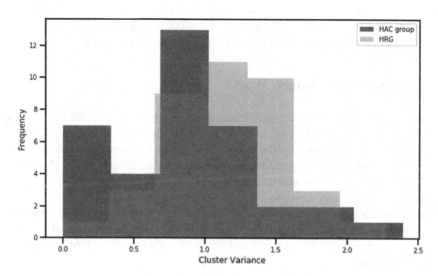

Fig. 5. Within cluster variation of patient-level cost in HRG vs HAC (Color figure online)

We further explore the patient characteristics of the HAC groups. Table 2 provides some illustrative examples that highlight the interpretability of some of the resulting clusters, based on patient characteristics. For example, when comparing the clusters Adult3 and Adult12, these have very similar average age, but Adult3 has the more severe burns (TBSA), higher LOS and cost, and so the necessity to have separate groups. In the child group, comparing clusters Child2 and Child11, we see a similar pattern, i.e. similarity with respect to average age but differentiation by severity of burn, LOS and cost.

Table 2. A sample* of HAC groups by patient characteristics (average).

Cluster Label	Age	TBSA	Adjusted LOS	Existing disorder (Significance)	Patient-level cost (£)
Adult3	41.02	50.59	13.71	0.27	99184.74
Adult12	42.99	31.30	9.08	0.49	56979.04
Adult20	46.61	19.49	9.28	0.79	39006.15
Adult11	67.64	5.28	7.36	2.68	19037.41
Child5	5.13	34.86	5.77	0.10	45585.72
Child12	4.85	25.10	5.08	0.07	32656.89
Child2	3.87	10.56	2.63	0.08	10595.17
Child11	3.87	2.56	0.88	0.03	1921.42

*Purposive sampling

Meanwhile comparing Adult20 and Adult11, there have different patient characteristics and thus different average patient-level cost. Child5 and Child10 though with similar adjusted LOS, Child5 has a higher TBSA, higher score with respect to the severity of existing disorders and thus a higher average patient-level cost. The observed similarity in adjusted LOS but higher difference in patient-level cost in this child5 and Child10 also supports the need for using more than one feature space in deriving the LDA target feature. These results highlight the effectiveness of the data-driven HAC grouper in generating groups with homogenous patient characteristics.

4   Conclusion and Future Work

HRGs are used as the basis for cost reimbursement in health systems. When designing HRGs, the accurate reflection of resource usage is imperative in ensuring different settings get fair reimbursements. The collection of patient-level cost, at a national scale, has created the possibility of generating improved data-driven groups. We have been able to highlight that improvements can be made in identifying patient case mix suitable for payment rate derivation. This was done with the implementation of a data driven process on burn care data to identify significant features and the subsequent use of a clustering algorithm. With the adopted analysis pipeline, we showed a visible difference in the distribution of patient-level cost variance when HAC groups are compared to HRGs.

There could be further reduction in within cluster variance with the use of state-of-the-art clustering algorithms that simultaneously consider Step 2, 3 and 4 of our analysis. A further limitation of this paper is the lack of certain features in the patient-level data such as current patient level HRG labels, in-hospital complications and interventions. The unavailability of patient-level data labelled with current HRGs, limited our ability to evaluate performance of current HRGs in terms of patient characteristics. Finally, the lack of complications and interventions data could limit the homogeneity by patient characteristics in the HAC identified groups.

Future work will be aimed at exploring changes to our analytical model, including the consideration of different approaches to dimensionality reduction and cluster analysis, as well as the inclusion of expert opinion in feature selection and group validation. This will be done using comprehensive anonymized patient-level data set. An improved analytical pipeline developed and validated using burn care data can be applied to other injury type. The adoption in other injury type would include the use of expert and data-driven feature selection phase, but with the aim of a standardised pipeline, the same dimensionality reduction and cluster analysis technique would be implemented.

References

1. Hopfe, M., Stucki, G., Marshall, R., Twomey, C.D., Üstün, T.B., Prodinger, B.: Capturing patients' needs in casemix: a systematic literature review on the value of adding functioning information in reimbursement systems. BMC Health Serv. Res. **16**(1), 40 (2015)
2. Jegers, M., Kesteloot, K., De Graeve, D., Gilles, W.: A typology for provider payment systems in health care. Health Pol. (New York) **60**(3), 255–273 (2002)
3. Mathauer, I., Wittenbecher, F.: Hospital payment systems based on diagnosis-related groups: experiences in low- and middle-income countries. Bull. World Health Organ. **91**(10), 746–756A (2013)
4. Fischer, W.: The DRG Family: State of affairs (2007)
5. Street, A., Kobel, C., Renaud, T., Thuilliez, J.: How well do diagnosis- related groups explain variations in costs or length of stay among patients and across hospitals? methods for analysing routine patient data. Health Econ. **21**, 6–18 (2012)
6. Sembiring, R.W., Zain, J.M., Embong, A.: Dimension reduction of health data clustering. Int. J. New Comput. Archit. Appl. **1**(3), 1041–1050 (2011)
7. National Burn Care Review Committee, National Burn Care Review Committee Report Standards and Strategy for Burn Care A Review of Burn Care in The British Isles Contents (2001)
8. Pedregosa, F., et al.: Scikit-learn: machine learning in Python. J. Mach. Learn. Res., **12**, 2825–2830 (2011)

A Comparative Assessment of Feed-Forward and Convolutional Neural Networks for the Classification of Prostate Lesions

Sabrina Marnell[1] , Patrick Riley[1(✉)] , Ivan Olier[1] ,
Marc Rea[2] , and Sandra Ortega-Martorell[1]

[1] Department of Applied Mathematics, Liverpool John Moores University,
Liverpool L3 3AF, UK
S.Marnell@2016.ljmu.ac.uk, P.J.Riley@2014.ljmu.ac.uk,
{I.A.OlierCaparroso,S.OrtegaMartorell}@ljmu.ac.uk
[2] Clatterbridge Cancer Centre NHS Foundation Trust, Wirral CH63 4JY, UK
m.rea@nhs.net

Abstract. Prostate cancer is the most common cancer in men in the UK. An accurate diagnosis at the earliest stage possible is critical in its treatment. Multiparametric Magnetic Resonance Imaging is gaining popularity in prostate cancer diagnosis, it can be used to actively monitor low-risk patients, and it is convenient due to its non-invasive nature. However, it requires specialist knowledge to review the abundance of available data, which has motivated the use of machine learning techniques to speed up the analysis of these many and complex images. This paper focuses on assessing the capabilities of two neural network approaches to accurately discriminate between three tissue types: significant prostate cancer lesions, non-significant lesions, and healthy tissue. For this, we used data from a previous SPIE ProstateX challenge that included significant and non-significant lesions, and we extended the dataset to include healthy prostate tissue due to clinical interest. Feed-Forward and Convolutional Neural Networks have been used, and their performances were evaluated using 80/20 training/test splits. Several combinations of the data were tested under different conditions and summarised results are presented. Using all available imaging data, a Convolutional Neural Network three-class classifier comparing prostate lesions and healthy tissue attains an Area Under the Curve of 0.892.

Keywords: Feed-forward neural networks · Convolutional neural networks · SPIE ProstateX · mpMRI · Prostate cancer

1 Introduction

The tools for the diagnosis of prostate cancer (PCa) are seeing change in recent times. In the UK, trials for diagnosis by Magnetic Resonance Imaging (MRI) scans, as a non-invasive test, are proving to provide benefit in various areas that the current first line of diagnosis (a PSA blood test and biopsy) lack [1]. In addition, active surveillance using multiparametric MRI (mpMRI) has gained popularity as an acceptable management option for low-risk prostate cancer patients, as it can delay or prevent unnecessary

© Springer Nature Switzerland AG 2019
H. Yin et al. (Eds.): IDEAL 2019, LNCS 11872, pp. 132–138, 2019.
https://doi.org/10.1007/978-3-030-33617-2_15

interventions - thereby reducing morbidity associated with overtreatment [2]. However, the volumetric analysis of mpMRI scans remains challenging as it requires specialist knowledge and is time consuming. This has motivated the use of machine learning techniques to assist with the analysis of these many and complex images, with the aim of increasing accuracy (especially relevant in places without access to specialist knowledge) and speed up the process (also reducing costs).

This study conducts a comparative analysis with both Feed-Forward Neural Network (FFNN) [3] and Convolutional Neural Network (CNN) [4] architectures as methods for utilising machine learning methodologies for the classification of prostate cancers and healthy tissue on mpMRI scans. It extends on work conducted by the authors previously [5], where the healthy prostate class was added to the SPIE ProstateX challenge dataset [6]. Using advice from collaborating clinicians, the contralateral of the lesion location was taken as healthy prostate tissue, extending from the two classes available in the dataset – clinically significant lesions and non-significant lesions. The latter, "non-significant" prostate lesions do not always require treatment as they hold a lower Gleason score [7].

Previously, various methods were applied to the original SPIE ProstateX challenge problem for classification against the two lesion classes – including transfer learning [8], SVM [9] and convolutional neural networks [10], but not including the healthy class, which is of clinical interest. Only the authors' previous work included the healthy class, in which SVM was used for binary classification and a voting ensemble system was implemented for the diagnosis of individual cases [5]. Hence, the natural next step was to test more sophisticated approaches, which led us to the use of neural networks, and perform a comparative assessment of both FFNNs and CNNs to model classification of prostate lesions against healthy tissue using SPIE ProstateX mpMRI data.

The structure of the rest of the paper is as follows: the Data section details a description of the SPIE ProstateX dataset used in the study, with the Classification Methods section describing the setup of the FFNN and CNN applications. The Results and Discussion section provides insights into the comparative assessment of the two machine learning algorithms to model the diagnosis of prostate cancer through mpMRI.

2 Data

The SPIE ProstateX challenge training data was attained for this study and was extended for clinical use. The data was distributed in DICOM format. Table 1 provides class label information for the dataset used for this study, with a more extensive description available in [5]. The Gleason score (GS) determines the aggressiveness of PCa. At least one lesion was found in every patient. The lesion significance level was stored in the metadata for the respective DICOM files; however, the Gleason Score was not provided.

Various parameters of MRI were provided; Apparent Diffusion Coefficient (ADC), K^{trans} and T2-weighted imaging are used in this paper due to their link in detecting clinical significance [11]. ADC is a measure of the magnitude of diffusion (of water molecules) within tissue and is calculated from diffusion weighted imaging. K^{trans}, a type of perfusion imaging, represents a measure of capillary permeability, calculated from dynamic contrast-enhanced imaging. T2-weighted imaging is a form of spin-echo pulse sequencing, showing fatty tissue and fluid brightly.

The contralateral was taken from 54 patients as described in [5]. This dataset was the training dataset of the challenge, however in this paper it has been used as the sole dataset – used for training and testing; at the time of writing, the challenge dataset test labels have not been released. Different planes were used whilst scanning the patients to create a 3D view around the region of interest (ROI): coronal, sagittal and transverse.

Table 1. Class label information on the extended SPIE ProstateX dataset used in this study.

Class	Gleason score	Available (N = 384)	Under-sampled data	Over-sampled data
Clinically significant	≥ 7	76	76	228
Non-significant	≤ 6	251	75	251
Healthy tissue	N/A	54	54	216

Both under-sampling and over-sampling techniques have been applied to the dataset and compared. Under-sampling is the practice of randomly deleting observations from the larger class, ensuring a good comparative ratio. Synthetic Minority Oversampling Technique (SMOTE) has been utilised for over-sampling, which synthetically manufactures observations of the unbalanced class which share a likeness with the said class; using the k-Nearest Neighbours technique. These methods were tested on the data as provided. Table 1 shows the class sizes for each sampling method, as well as the original class sizes.

3   Classification Methods

For creating the FFNN and CNN classifiers, we followed the steps detailed below:

(a) Pre-processing, region of interest extraction, vectorisation and standardisation:
Various rules were required for pre-processing this large data set; an extensive description, including a description of the contralateral for healthy prostate tissue, is available in [5]. For this study, the FFNNs utilised a 5 * 5 mm centred patch extracted at 1 px/mm around the ROI, while and the CNNs utilised 25 * 25 mm centred patch extracted at 1 px/mm around the ROI, with the input in image format i.e. [25 25 1]. All data was standardised by subtracting the mean and dividing by the standard deviation, ensuring that each dimension was approximately normal.

(b) Lesion classification using FFNNs: The data for FFNNs were inputted as flattened vectors. Both binary and three-class classifiers were tackled in this study. The network architecture for this model FFNN consisted of input layers, hidden layers and output layers, utilising both dense and dropout layers. The dropout layer is used for regularisation. *ReLU* and *Softmax* activation functions were utilised, the loss function was *sparse categorical cross-entropy* and the *Adam* optimiser was utilised.

(c) <u>Lesion classification using CNNs</u>: CNNs were used for both binary and three-class classification. The CNN architecture in this work used an array of different layers – two-dimensional convolutional layers, pooling layers, dropout layers and a dense layer. The CNN employed *ReLU* and *Softmax* activation functions, *categorical cross-entropy* loss function and a *stochastic gradient descent* optimizer.

(d) <u>Hyperparameter selection and finetuning</u>: Hyperparameter selection has been utilised upon a large selection of combinations of parameters for model training. Table 2 lists the values for hyperparameter selection.

(e) <u>Validation</u>: For all models, an 80/20 out-of-bag sampling method for training/test has been utilised.

Table 2. Hyper-parameter tuning values performed (six for FFNN and twelve for CNN).

	Hyper-parameter	Tested values	Hyper-parameter	Tested values
FFNN	Dense units 1st layer	25, 50, 100, 150, 200, 250	Dropout units 4th layer	0.05, 0.1
	Dropout rate 2nd layer	0.05, 0.1, 0.2	Epochs	25, 50, 100
	Dense units 3rd layer	50, 100, 150	Batch size	32, 64
CNN	Filters 1st layer	16, 32, 64	Filters 5th layer	32, 64
	Kernel size 1st layer	2, 3	Kernel size 5th layer	2, 3
	Filters 2nd layer	16, 32, 64	Dropout rate 6th layer	0.05, 0.1, 0.2, 0.3
	Kernel size 2nd layer	2, 3	Dense units 7th layer	50, 100, 200
	Dropout rate 3rd layer	0.1, 0.2	Dropout rate 8th layer	0.1, 0.2, 0.3
	Filters 4th layer	32, 64	Epochs	15, 25, 50
	Kernel size 4th layer	2, 3	Batch size	32, 64

The total number of models developed with the FFNN architecture was 864, and the total of models with CNN was 994. This allowed us to identify the best set of hyper-parameter values for each architecture and the data at hand. Classification was conducted against the clinical significance label denoted to the prostate lesion, or healthy tissue. Binary classifiers have been tested as well as the three-class classification problem, with the Area Under the Curve (AUC) reported for each test. A McNemar test was used to determine whether the results between the two architectures were statistically significant.

4 Results and Discussion

Various comparative experiments of FFNNs and CNNs were performed. Tests for the binary classifiers are averaged over single modality tests – for example, for the Significant vs. Non-Significant classifier, the classification was conducted for T2-weighted only, for ADC only and for K^{trans} only, and averaged. Results are summarised in Table 3.

Table 3. Summarised results for the comparative assessment of the FFNN and CNN application to the extended SPIE ProstateX dataset. The binary classifiers were performed on each of the mpMRI scan parameters alone and are then averaged for each separate classifier. S: Significant; N: Non-Significant; H: Healthy.

	Under-sampled data		Over-sampled data	
	FFNN	CNN	FFNN	CNN
Single mpMRI, binary: S vs. N	0.703 ± 0.025	0.619 ± 0.059	0.831 ± 0.023	0.837 ± 0.047
Single mpMRI, binary: S vs. H	0.871 ± 0.033	0.583 ± 0.074	0.954 ± 0.011	0.905 ± 0.054
Single mpMRI, binary: N vs. H	0.645 ± 0.091	0.720 ± 0.048	0.873 ± 0.020	0.960 ± 0.031
All mpMRI, three-classes: S vs. N vs. H	0.628	0.648	0.824	0.892

The first thing to notice from the results is that always the over-sampling strategy outperformed the under-sampling, regardless of the neural network architecture used. This is not surprising as during the under-sampling process we are bound to lose what can turn out to be valuable information from the cases left out, whilst the over-sampling process would benefit from keeping those cases in the dataset. This is even more the case since the size of the dataset is not very large; hence, the reduction of a number of observations may limit the generalisation capabilities of the models since the data used would not be properly representing the population.

Focusing the attention from this point onwards on the results obtained with the over-sampled data for the binary classifiers, using a single mpMRI, we can see that both neural network architectures have produced more competitive results than the previous ones in [5] using SVM, where the AUC for Significant vs. Non-significant was 0.72 (in this study, FFNN: 0.83, CNN: 0.84), Significant vs. Healthy was 0.87 (in this study, FFNN: 0.95, CNN: 0.91), and Non-significant vs. Healthy was 0.71 (in this study, FFNN: 0.87, CNN: 0.96). We compared the obtained results with the ones in [5] since the dataset used is the same. One of the reasons the results are so much improved in the current study can be explained by the use of the SMOTE over-sampling (which was not the case in the previous study), leading to a better informed dataset of which both neural network methods made the most of.

Previous works that looked at the discrimination of the Clinically Significant lesions from the Non-significant ones, reported AUCs of 0.83 in [8], 0.811 in [9] and 0.84 in [10]. This compares with AUCs of approximately 0.83 (when using FFNN) and 0.84 (when using CNN) in our results for the same discrimination problem, hence we can conclude from here that we have matched the results from these previous study, whilst adding extra value from the inclusion of the Healthy class. In the case of the comparison of the other two binary classifiers (i.e. Significant vs. Healthy and Non-Significant vs. Healthy) with the rest of the literature was not possible since the Healthy class was introduced to this dataset following the request of clinical collaborators.

After being reassured that the results for the Significant vs. Non-Significant problem are in line with the literature, we can focus the attention on the comparison between FFNN and CNN. Starting with the binary classifiers: In the case of the Significant vs. Non-Significant classifier, the McNemar test resulted in a p-value > 0.05 (any p-value > 0.05 is deemed as not statistically significant), hence we conclude that both FFNN and CNN models are equivalent. In the case of the other two binary classifiers, FFNN was better for the Significant vs. Healthy classifier, whilst CNN was better for the Non-significant vs. Healthy. Hence, in the case of these binary classification problems, we can only advice that both architectures should be considered, as each of them will have something to offer. However, if only one architecture was going to be used, and more weight was given to the accurate identification of the clinically significant class using binary classifiers, we would recommend the use of FFNN, since it produced a better outcome in the separation of this class from healthy tissue.

When looking at the three-class classifier that utilises all three mpMRI data available to the study, the CNN performed the best, with an AUC of 0.89. A McNemar test showed that between the FFNN and the CNN over all three classes, the differences in predictions observed are statistically significant (p-value < 0.05). This corroborates with [11], which denotes there is a relationship in clinical significance between the utilised MRI scan parameters in this study – T2-weighted, ADC and K^{trans}. Therefore, we would recommend the use of CNN in the scenario where a three-class classifier is implemented using these MR images for the simultaneous separation of the Significant, Non-significant, and Healthy classes.

5 Conclusions and Further Work

This study looked at a comparative assessment of FFNNs and CNNs, both in the context of binary and a three-class classifier for the separation of prostate lesions against healthy tissue. The inclusion of the healthy class to a publicly available dataset was motivated by the interest of clinical collaborators. The use of both FFNN and CNN architectures proved successful for both binary and three-class classifiers, leading to clinical impact.

Future work will look at interpreting the CNN features through sensitivity analysis. As opposed to FFNNs, CNN features are by nature sparse. This would provide insights into tissue relevance discrimination.

Acknowledgements. This work has been funded by the LJMU Scholarship Fund.

References

1. Prostate cancer screening scan hope - BBC News
2. An, J.Y., Sidana, A., Choyke, P.L., Wood, B.J., Pinto, P.A., Türkbey, İ.B.: Multiparametric magnetic resonance imaging for active surveillance of prostate cancer. Balkan Med. J. **34**, 388–396 (2017). https://doi.org/10.4274/balkanmedj.2017.0708
3. McCulloch, W.S., Pitts, W.: A logical calculus of the ideas immanent in nervous activity. Bull. Math. Biophys. **5**, 115–133 (1943). https://doi.org/10.1007/BF02478259
4. Srivastava, N., Hinton, G., Krizhevsky, A., Salakhutdinov, R.: Dropout: A Simple Way to Prevent Neural Networks from Overfitting (2014)
5. Riley, P., Olier, I., Rea, M., Lisboa, P., Ortega-Martorell, S.: A voting ensemble method to assist the diagnosis of prostate cancer using multiparametric MRI. In: Vellido, A., Gibert, K., Angulo, C., Martín Guerrero, J.D. (eds.) WSOM 2019. AISC, vol. 976, pp. 294–303. Springer, Cham (2020). https://doi.org/10.1007/978-3-030-19642-4_29
6. Litjens, G., Debats, O., Barentsz, J., Karssemeijer, N., Huisman, H.: Computer-aided detection of prostate cancer in MRI. IEEE Trans. Med. Imaging **33**, 1083–1092 (2014). https://doi.org/10.1109/TMI.2014.2303821
7. Gallagher, J.: Prostate cancer treatment "not always needed" - BBC News/ Health (2016). https://www.bbc.co.uk/news/health-37362572
8. Chen, Q., Xu, X., Hu, S., Li, X., Zou, Q., Li, Y.: A transfer learning approach for classification of clinical significant prostate cancers from mpMRI scans. In: Armato III, S.G., Petrick, N.A. (eds.) SPIE Medical Imaging 2017: Computer-Aided Diagnosis, p. 101344F. International Society for Optics and Photonics, Orlando (2017)
9. Kitchen, A., Seah, J.: Support vector machines for prostate lesion classification. In: Armato III, S.G., Petrick, N.A. (eds.) SPIE Medical Imaging 2017: Computer-Aided Diagnosis, p. 1013427. International Society for Optics and Photonics, Orlando (2017)
10. Seah, J.C.Y., Tang, J.S.N., Kitchen, A.: Detection of prostate cancer on multiparametric MRI. In: Armato, S.G., Petrick, N.A. (eds.) SPIE Medical Imaging 2017: Computer-Aided Diagnosis. p. 1013429. International Society for Optics and Photonics (2017)
11. Langer, D.L., et al.: Prostate tissue composition and MR measurements: investigating the relationships between ADC, T2, Ktrans, ve, and corresponding histologic features. Radiology **255**, 485–494 (2010). https://doi.org/10.1148/radiol.10091343

Special Session on Machine Learning in Automatic Control

A Method Based on Filter Bank Common Spatial Pattern for Multiclass Motor Imagery BCI

Ziqing Xia[1,2], Likun Xia[1,2,3,4(✉)], and Ming Ma[5]

[1] College of Information Engineering,
Capital Normal University, Beijing, China
xlk@cnu.edu.cn
[2] Laboratory of Neural Computing and Intelligent Perception (NCIP),
New York, USA
[3] International Science and Technology Cooperation Base of Electronic System
Reliability and Mathematical Interdisciplinary, Beijing, China
[4] Beijing Advanced Innovation Center for Imaging Technology, Beijing, China
[5] School of Medicine, Stanford University, Stanford, USA

Abstract. The Common Spatial Pattern (CSP) algorithm is capable of solving the binary classification problem for the motor image task brain-computer interface (BCI). This paper proposes a novel method based on the Filter Bank Common Spatial Pattern (FBCSP) termed the Multiscale and Overlapping FBCSP (MO-FBCSP). We extend the CSP algorithm for multiclass by using the one-versus-one (OvO) strategy. Multiple periods are selected and combined with the overlapping spectrum of the filter bank which contains useful information. This method is evaluated on the benchmark BCI Competition IV dataset 2a with 9 subjects. An average accuracy of 80% was achieved with the random forest (RF) classifier, and the corresponding kappa value was 0.734. Quantitative results have shown that the proposed scheme outperforms the classical FBCSP algorithm by over 12%.

Keywords: Brain-computer interface · Motor imagery · Machine learning · EEG

1 Introduction

Brain-computer interface (BCI) is a technology which can directly connect the brain to external devices. The devices are controlled by a general brain-like model that is built with machine learning (ML) techniques based on Electroencephalogram (EEG) signals [1], whereas motor imagery BCI (MI-BCI) is committed to using ML to recognize EEG signal in different patterns. The MI-BCI can be widely applied in many fields including medicine, education, military, and entertainment [2]. However, the application of BCI in reality is often restricted by low bit-transfer rates that barely meet requirements. In order to improve the situation, one way is to allow subjects to generate multiple states of motor imagery [3].

© Springer Nature Switzerland AG 2019
H. Yin et al. (Eds.): IDEAL 2019, LNCS 11872, pp. 141–149, 2019.
https://doi.org/10.1007/978-3-030-33617-2_16

The event-related synchronization/desynchronization (ERS/ERD) phenomenon is very obvious in the motion imagination of the left and right hands. However, the phenomenon of tongue and feet is weak and only found from certain subjects [4]. It has resulted in low accuracy for the classification of multiclass motor imagery BCI (MI-BCI). In addition, only a small size sample of the motor imagery EEG signal is produced.

In this paper, we propose a novel method termed Multiscale and Overlapping Filter Bank Common Spatial Pattern (MO-FBCSP) based on the Filter Bank Common Spatial Pattern (FBCSP) algorithm. The CSP can maximize the difference between two types of EEG. But if there are multiple types of signals, methods such as One-versus-One (OvO) and One-versus Rest (OvR) are often used. For four classes of tasks, the OvO strategy needs six CSP filters, whereas OvR strategy requires four CSP filters. Since the OvR algorithm uses all samples at a time, whereas the OVO only needs two types of signals, the OvO strategy is selected in this paper. In addition to 4–40 Hz, the frequency band covered by 6–42 Hz is selected. In addition to the overlapping filter bank, the multiple time segmentation as an alternative selection method is also taken into consideration.

The contribution of our work is that the appropriate number of time windows and the filter banks of overlapping frequency band can fully capture the feature, while the computational cost is taken into account. Moreover, the proposed method can extract more features than the classical FBCSP algorithm. Improvement of over 12% in accuracy is achieved as compared to the classical CSP and FBCSP algorithm. The training and testing time is also reduced significantly.

2   Related Work

CSP was originally used to identify people with neurological disorders and healthy people [5]. In the brain activity of motor imagery, the ERS/ERD phenomenon occurs, which can construct a feature vector with significant difference. The classical CSP algorithm is applied for the two-class task. Given a signal $X = \mathbb{R}^{N_c \times N_T}$, which is then subtracted by the mean value of each channel, the norm covariance matrix C is:

$$C = \frac{1}{\text{trace}(XX^T)} \left(XX^T \right) \tag{1}$$

where N_C denotes the number of channels, N_T indicates the number of samples, and X^T is the transpose of the matrix X.

The basic idea of the CSP algorithm is to look for a spatial filter $W \in \mathbb{R}^{N_c}$ which can maximize the differences of signals of two classes. This is considered to be equivalent to maximizing the Rayleigh quotient, where C_k denotes the mean of the norm covariance matrix which belongs to class k.

$$R(w) = \frac{w^T C_1 w}{w^T C_2 w} \tag{2}$$

$$f_i = \log\left(\frac{w_i^T X X^T w}{\sum_{j=1}^{2m} w_j^T X X^T w}\right) \tag{3}$$

We take several pairs of the column vectors corresponding to relatively large and small eigenvalues. These eigenvectors are the spatial filters as expected, where f indicates the eigenvectors of EEG signal and w is the spatial filters.

The classical CSP algorithm relies on manual selection of frequency bands. Several methods were proposed to select the optimal temporal frequency band for the CSP algorithm, such as the Separable Common Spatio-Spectral Pattern (CSSP) [6], the Common Sparse Spectral-Spatial Pattern (CSSSP) [7], and Filter Bank Common Spatial Pattern (FBCSP) [8]. Besides, wavelet decomposition is often used for feature extraction [9]. The FBCSP performs the CSP algorithm on every frequency band and selects features by mutual information-based best individual feature algorithm [8]. Combining the OvR expansion strategy with Bayesian classifiers, they obtained a champion in the BCI Competition IV dataset 2a by achieving a mean kappa value of 0.57.

Deep learning methods achieve state-of-the-art level in computer vision and natural language processing. In order to deal with the problem, Lu et al. proposed a deep learning scheme based on restricted Boltzmann machine [10]. Sakhavi et al. used the FBCSP algorithm as their data preparation method and proposed a novel envelope representation for the MI-EEG [11]. They combined envelope representation with Convolutional Neural Networks (CNN) and achieved a mean kappa value of 0.659. Amin et al. extracted and fused multilevel convolutional features from different CNN layers. The features are abstract representations of the EEG signal at various levels. Their method achieved a mean accuracy of 0.754 and the kappa value of 0.672 [12].

Riemannian approaches, spearheaded by the use of covariance matrices, are such a promising tool that a growing number of researchers have started to use it. Barachant et al. [13] started to use Riemannian approaches by mapping the covariance matrices onto the Riemannian tangent space and achieved a mean kappa value of 0.567. Gaur et al. [14] proposed a subject specific multivariate empirical mode decomposition (MEMD) based on filtering method, and achieved a mean kappa value of 0.6. The study [15] used the Large Multiscale Temporal and Spectral Features based on Riemannian covariance features, and achieved a mean accuracy of 0.755 and the kappa value of 0.673.

3   Methodology

The MO-FBCSP is structured as shown in Fig. 1. Initially, it takes the multi-channel EEG signals as inputs, and then processes them in order to obtain the sufficient features. Finally, classification is performed using the random forest (RF). In order to capture more robust features, two filter banks with overlapping frequency band are selected. Both individual reaction time and the time windows for performing motor imagery may be different. If more time periods are selected, longer training time and testing time are taken due to the large size of feature vector. In this case, five overlapping time windows are selected. We record the filter bank of 4–40 Hz as B1 and the filter bank of 6–42 Hz as B2. Each of the banks contains 9 filters, where the pass band of each filter is 4 Hz. Such selection of redundant time windows and frequency bands enhances the robustness of the feature extractor.

A crucial step in the MO-FBCSP is z-score standardization. It can dramatically improve the accuracy of the EEG classification. This is performed after extracting the feature vector. Normalization is another method of feature preprocessing which changes the sum of the squares of the features in a sample to 1.

In view of the superior classification effect and easiness to tuning the hyper-parameters, the feature vectors generated by all binary CSP filters are placed into one row as the feature vector. The RF is used for feature classification with Gini coefficient as the criterion, where each decision tree is allowed to randomly divide the number of the square root of the number of total attributes as a subset sqrt(n), and then select the best attribute from the subset for classification. There are 1,000 decision trees in the RF classifier used in this study. The parameters are obtained based on 5-fold cross validation (CV).

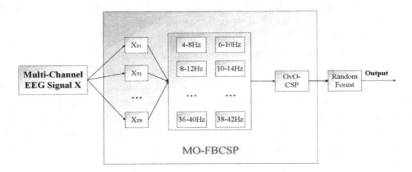

Fig. 1. The architecture of MO-FBCSP

4   Experiments

4.1   Dataset Acquisition

This paper evaluates the proposed method on the BCI Competition IV 2a. The EEG data was collected from 9 subjects (4 females and 5 males with age range between 22 and 26 years) with four different motor imaging tasks, namely left hand (class 1), right hand (class 2), feet (class 3) and tongue (class 4) [16]. Each subject was required six runs. Each run contained 48 trials, yielding a total of 288 trials per session. Each subject completed two sessions on different days. One session is the training set, and the other session is the testing set. The number of trials per class is balanced. At the beginning of a trial (T = 0 s), a fixation cross appeared on the screen, and a short transitory warning tone was presented. At T = 2 s, a cue in the form of an arrow pointed either to the left, right, up or down which corresponds to the one of four classes left hand, right hand, tongue or foot. This prompts subjects to perform the desired task. At T = 6 s, the fixation cross disappeared from the screen and the subjects stopped their motor imagery. Finally, a short break was followed where the screen was black again. The signals were sampled with 250 Hz and bandpass-filtered between 0.5 Hz and 100 Hz. The sensitivity of the amplifier was set to 100 μV. Moreover, one 50 Hz notch filter was enabled to suppress line noise.

4.2 Parameter Setup

As illustrated in Table 1, five time windows (T1–T5) and two filter banks (B1, B2) are selected. Each of the banks includes nine filters and each filter selected is a 2nd order Butterworth filter with the bandwidth of 4 Hz. As is mentioned above, this paper uses the OvO strategy to extend the CSP algorithm for the four-class problem. A total of six CSP filters are needed for a time window and one frequency band. In each binary CSP filter, two pairs of eigenvectors are selected.

Table 1. Time window and filter bank selection

T1	T2	T3	T4	T5	B1	B2
2–5 s	2.5–5.5 s	3–6 s	2.5–4.5 s	4.5–6.5 s	4–40 Hz	6–42 Hz

The importance of z-score standardization of feature vectors has been repeatedly mentioned above. In fact, standardization is a generally required step before classification. For many machine learning estimators, if features are very far from standard Gaussian distribution, the performance may be very poor. Normalization mentioned above is another standardization method. In order to verify the performance of different feature preprocessing methods, a comparison is conducted, where B1, T3 and one pair of features in a binary CSP filter are chosen. The results are shown in Fig. 2. It can be seen that the performance of normalization is worse than that of z-score standardization, and the performance can be improved by using z-score standardization after normalization. Therefore, in this case, the preprocessing method of using z-score standardization after normalization is selected, i.e., normalization + z-score (n + z).

The parameters selected in the case are: Time window: T1–T5; Filter bank: B1, B2; Two pairs. The decision can be found in Table 2.

Table 2. Experimental results for parameter selection

Time window	T3	T3	T1–T3	T1–T3	T1–T5	T1–T5	T1–T5
Filter Bank	B1	B1, B2	B1	B1, B2	B1	B1, B2	B1, B2
Pair number	1	1	1	1	1	1	2
Feature Number	108	216	324	648	540	1080	2160
Preprocess	n+z	n + z	n + z	n + z	n + z	n + z	n + z
Classifier	RF	RF	RF	RF	RF	RF	RF
S1	**0.854(0.806)**	0.854(0.806)	0.896(0.861)	0.889(0.852)	0.885(0.847)	0.892(0.856)	0.917(0.889)
S2	0.622(0.495)	0.608(0.477)	0.604(0.472)	0.594(0.458)	0.583(0.444)	0.604(0.472)	0.608(0.477)
S3	**0.920(0.894)**	0.920(0.894)	0.931(0.907)	0.934(0.912)	0.955(0.940)	0.948(0.931)	0.941(0.921)

(*continued*)

Table 2. (*continued*)

Time window	T3	T3	T1–T3	T1–T3	T1–T5	T1–T5	T1–T5
S4	0.656(0.542)	0.688(0.583)	0.705(0.606)	0.733(0.644)	0.705(0.606)	0.726(0.634)	0.743(0.657)
S5	0.667(0.556)	0.694(0.593)	0.684(0.579)	0.701(0.602)	0.670(0.560)	0.684(0.579)	0.708(0.611)
S6	**0.517(0.356)**	0.542(0.389)	0.569(0.426)	0.590(0.454)	0.562(0.417)	0.597(0.463)	0.642(0.523)
S7	0.892(0.856)	0.899(0.866)	0.868(0.824)	0.899(0.866)	0.875(0.833)	0.896(0.861)	0.903(0.870)
S8	0.844(0.792)	0.819(0.759)	0.865(0.819)	0.875(0.833)	0.872(0.829)	0.872(0.829)	0.878(0.838)
S9	**0.788(0.718)**	0.781(0.708)	0.847(0.796)	0.847(0.796)	0.847(0.796)	0.868(0.824)	0.861(0.815)
Avg	0.751(0.668)	0.756(0.675)	0.774(0.699)	0.785(0.713)	0.773(0.697)	0.787(0.717)	**0.800(0.734)**
Avg-Training-Time	1.013	1.381	1.380	1.496	1.499	1.591	1.793
Avg-Testing-Time	0.204	0.226	0.238	0.227	0.238	0.205	0.217

4.3   Quantitative Results

Given the confusion matrix L of a classification result, the accuracy is defined blow,

$$accuracy = \frac{\sum_i L_{ii}}{N}, \tag{4}$$

where N denotes the total number of samples. The kappa value is defined as,

$$kappa = \frac{P_0 - P_e}{1 - P_e}, \tag{5}$$

where P_0 corresponds to accuracy and the formula of P_e is as follows,

$$P_e = \frac{\sum_{i=1}^{m} L_{:,i} \times L_{i,:}}{L \times L}, \tag{6}$$

where m denotes the total number of class, $L_{:,i}$ and $L_{i,:}$ denote the sum of ith column and row of the confusion matrix, respectively.

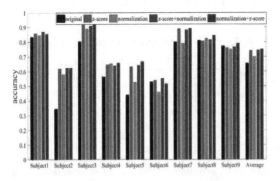

Fig. 2. Comparison of various feature preprocessing methods

The parameters are evaluated and decided in Table 2, as mentioned in Sect. 4.2. We compared the MO-FBCSP with existing methods by observing the kappa value. The comparison results are highlighted in Table 3.

Table 3. Comparison between MO-FBCSP and other related work

Method	S1	S2	S3	S4	S5	S6	S7	S8	S9	Avg
Ang et al. [8]	0.68	0.42	0.75	0.48	0.40	0.27	0.77	0.75	0.61	0.57
Wang [18]	0.67	0.49	0.77	0.59	0.52	0.31	0.48	0.75	0.65	0.58
Kam et al. [19]	0.74	0.35	0.76	0.53	0.38	0.31	0.84	0.74	0.74	0.60
Nicolas-Alonso et al. [20]	0.83	0.51	0.88	0.68	0.56	0.35	**0.90**	0.84	0.75	0.70
He et al. [21]	0.69	0.51	0.87	**0.85**	**0.78**	0.42	0.54	**0.97**	0.45	0.66
Gaur et al. [14]	0.86	0.24	0.70	0.68	0.36	0.34	0.66	0.75	0.82	0.60
Xie et al. [22]	0.77	0.43	0.78	0.52	0.51	0.32	0.78	0.78	0.77	0.63
Sakhavi et al. [11]	0.83	0.54	0.87	0.56	0.50	0.27	0.86	0.78	0.73	0.66
Hersche et al. [15]	0.87	0.41	0.75	0.63	0.59	0.42	0.81	0.78	0.80	0.67
Razi et al. [17]	0.78	**0.59**	0.85	0.72	0.67	**0.57**	0.81	0.86	**0.88**	**0.75**
MO-FBCSP	**0.89**	0.48	**0.92**	0.66	0.61	0.52	0.87	0.84	0.82	0.734

It is observed that our method has achieved kappa value of 0.73, which is much higher than most of the state-of-art methods in the literature except for one from [17] (kappa value = 0.75), which is about 2.7% less. However, in the case, only one classifier is used in our method, whereas the method in [17] used thirteen classifiers, which should be considerably slower than our method. In the real-time BCI system, the testing time is one of the most important factors. In this case, the average test time is 0.217 s with 288 samples, so the mean testing time is about 0.753 ms for each sample. The time costs are obtained using a system with Intel(R) Core i7 2.80 GHz CPU and 8 GB RAM. The test sample in [17] using 34.3 ms by a system with Intel(R) Core (TM) 2.50 GHz CPU and 12 GB RAM. Therefore, despite under the different conditions, the training and testing time of the proposed method is much faster than the one in [17].

5 Conclusion

In this study, we proposed a supervised learning method termed MO-FBCSP for multiclass motor imagery EEG classification. It integrates normalization with z-score standardization after extracting the feature vectors in order to improve the classification accuracy. The Mutual Information Based Individual Feature (MIBIF) method was used to reduce the number of feature vectors and also reduce the training time. As compared with the existing method [21], the MO-FBCSP is capable of greatly reducing the number of feature vectors while improving the accuracy. Moreover, the training time and testing time is quick enough to meet real-time BCI applications. So far, the MO-FBCSP has achieved reasonable accuracy and high speed. In the future work, more effective classifiers will be taken into consideration.

Acknowledgement. This work was partially supported by a Natural Science Foundation of China (NSFC) grant (61572076) and Beijing Advanced Innovation Center for Imaging Technology.

References

1. Zhang, Y., Wang, Y., Zhou, G., et al.: Multi-kernel extreme learning machine for EEG classification in brain-computer interfaces. Expert Syst. Appl. **96**, 302–310 (2018)
2. Rajkomar, A., Dean, J., Kohane, I.: Machine learning in medicine. N. Engl. J. Med. **380**, 1347–1358 (2019)
3. Dornhege, G., Blankertz, B., Curio, G., Muller, K.R.: Boosting bit rates in noninvasive EEG single-trial classifications by feature combination and multiclass paradigms. IEEE Trans. Biomed. Eng. **51**(6), 993–1002 (2004)
4. Pfurtscheller, G., Brunner, C., Schlögl, A., Silva, F.H.: Mu rhythm (de)synchronization and EEG single-trial classification of different motor imagery tasks. NeuroImage **31**(1), 153–159 (2006)
5. Koles, Z.J., Lazar, M.S., Zhou, S.Z.: Spatial patterns underlying population differences in the background EEG. Brain Topogr. **2**(4), 275–284 (1990)
6. Lemm, S., Blankertz, B., Curio, G., et al.: Spatio-spectral filters for improving the classification of single trial EEG. IEEE Trans. Biomed. Eng. **52**(9), 1541–1548 (2005)
7. Dornhege, G., Blankertz, B., Krauledat, M., et al.: Combined optimization of spatial and temporal filters for improving brain-computer interfacing. IEEE Trans. Biomed. Eng. **53**(11), 2274–2281 (2006)
8. Ang, K.K., Chin, Z.Y., Wang, C., Guan, C., Zhang, H.: Filter bank common spatial pattern algorithm on BCI competition IV Datasets 2a and 2b. Front. Neurosci. **6**, 39. https://doi.org/10.3389/fnins.2012.00039
9. Iacoviello, D., Petracca, A., Spezialetti, M., et al.: A classification algorithm for electroencephalography signals by self-induced emotional stimuli. IEEE Trans. Cybern. **46**(12), 3171–3180 (2016)
10. Lu, N., Li, T., Ren, X., et al.: A deep learning scheme for motor imagery classification based on restricted Boltzmann machines. IEEE Trans. Neural Syst. Rehabil. Eng. **25**(6), 566–576 (2017)
11. Sakhavi, S., Guan, G., Yan, S.C.: Learning temporal information for brain-computer interface using convolutional neural networks. IEEE Trans. Neural Netw. Learn. Syst. **29**(11), 5619–5629 (2018)
12. Amin, S.U., Alsulaiman, M., Muhammad, G., et al.: Multilevel weighted feature fusion using convolutional neural networks for EEG motor imagery classification. IEEE Access **7**, 18940–18950 (2019)
13. Barachan, A., Bonnet, S., Congedp, M.: Multiclass brain-computer interface classification by riemannian geometry. IEEE Trans. Biomed. Eng. **59**(4), 920–928 (2011)
14. Gaur, P., Pachori, R.B., Wang, H., et al.: A multi-class EEG-based BCI classification using multivariate empirical mode decomposition based filtering and Riemannian geometry. Expert Syst. Appl. **95**(1), 201–211 (2018)
15. Hersche, M., Rellstab, T., Schiavone, P.D., et al.: Fast and accurate multiclass inference for mi-bcis using large multiscale temporal and spectral features. In: 26th European Signal Processing Conference (EUSIPCO), Rome, Italy, pp. 1690–1694. IEEE (2018)
16. Brunner, C., Leeb, R., Muller-Putz, R., Schlogl, A., Pfurtscheller, G.: BCI competition 2008 - Graz data set A. http://www.bbci.de/competition/iv/desc_2a.pdf. Accessed 10 July 2019

17. Razi, S., Mollaei, M.R.K., Ghasemi, J.: A novel method for classification of BCI mul-ti-class motor imagery task based on Dempster-Shafer theory. Inf. Sci. **484**, 14–26 (2019)
18. Wang, H.X.: Multiclass filters by a weighted pairwise criterion for EEG single-trial classification. IEEE Trans. Biomed. Eng. **58**(5), 1412–1420 (2011)
19. Kam, T.E., Suk, H.I., Lee, S.W.: Non-homogeneous spatial filter optimization for EEG-based brain-computer interfaces. In: International Winter Workshop on Brain-Computer Interface, Gangwo, South Korea, pp. 26–28. IEEE (2013)
20. Nicolas-Alonso, L.F., Corralejo, R., Gomez-Pilar, J., et al.: Adaptive semi-supervised classification to reduce intersession non-stationarity in multiclass motor imagery-based brain computer interfaces. Neurocomputing **159**, 186–196 (2015)
21. He, L.H., Hu, D., Wan, M., et al.: Common Bayesian network for classification of EEG-Based multiclass motor imagery BCI. IEEE Trans. Syst. Man Cybern. Syst. **46**(6), 843–854 (2016)
22. Xie, X.F., Yu, Z.L., Gu, Z.H., et al.: Bilinear regularized locality preserving learning on riemannian graph for motor imagery BCI. IEEE Trans. Neural Syst. Rehabil. Eng. **26**(3), 698–708 (2018)

Safe Deep Neural Network-Driven Autonomous Vehicles Using Software Safety Cages

Sampo Kuutti[1]([⊠]) [iD], Richard Bowden[2] [iD], Harita Joshi[3], Robert de Temple[3], and Saber Fallah[1] [iD]

[1] Connected Autonomous Vehicles Lab, University of Surrey, Guildford, UK
{s.j.kuutti,s.fallah}@surrey.ac.uk
[2] Centre for Vision, Speech and Signal Processing,
University of Surrey, Guildford, UK
r.bowden@surrey.ac.uk
[3] Jaguar Land Rover, Coventry, UK
{hjoshi3,rdetempl}@jaguarlandrover.com

Abstract. Deep learning is a promising class of techniques for controlling an autonomous vehicle. However, functional safety validation is seen as a critical issue for these systems due to the lack of transparency in deep neural networks and the safety-critical nature of autonomous vehicles. The black box nature of deep neural networks limits the effectiveness of traditional verification and validation methods. In this paper, we propose two software safety cages, which aim to limit the control action of the neural network to a safe operational envelope. The safety cages impose limits on the control action during critical scenarios, which if breached, change the control action to a more conservative value. This has the benefit that the behaviour of the safety cages is interpretable, and therefore traditional functional safety validation techniques can be applied. The work here presents a deep neural network trained for longitudinal vehicle control, with safety cages designed to prevent forward collisions. Simulated testing in critical scenarios shows the effectiveness of the safety cages in preventing forward collisions whilst under normal highway driving unnecessary interruptions are eliminated, and the deep learning control policy is able to perform unhindered. Interventions by the safety cages are also used to re-train the network, resulting in a more robust control policy.

Keywords: Automatic control · Autonomous vehicles ·
Cyber-physical systems · Deep learning · Safety

1 Introduction

Autonomous vehicles are proposed as the future of intelligent transportation systems to address problems such as traffic congestion, pollution, and road

© Springer Nature Switzerland AG 2019
H. Yin et al. (Eds.): IDEAL 2019, LNCS 11872, pp. 150–160, 2019.
https://doi.org/10.1007/978-3-030-33617-2_17

safety [4,14,16,19]. Deep learning has emerged as a popular artificial intelligence technique for autonomous vehicles, and has been proposed for many uses in autonomous vehicles, including vehicle control [13]. The downside of deep learning techniques is the opaqueness of the learned systems. In a safety-critical system, such as an autonomous vehicle, the safety of all sub-components as well as the overall system must be validated to a high level of safety assurance. The lack of interpretability in deep neural networks currently prevents effective safety validation. Moreover, the complex nature of the driving task and the operational environment mean that targeted testing methods are less useful due to the inability to test the autonomous vehicle in all possible use cases [3,10]. In order to introduce deep learning solutions to the next generation of intelligent vehicles, new safety validation techniques must be found to address them specifically [13].

Safety cages have been proposed for black box systems, where the safety of the system must be ensured without necessarily having full understanding of how the system works. Given the opaqueness of deep neural networks, such safety systems show great promise to improve the safety of the overall system. Safety cages eliminate unsafe actions by imposing limits on possible control actions. Utilising run-time monitoring, the safety cages can change dynamically based on the state of the system and its environment [7]. For instance, the safety cages in an autonomous vehicle can change the limits on acceleration based on the relative distance to nearby vehicles or the current speed of the vehicle. However, given that the safety cages intervene on the control output, a well designed safety cage must minimise unnecessary interventions. Safety cages have been used in cyber-physical systems where full offline safety validation is not possible, such as in robotics [6,12] or aerospace [17] applications, to intervene on the controller outputs in the presence of faults or dangerous control outputs. For autonomous vehicles, Heckemann et al. [7] suggested that these techniques could be useful for ensuring the safety of complex and adaptive machine learning systems in autonomous vehicles. Adler et al. [1] proposed safety cages based on five constraints such as "accelerating if a slower vehicle is closely in front" to meet the five Automotive Safety Integrity Levels (ASIL) defined in ISO26262 [8].

The contributions of this paper are three-fold. First, we present an imitation learning method for longitudinal control of an autonomous vehicle. The model is trained on data collected in IPG CarMaker [9], where the default driver demonstrates the desired driving policy. Second, we present two software safety cages designed to prevent forward collisions in an autonomous vehicle during highway driving. The safety cages are used to intervene on the control output of the neural network in safety critical scenarios, only when the network output is not reacting adequately to the current danger. Unnecessary interventions on the control output, which would degrade the performance of the controller and the comfort of the passengers, are minimised. Extensive testing under normal driving scenarios and in critical scenarios on two different neural networks demonstrate the effectiveness of the approach. Third, we demonstrate that the interventions by the safety cages can be used for re-training the network, improving the robustness of the learned policy to mistakes in novel states. The remainder of the paper is structured as follows. Section 2 describes the neural network algorithm developed for longitudinal control of an autonomous vehicle. Section 3 describes the developed safety cages to prevent forward collisions. Testing of the safety cages in a simulated car following scenario are presented in Sect. 4. Finally, concluding remarks and potential future work is given in Sect. 5.

2 Longitudinal Control Algorithm

2.1 Data Collection

The collection of training data was carried out by defining various highway driving scenarios in IPG CarMaker [9] where the lead vehicle varies its velocity over time. In these scenarios, CarMaker's pre-defined driver, IPG Driver, was used to control the host vehicle and was set to maintain a 2 s time headway from the lead vehicle. The velocity of the lead vehicle was in the interval [17, 30] m/s, whilst the acceleration was limited to the interval $[-2, 2]$ m/s^2, since higher acceleration magnitudes would be uncomfortable for passengers [20]. All data was collected under dry road conditions with a road friction coefficient $\mu = 1.0$. The combined scenarios amount a total of 2 h of driving data. Sampling the simulation at 50 Hz, this amounts to 375,000 data points. The collected data set was then split into the training and validation data sets, with 80% of data in the training set and 20% in the validation set. This data was solely used for training and validating the deep learning control policy. A further 10 h of simulation, for each control policy, was used in live testing of these policies in conjunction with the safety cages in a variety of scenarios and road conditions that were not seen during training, and hence would be expected to prove challenging for the control policy to generalise to.

2.2 Learning Algorithm

The inputs to the network were defined as time headway t_{hw}, relative velocity v_{rel}, host vehicle velocity v, and host vehicle acceleration a. The output of the network y was defined as a single parameter based on the gas pedal and brake pedal values such that $y \in [-1, 1]$, where positive values signal the use of the gas pedal whereas negative values correspond to the use of the brake pedal. The activation function at each hidden layer neuron is the Rectified Linear Unit (ReLU) function, whilst the output layer uses a tanh activation. The network output is then compared to the output of IPG Driver, $\hat{y}_k$, in the training data. The model is then trained through imitation learning, using the mean square error between predicted output and ground truth as the loss function. After performing a grid search for the network hyperparameters, the final network architecture has 3 hidden layers with 50 neurons each, trained with a learning rate of 1×10^{-2} and batch size of 100. The final model was trained for 1×10^{6} training steps, resulting in a final validation loss of 0.0150463. For the purposes of testing the effectiveness of the safety cages on an unsafe controller, a second smaller neural network was also trained for the same task. The second suboptimal network has the same parameters as the above-mentioned network, except it only has one hidden layer with 10 neurons. Note, this network's parameters were not optimised for performance, as a suboptimal model gives a better opportunity to investigate the safety benefit offered by the proposed safety cages. To avoid confusion between the two trained networks, the deeper network with optimised hyperparameters will henceforth be referred to as the *deep network* and the smaller suboptimal network will be referred to as the *shallow network*. The losses during training for both networks can be seen in Fig. 1.

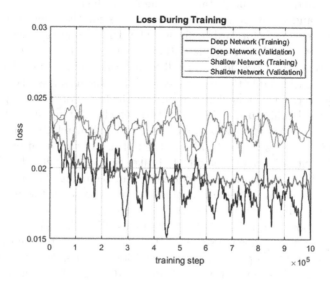

Fig. 1. Smoothed loss curves for training and validation sets.

3   Safety Cages

Software safety cages have been used in many applications as a means to improve safety by limiting outputs to a safe operational envelope. In their simplest forms, the safety cages can be hard upper/lower limits on the output. By using run-time monitoring to observe the state of the system and its environment, context-aware safety cages can use the observed states to dynamically limit the control output as the situation requires. For instance, in the problem of autonomous driving, a control output calling for full acceleration may be safe when there are no vehicles ahead but would be unsafe if the host vehicle is already close to the lead vehicle. Using such situational awareness, the potentially dangerous outputs of the neural network can be prevented by limiting their outputs during critical scenarios. Therefore, the safety validation requirements on the neural network can be relaxed, given that the software safety cages can be validated with high assurance [11].

The safety cages developed here focus on the longitudinal control of an autonomous vehicle described in previous sections, attempting to prevent forward collisions in highway driving scenarios. The safety cages observe the time headway (t_{hw}) and time-to-collision (TTC) to the lead vehicle as given by:

$$t_{hw} = \frac{x_{rel}}{v} \tag{1}$$

$$TTC = \frac{x_{rel}}{v_{rel}} \tag{2}$$

where v is the host vehicle velocity, v_{rel} is the relative velocity of the host and lead vehicles, and x_{rel} is the distance to the lead vehicle.

The TTC and t_{hw} were chosen as the metrics for the safety cages as they represent the risk of a forward collision. The TTC value represents the time required for two vehicles to collide if they continue at their current velocities and trajectories. Therefore, for a car following scenario in a single lane, such as the scenarios considered in this paper, a low TTC value means a forward collision is likely. However, TTC alone does not provide the full information regarding the risk of a forward collision. For instance, two vehicles driving at high speeds on the highway may be very close to each other, but if their relative speed is low or zero the TTC metric would not indicate the full risk of the situation. If the lead vehicle in this situation had to suddenly brake, the vehicle behind would be too close to react in time and prevent a collision from occurring. Therefore, a metric such as time headway is useful. Time headway represents the intervehicular distance in time, based on the host vehicle's velocity. Since t_{hw} does not make assumptions about the lead vehicle's actions as TTC does, it acts as a good safety metric in a car following scenario. These observed states are then used to identify a risk level for a possible forward collision. The risk levels were based on the TTC and t_{hw} based risk threshold presented in [2,5,15] with the final risk thresholds tuned to avoid collisions whilst minimising unnecessary interventions by the safety cages, as shown in Fig. 2. The safety cages impose a minimum brake pedal value when the risk of forward collision exceeds the given threshold. The safety cages intervene on the control action, if the neural network outputs a control action with less than the minimum required braking. Therefore, the safety cages do not decide the correct action for each scenario, but only intervene on the control action as a fall-back safety mechanism when the neural network is not responding to the level of risk adequately. Using the notation from the previous section, where a negative output y represents braking, the final control action is given by

$$y = min(y_{nn}, y_{sc}) \tag{3}$$

where y_{nn} is the output of the neural network and y_{sc} is the minimum required braking imposed by the safety cages.

4   Simulation Results

In order to investigate the effectiveness of the proposed safety cages, the two trained neural networks were tested under various car following scenarios. All

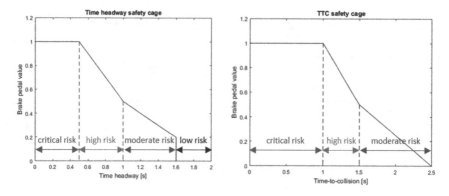

Fig. 2. Safety Cages with Time headway and Time-to-Collision based safety envelopes.

testing was done in the IPG CarMaker simulation platform. The scenarios were broken down to 5-min episodes, where the episode ends after the 5 min or if a crash occurred. At the start of each episode a road friction coefficient value between 0.4 and 1.0 was chosen. The lead vehicle performed randomly chosen manoeuvrers, with the velocity limited to $v_{lead} \in [17, 40]\,\mathrm{m/s}$, and the acceleration limited to $a_{lead} \in [-2, 2]\,\mathrm{m/s^2}$. The exception to this was emergency braking manoeuvrers, which the lead vehicle performed, on average, once an hour. During emergency braking the deceleration was chosen between $a_{lead} \in [-6, -3]\,\mathrm{m/s^2}$. The combined testing includes 40 h of driving overall, with various road conditions and different manoeuvrers performed by the lead vehicle. This includes testing the deep network with safety cages, shallow network with safety cages, shallow network without safety cages, and re-trained shallow network, for 10 h each. We start the section by presenting results of the key types of scenarios (e.g. normal highway driving, emergency braking, wet road conditions) to investigate what the networks have learned and the effectiveness of the safety cages. Finally, we present and discuss the overall results of each 10 h test run.

The first tests validated the performance of the networks under typical highway driving scenarios similar to those seen in the training data (Fig. 3a). Both networks show that they have learned a reasonable driving policy, keeping a safe headway close to the target headway of 2 s. Moreover, no safety cage interventions are required in these scenarios for either network. This is not surprising as we are asking the networks to predict vehicle control actions for scenarios similar to those they were trained to operate in. Following this, the generalisation capability to completely new scenarios was tested. Firstly, the networks were tested under different road conditions. Since all training data was from dry road conditions (road friction coefficient $\mu = 1.0$), the performance was tested under various μ values ranging from 0.4 to 1.0. Vehicle following scenario at $\mu = 0.55$ can be seen in Fig. 3b. Both networks have learned to keep a safe time headway and adjust well to different road conditions, even if they were trained only on dry road condition. Again, no interventions by the safety cages occur in these scenarios. This in itself is impressive and demonstrates the power of deep learning.

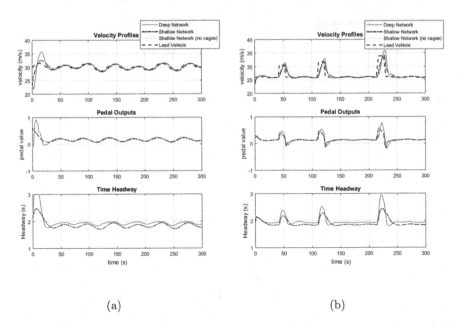

Fig. 3. Neural network controllers following a lead vehicle at (a) $\mu = 1.0$ and (b) $\mu = 0.55$.

The second generalisation tests involved emergency braking performed by the lead vehicle. Since no emergency manoeuvrers were included in the training data, the networks' response to emergency braking by the lead vehicle is more interesting. Figure 4a and b show emergency braking manoeuvrers performed by the lead vehicle with a deceleration of $5\,\mathrm{m/s^2}$ at μ values of 0.9 and 0.5, respectively. Here, the deep network generalises its previously learned rules to perform well at the emergency manoeuvrer, whilst the shallow network fails to generalise. Both emergency braking scenarios lead to the shallow network causing a forward collision with the lead vehicle when no safety cages are used. The shallow network initially starts to decelerate as the lead vehicle decelerates, but when seeing inputs to the network not seen during training, it cannot generalise to the new situation due to insufficient amount of parameters compared to the deep network and begins to accelerate until it crashes. However, when safety cages intervene on the shallow network's control action, the braking by the safety cages brings the vehicle back from the low time headway to a safe one, where the shallow network resumes control of the vehicle. Moreover, these scenarios show that the safety cages can respond adequately to an emergency scenario by keeping the vehicle at a safe distance without excessive braking.

The networks were each tested in 10 h of overall driving, for which the results can be seen in Table 1. The results from the IPG Driver demonstrator are also shown as a baseline for comparison. It can be seen from the results that the deep network has learned a safe driving policy, which keeps the vehicle at a safe

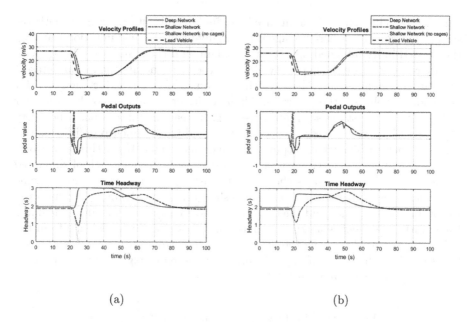

(a)   (b)

Fig. 4. Neural network controllers with the lead vehicle performing an emergency braking manoeuvrer (a) at $\mu = 0.9$, where the shallow network without safety cages crashes at t = 24.72 s and (b) at $\mu = 0.5$, where the shallow network without safety cages crashes at t = 21.96 s.

distance from the lead vehicle without requiring the safety cages to intervene. The results show that the deep network has learned to outperform the IPG Driver, showing it generalises better to new scenarios (e.g. emergency braking) compared to the rule-based system. Also, given that the deep network can safely operate the vehicle, it can be seen that the safety cages never unnecessarily intervened on the control output, which could degrade the performance of the controller and lead to discomfort for the passengers. In comparison, the shallow network showed adequate performance in situations where the network inputs were in the same region as those seen during training. However, under emergency braking scenarios the shallow network fails to generalise and unexpectedly begins to accelerate until it crashes. This type of unexpected behaviour in new scenarios shows the efficacy of the proposed safety cages. The safety cages intervene on the network outputs, by decelerating the vehicle to a low risk region before handing the control back to the neural network again. Without the safety cages, the shallow network crashed 6 times in 10 h, whilst using the safety cages caused the cages to intervene on 360 control actions (equal to only 14.40 s total duration) which prevented all collisions from occurring.

As an additional experiment for improving the robustness of the neural network, the interventions by the safety cages were used to augment the original training set. The shallow network was then re-trained for a further 100,000 training steps using the new augmented training set and previously stated training parameters. This approach is inspired by similar multi-stage training methods, such as DAgger [18], where the agent is initially trained using imitation learning, then tested in its intended operational domain, and the states seen during testing are re-labelled by the expert (which was used to generate the initial dataset) to create an augmented dataset, which is then used for re-training. However, for autonomous vehicles, the expert (i.e. human driver) would be costly to use for re-labelling all the states seen during testing. Instead, in our approach the safety cages provide a more accessible signal for ground truth, albeit only in scenarios where the neural network is outputting dangerous actions. Using this framework, the re-trained shallow network was then tested again through 10 h of simulated driving, with the new results shown in the last row of Table 1. Although some interventions are still required, the results show that the re-training has improved the overall performance and safety of the shallow network. Thus, in our training framework, the agent is allowed to make mistakes and the knowledge from the safety cages is used to teach the network how to correct these mistakes, thereby boosting the robustness of the network without requiring costly queries to a human expert for ground truth labels.

Table 1. Neural network performance with 10 h of testing.

Network	x_{rel} min./mean (m)	v_{rel} max./mean (m/s)	t_{hw} min./mean (s)	Collisions	Safety cage violations
IPG Driver	10.7372/75.1552	13.8896/0.1866	1.0459/2.5471	0	–
Deep	23.8440/57.3687	8.8781/0.0197	1.7383/1.9895	0	0
Shallow	7.2504/54.4619	13.4619/0.0096	0.7765/1.8836	0	360
Shallow (no safety cages)	0/54.4530	20.8884/0.0005	0/1.8789	6	–
Shallow (re-trained)	13.1289/57.5260	11.2086/0.0222	1.1527/1.9881	0	198

5   Concluding Remarks

In this paper, two safety cages were proposed for prevention of forward collisions in an autonomous vehicle. The safety cages aim to improve the safety and reliability of the system, without requiring any white-box knowledge of the machine learning system. This is achieved by limiting the control output of the system to a safe envelope, which is defined by the time headway and time-to-collision to the vehicle in front. Therefore, by using run-time monitoring to observe the state of the host vehicle and the lead vehicle, the control action can be limited dynamically by a context-aware software safety cage. The results presented in Sect. 4 demonstrated the efficacy of the proposed safety cages. The safety cages were shown to correctly identify unsafe scenarios where control interventions were required and bring the vehicle back to a low risk region.

Moreover, the results under normal highway driving indicate that the safety cages do not unnecessarily intervene on the controllers actions and degrade the overall performance of the system. Instead, the safety cages step in when the neural network shows unexpected behaviour (e.g. by failing to generalise to a completely new scenario not included in the training data set) and ensure safe control actions are used. Therefore, the safety cages increase the confidence in the safety of the autonomous vehicle, by ensuring that incorrect outputs can be eliminated and giving an idea of what the worst case performance of the vehicle would be in safety-critical scenarios. Furthermore, it was shown that interventions by the safety cages could be used for re-training the network, which improved the performance and safety of the learned policy.

This work opens multiple potential avenues for future work. The safety cages presented here mitigate forward collisions under highway driving. The presented techniques could be used to further develop safety cages to account for lateral control, urban driving, etc. To improve the safety offered by the safety cages, fault identification and mitigation could be used to identify the effect of faulty measurements on the effectiveness of the safety cages. Furthermore, the interventions by the safety cages could be leveraged to better understand what the network has learned (or has not learned) by identifying the edge cases where the network fails. Further improvements to the re-training framework could also be investigated, for example, by using iterative testing and re-training, or by addressing the imbalance between the scarce interventions of the safety cages relative to the significantly larger original dataset.

Acknowledgments. This work was supported by the UK-EPSRC grant EP/R512217/1 and Jaguar Land Rover.

References

1. Adler, R., Feth, P., Schneider, D.: Safety engineering for autonomous vehicles. In: 2016 46th Annual IEEE/IFIP International Conference on Dependable Systems and Networks Workshop (DSN-W), pp. 200–205. IEEE (2016)
2. Archer, J.: Indicators for traffic safety assessment and prediction and their application in micro-simulation modelling: a study of urban and suburban intersections. Ph.D. thesis, KTH (2005)
3. Burton, S., Gauerhof, L., Heinzemann, C.: Making the case for safety of machine learning in highly automated driving. In: Tonetta, S., Schoitsch, E., Bitsch, F. (eds.) SAFECOMP 2017. LNCS, vol. 10489, pp. 5–16. Springer, Cham (2017). https://doi.org/10.1007/978-3-319-66284-8_1
4. Department for Transport: Research on the Impacts of Connected and Autonomous Vehicles (CAVs) on Traffic Flow: Summary Report (2017)
5. Glaser, S., Vanholme, B., Mammar, S., Gruyer, D., Nouveliere, L.: Maneuver-based trajectory planning for highly autonomous vehicles on real road with traffic and driver interaction. IEEE Trans. Intell. Transp. Syst. **11**(3), 589–606 (2010)
6. Haddadin, S., et al.: On making robots understand safety: embedding injury knowledge into control. Int. J. Robot. Res. **31**(13), 1578–1602 (2012)

7. Heckemann, K., Gesell, M., Pfister, T., Berns, K., Schneider, K., Trapp, M.: Safe automotive software. In: König, A., Dengel, A., Hinkelmann, K., Kise, K., Howlett, R.J., Jain, L.C. (eds.) KES 2011. LNCS (LNAI), vol. 6884, pp. 167–176. Springer, Heidelberg (2011). https://doi.org/10.1007/978-3-642-23866-6_18

8. International Organization for Standardization: ISO 26262: Road vehicles-functional safety. International Standard ISO/FDIS (2011)

9. IPG Automotive GmbH: Carmaker: Virtual testing of automobiles and light-duty vehicles (2017). https://ipg-automotive.com/products-services/simulation-software/carmaker/

10. Kalra, N., Paddock, S.M.: Driving to safety: how many miles of driving would it take to demonstrate autonomous vehicle reliability? Transp. Res. Part A Policy Pract. **94**, 182–193 (2016)

11. Koopman, P., Wagner, M.: Challenges in autonomous vehicle testing and validation. SAE Int. J. Transp. Saf. **4**(1), 15–24 (2016)

12. Kuffner Jr., J.J., Anderson-Sprecher, P.E.: Virtual safety cages for robotic devices, US Patent 9,522,471, 20 December 2016

13. Kuutti, S., Fallah, S., Bowden, R., Barber, P.: Deep Learning for Autonomous Vehicle Control: Algorithms, State-of-the-Art, and Future Prospects. Morgan & Claypool Publishers, San Rafael (2019)

14. Kuutti, S., Fallah, S., Katsaros, K., Dianati, M., Mccullough, F., Mouzakitis, A.: A survey of the state-of-the-art localization techniques and their potentials for autonomous vehicle applications. IEEE Internet Things J. **5**(2), 829–846 (2018)

15. Lu, J., Dissanayake, S., Xu, L., Williams, K.: Safety evaluation of right-turns followed by u-turns as an alternative to direct left turns: Crash data analysis. Florida Department of Transportation (2001)

16. Montanaro, U., et al.: Towards connected autonomous driving: review of use-cases. Veh. Syst. Dyn. **57**, 1–36 (2018)

17. Polycarpou, M., Zhang, X., Xu, R., Yang, Y., Kwan, C.: A neural network based approach to adaptive fault tolerant flight control. In: Proceedings of the 2004 IEEE International Symposium on Intelligent Control, pp. 61–66. IEEE (2004)

18. Ross, S., Gordon, G., Bagnell, D.: A reduction of imitation learning and structured prediction to no-regret online learning. In: Proceedings of the Fourteenth International Conference on Artificial Intelligence and Statistics, pp. 627–635 (2011)

19. Thrun, S.: Toward robotic cars. Commun. ACM **53**(4), 99 (2010)

20. Treiber, M., Kesting, A.: Car-following models based on driving strategies. In: Traffic Flow Dynamics, pp. 181–204. Springer, Heidelberg (2013). https://doi.org/10.1007/978-3-642-32460-4_11

Wave and Viscous Resistance Estimation by NN

D. Marón[1] and M. Santos[2(✉)] (iD)

[1] Allseas Engineering B.V., Delft, The Netherlands
d.maronblanco@gmail.com
[2] University Complutense of Madrid,
Profesor Gª Santesmases 9, 28040 Madrid, Spain
msantos@ucm.es

Abstract. Ship resistance estimation is one of the most important problems to be solved by naval architects at the early stages of the ship design project. This paper presents a comparison of methods that are used to estimate the resistance value of a vessel, studying the two terms that are the most relevant, the viscous resistance (depending on the form factor) and the resistance to waves, that appears in any floating device. This work focuses on the estimation of the form factor since it is a parameter difficult to estimate in the design early phases, and it is not always available in the measurements provided by real experiments with ship prototypes in towing tanks. Different estimation methods are applied and they are compared with the direct estimation and with the prediction obtained with a feedforward neural network. The results support the suitability of the neural networks to identify these vessel shape and wave related variables.

Keywords: Ship resistance · Neural networks · Form factor · Floating devices · Wave coefficient

1 Introduction

One of the main problems to be solved by naval architects when designing a new vessel is the estimation of the water resistance that depends on the ship shape. An accurate estimation of the ship resistance at the very early stages of the project will allow a good design regarding such important aspects as the propulsive system, the power system and the fuel consumption, among others. Even more, this resistance has been proved to be a key factor to any floating device [1].

There are different methods to estimate this resistance, namely: empirical methods, computational fluid dynamics simulation (CFD) and model test [2]. Empirical methods have been traditionally used but they require some information such as the hull dimensions and shape characteristics (length, beam or the displacement). CFD simulation is more time consuming. To obtain more reliable estimations, model test campaigns are usually performed with scaled replicas at towing tank facilities.

The main objective of this paper is to obtain a preliminary estimation of the ship resistance at the early stages of the design project by applying artificial intelligence techniques, particularly Neural Networks (NN), for mono-hull vessels.

H. Yin et al. (Eds.): IDEAL 2019, LNCS 11872, pp. 161–168, 2019.
https://doi.org/10.1007/978-3-030-33617-2_18

Results have been compared to those obtained by traditional methods and show a good agreement between the estimated parameters and the real data.

The structure of the paper is as follows. Definition of ship resistance is given in Sect. 2. Materials and methods are described in Sect. 3. The direct estimation of the form factor is explained in Sect. 4 and compared with the estimation by NN in Sect. 5. Conclusions and future works end the paper.

2   Ship Resistance

The ship resistance is the force required to tow a ship at a certain speed in calm water. It can be computed as the sum of two components [3], the viscous resistance, R_V, and the wave resistance, R_W.

$$R = R_V + R_W \tag{1}$$

The viscous resistance is due to the fluid viscosity involving the hull friction and the pressure changes as a result of the hull shape. It is composed of frictional and residual components arising from pressure gradients along the hull. The viscous resistance can be defined as follows:

$$R_V = C_{FP} \cdot (1 + k) \tag{2}$$

Being C_{FP} the resistance associated to a flat platform with equivalent wetted surface, and $(1 + k)$ the form factor.

Regarding the wave resistance, when a body is floating in a free surface between two fluids (water-air), pressure fields due to its surroundings generate a wave pattern that accompanies the body in its movement. A new force is generated when integrating the pressure field, named wave making resistance, R_W, that opposes to the movement.

The wave resistance, in a vessel, is strongly dependent on the speed, the hull shape and the dimensions.

Thus, the total resistance is obtained adding the viscous and the wave resistance components.

$$R = R_F(1 + k) + R_W \tag{3}$$

The total ship resistance at full-scale can increase if other less important components are considered such as the hull appendages, the hull rugosity, etc. These additional resistance components strongly depend on each particular vessel and their magnitude is usually smaller than the main resistance components. Therefore, the analysis here presented is focused on the wave and viscous resistance.

3 Materials and Methods

3.1 Material

The research centre INTA-CEHIPAR [4] provided historic test data from its towing tank facility with information of model test campaigns. The data base contains the following "text files":

- Carenas: with the main dimensions of each scaled model, such as the length, beam, or the block coefficient and the type of ship.
- Sit: loading conditions for each test such as the waterline length or the wetted surface.
- Remolques: model tests run for a specific hull and a specific loading condition, such as the different vessel speed values tested, the model scale, the estimation of the form factor (not always available) and the temperature of the tank water during the tests.
- Tab_remol: from these data the different resistance components for each full-scale hull and loading condition can be obtained.

The data base contains more than two thousand model tests and more than thirty thousand velocities measured. This data has been manually pre-processed to delete unnecessary data and correct duplicated values.

As a result, a binary MATLAB file for each ship type and scaled hull shape has been generated. These data will be used to train the neural network.

3.2 Methods

There are two different types of empirical methods used to estimate the resistance: systematic series and statistical formulas. The series are based on the results of several ship model tests performed with realistic hulls in the towing tank, varying the characteristics in a systematic way. On the other hand, there are some statistical formulas developed by different authors that use model test measurements from different vessels. The most well-known wave resistance method is the Holtrop and Mennen one (1978). This one and other methods will be applied and compared to the estimations obtain by neural networks.

Artificial neural networks are suitable for this application due to the following characteristics [5]:

- The learning capacity allows the NN to be applied to different types of vessels and hull shapes.
- The strong non-linear relation between the Froude Number and the wave resistance makes easier the estimation of the ship resistance.
- Easy implementation. The Neural Network toolbox of Matlab software has been used.
- Continuous learning and thus, adaptation to other types of ship or data from other towing tanks.

4 Form Factor Empirical Estimation

Unfortunately, the form factor is not always given as a result of the towing tank experiments, as it happens in more than half of the available tests. Thus, the form factor has been estimated by several methods and compared with the test data when this value is given by CEHIPAR. The most suitable method has been selected based on an error analysis and the recommended procedures of the ITTC [6].

The following methods have been applied and compared.

- Alaez method [7].
- Sasajima and Tanaka [8].
- Holtrop and Mennen method [9].
- Prohaska method [10].

The error is calculated as the difference between the estimated form factor and the real value when available in the experiment data base. The number of available form factor for the hull model tests is 650.

Table 1 shows the results. For each method, and depending on the range of application, the number of estimations is different. The two simplest methods, Alaez and Sasajima, are applicable for less than 200 hull shapes. Regarding the mean square error, the Prohaska method is the one that gives the best performance.

Table 1. Form factor estimation by several methods. Error analysis.

Method	Alaez	Sasajima & Tanaka	Holtrop & Mennen	Prohaska
No. data	193	124	646	445
Mean square error	0.0216	0.0151	0.0273	0.0100
Max square error	0.0755	0.0385	0.0994	0.0460

Figure 1 shows the error histograms for each analysed method. Alaez and Sasajima give the smallest number of estimations, and the errors are between ±0.2. The Holtrop method shows a larger number of estimations, with errors within the same range. Prohaska method gives also many estimations within a smaller range, ±0.1. Therefore, Holtrop and Prohaska methods give a good performance in a wider range and Prohaska's gives the best performance.

According to these results and bearing in mind that the ITTC recommends the use of Prohaska method, this one is considered to increase the number of points in the data base that we will be used to train the network for the estimation of the form factor.

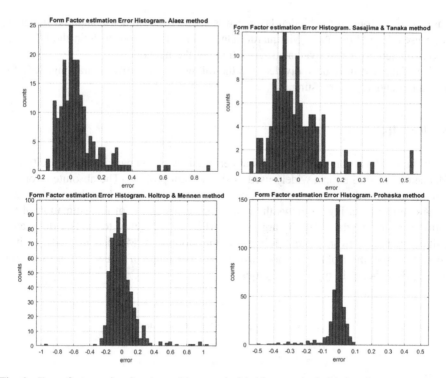

Fig. 1. Form factor estimation (error histogram). (a) Alaez method; (b) Sasajima and Tanaka; (c) Holtrop & Mennen; (d) Prohaska, respectively.

5 Estimation of the Form Factor by Neural Networks

In the design phase of a ship, the hull shapes are not defined in detail and are still subject to changes. Consequently, the NN inputs should be preliminary hull dimensions and parameters defining the geometry. Besides, the inputs have been included considering their influence on the output, trying to minimize them as much as possible.

Taking into account these considerations, the following input parameters have been used for the estimation of the form factor:

1. Waterline length (L_{wl}).
2. Beam (B).
3. Mean draft (T).
4. Wetted surface (S_w).
5. Block coefficient (C_B).
6. Prismatic coefficient (C_P).
7. Midship section coefficient (C_M).
8. Waterplane coefficient (C_{WP}).

A feedforward network with two hidden layers has been used to estimate the form factor. The NN has 8 inputs and one output. After a sensitivity analysis to select the number of neurons in the hidden layer it was set to 10. Symmetric sigmoid functions were used. The "Levenberg-Marquardt" algorithm was selected to train the NN. The mean squared error was evaluated when training the network.

The data base have been divided into the following sets: 70% for training, 15% for validation and 15% for testing.

6   Results

Once the NN has been trained, its global performance is compared with the Holtrop & Mennen method, the most widely used in ship design projects. The absolute error and root mean squared error with respect to the model test data bases have been calculated.

As an example, Fig. 2 shows the relative error histogram of the estimation of the form factor for the data base of cargo ships, where "rann" refers to the NN (resistance artificial neural network) and "Holtrop" is the Holtrop & Mennen method. As it is possible to see, the Holtrop method covers a smaller range of values and the accuracy of the NN is better in most of the cases.

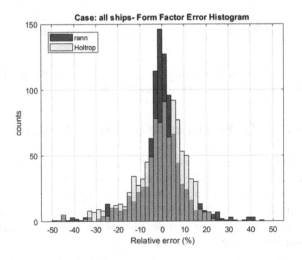

Fig. 2. Example of relative error histogram of form factor estimation

The main drawback of using the NN for the estimation is the reduced number of available data. Thus, the estimation of the form factor has been applied only to three different sets: all ships, cargo vessels and warships.

Table 2 presents a summary of the performance of the NN when estimating the form factor (rms: root mean square error).

In Fig. 3 we can see how the NN may not work well for other types of ships. On the other hand, this figure shows a better performance of the NN with respect to the Holtrop method for all cases.

Table 2. NN performance. Direct estimation of the form factor.

Case	Form factor	
	no. pts.	rms
All ships	973	0.144
Cargo ships	749	0.129
Warships	120	0.131

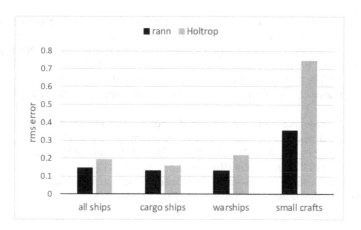

Fig. 3. Root mean squared error: NN vs Holtrop. Form factor estimation.

7   Conclusions and Future Works

In this work, the suitability of neural networks for the estimation of the resistance of floating devices has been proved. The results show a good agreement with the data base of the real experiments.

As future work, we propose to develop a complete simulation tool for ship resistance prediction based on neural networks, as an alternative to traditional empirical methods based on regression analysis, and trying different activation functions such as ELU/ReLU for the intermediate layers and no activation at the final layer [11]. Different optimizers and different identification configuration could also be used [12].

Acknowledgment. This work was partially supported by the Spanish Ministry of Science, Innovation and Universities under Project number RTI2018-094902-B-C21.

References

1. Herrero, E.R., et al.: Efficient parameter estimation and control based on a modified LOS guidance system of an underwater vehicle. Rev. Ibero Aut. Inf. Ind. **15**(1), 58–69 (2018)
2. Larsson, L., Raven, H.: Ship resistance and flow. Society of Naval Architects and Marine Engineers, New York (2010)

3. Baquero, A.: Resistencia al avance del buque. ETSIN, Madrid, Spain (2013)
4. INTA-CEHIPAR. http://www.inta.es/WEB/ICTS-CEHIPAR/en/inicio/. Accessed 31 July 2019
5. Santos, M., López, R., de la Cruz, J.M.: A neuro-fuzzy approach to fast ferry vertical motion modelling. Eng. Appl. Artif. Intell. **19**(3), 313–321 (2006)
6. ITTC. The Specialist Committee on Procedures for Resistance, Propulsion and Propeller Open Water Tests. In: 23rd International Towing Tank Conference Proceedings, vol. II, pp. 377–386 (2014)
7. Aláez-Zazurca, J.A.: Comportamiento del buque en la mar. Universidade da Coruña, Spain (1996)
8. Sasajima, H., Tanaka, I.: Form effects on viscous resistance and their estimation for full ships. In: 10th International Towing Tank Conference Proceedings, p. 122 (1963)
9. Holtrop, J., Mennen, G.G.J.: An approximate power prediction method. Int. Shipbuild. Prog. **29**(385), 166–170 (1982)
10. Prohaska, C.W.: A simple method for the evaluation of the form factor and the low speed wave resistance. In: 11th International Towing Tank Conference Proceedings, pp. 65–66 (1966)
11. Mira, J., Delgado, A.E., Alvarez, J.R., de Madrid, A.P., Santos, M.: Towards more realistic self contained models of neurons: High-order, recurrence and local learning. In: Mira, J., Cabestany, J., Prieto, A. (eds.) IWANN 1993. LNCS, vol. 686, pp. 55–62. Springer, Heidelberg (1993). https://doi.org/10.1007/3-540-56798-4_124
12. Sierra, J.E., Santos, M.: Modelling engineering systems using analytical and neural techniques: hybridization. Neurocomputing **271**, 70–83 (2018)

Neural Controller of UAVs with Inertia Variations

J. Enrique Sierra-Garcia[1]($\boxtimes$) (iD), Matilde Santos[2]($\boxtimes$) (iD),
and Juan G. Victores[3]($\boxtimes$) (iD)

[1] Department of Civil Engineering, University of Burgos, 09006 Burgos, Spain
jesierra@ubu.es
[2] Computer Science Faculty, Complutense University of Madrid,
28040 Madrid, Spain
msantos@ucm.es
[3] University Carlos III of Madrid, 28911 Madrid, Spain
jcgvicto@ing.uc3m.es

Abstract. Floating offshore wind turbines (FOWT) are exposed to hard environmental conditions which could impose expensive maintenance operations. These costs could be alleviated by monitoring these floating devices using UAVs. Given the FOWT location, UAVs are currently the only way to do this health monitoring. But this means that UAV should be well equipped and must be accurately controlled. Rotational inertia variation is a common disturbance that affect the aerial vehicles during these inspection tasks. To address this issue, in this work we propose a new neural controller based on adaptive neuro estimators. The approach is based on the hybridization of feedback linearization, PIDs and artificial neural networks. Online learning is used to help the network to improve the estimations while the system is working. The proposal is tested by simulation with several complex trajectories when the rotational inertia is multiplied by 10. Results show the proposed UAV neural controller gets a good tracking and the neuro estimators tackle the effect of the variations of the rotational inertia.

Keywords: Inertia variations · Neuro-estimator · UAV · Neural networks · FOWT

1 Introduction

In recent years, new and valuable applications of unmanned aerial vehicles (UAV) have emerged in different sectors such as defense, security, construction, agriculture, entertainment, shipping, etc. One of the most recent applications is the inspection and maintenance of offshore wind turbines and, particularly, floating devices (FOWT) [1]. Due to their location in deep seas, to take images and to get those places, efficient and robust UAV controllers are necessary. To address this complex task, soft computing has been proved an efficient approach [2–5], even more when internal parameters of the system changes during the operation.

H. Yin et al. (Eds.): IDEAL 2019, LNCS 11872, pp. 169–177, 2019.
https://doi.org/10.1007/978-3-030-33617-2_19

In this work we address the problem of UAV rotational inertia variations while the system is working. There are few works that study the effect of payload or inertia variation on the quadrotor dynamics. In [6], an adaptive control is used to estimate the parameter variation. In [7], adaptive parameter estimation is used to update the control action according to the current UAV mass and inertia moment. More recently, Wang [8] estimates the variations in the UAV payload and the effect of the wind gusts. But this topic demands further research.

We propose the design of a control strategy based on neural networks to cope with variations in the UAV mass and the rotational inertia. The controller combines feedback linearization, PIDs and adaptive neural estimators. This approach is tested by simulation for several trajectories giving good results.

The paper is organized as follows. Section 2 is focused on the description of the system and the modelling of the UAV inertia variation. The neural controller and the neuro estimators are studied in Sect. 3. Results are discussed in Sect. 4. Conclusions and future works end the paper.

2 System and Disturbances Models

A quadrotor is composed by four rotors which lift and propel it. Its dynamics is described by the position (x, y, z) and Euler's angles (ϕ, θ, ψ). The angular and translational dynamics of the system are given by the following equations [9]:

$$\tau = J\dot{\omega} + \omega \times J\omega, \quad J = \begin{pmatrix} I_x & 0 & 0 \\ 0 & I_y & 0 \\ 0 & 0 & I_z \end{pmatrix}, \quad m_Q \dot{v} = RT - m_Q g e_3 \quad (1)$$

Where τ is the torque vector (N.m), J is the inertia tensor (Nm^2), ω is the angular velocity (rad/s), m_Q is the UAV mass (Kg), R is the rotation matrix, g is the gravitational acceleration (m/s^2), T is a vector of forces (N). Vectors τ and T are related to the propeller speeds, $\Omega_1, \ldots, \Omega_4$, velocities of the rotors 1 to 4 (rad/s), the thrust coefficient b (N.s^2), the drag coefficient d (N.m.s^2), and the longitude of each arm l (m).

Instead of using the rotor speeds to control the UAV, it is a common practice to use the control signals u_1, u_2, u_3 and u_4, as defined by (2). This matrix is invertible, hence we can calculate velocity references for the rotors from a set of control signals.

$$\tau = \begin{pmatrix} bl(\Omega_4^2 - \Omega_2^2) \\ bl(\Omega_3^2 - \Omega_1^2) \\ d(\Omega_2^2 + \Omega_4^2 - \Omega_1^2 - \Omega_3^2) \end{pmatrix}, T = \begin{pmatrix} 0 \\ 0 \\ b(\Omega_1^2 + \Omega_2^2 + \Omega_3^2 + \Omega_4^2) \end{pmatrix}, \begin{bmatrix} u_1 \\ u_2 \\ u_3 \\ u_4 \end{bmatrix}$$

$$= \begin{bmatrix} 1 & 1 & 1 & 1 \\ 0 & -1 & 0 & 1 \\ -1 & 0 & 1 & 0 \\ 1 & -1 & 1 & -1 \end{bmatrix} \begin{bmatrix} \Omega_1^2 \\ \Omega_2^2 \\ \Omega_3^2 \\ \Omega_4^2 \end{bmatrix} \quad (2)$$

Finally, from Eqs. (1) and (2), the following system of equations is derived:

$$\ddot{\phi} = \dot{\theta}\dot{\psi}(I_y - I_z)/I_x + (lb/I_x)u_2, \ddot{\theta} = \dot{\phi}\dot{\psi}(I_z - I_x)/I_y + (lb/I_y)u_3 \qquad (3)$$

$$\ddot{\psi} = \dot{\phi}\dot{\theta}(I_x - I_y)/I_z + (d/I_z)u_4, \ddot{X} = -(sin\theta\,cos\phi)(b/m_Q)u_1 \qquad (4)$$

$$\ddot{Y} = (sin\phi)(b/m_Q)u_1, \ddot{Z} = -g + (cos\theta\,cos\phi)(b/m_Q)u_1 \qquad (5)$$

The values used to simulate the model, extracted from [9], are the following: $l = 0.232$ m; $m_Q = 0.52$ kg; $d = 7.5e^{-7}$ N.m.s^2; $b = 3.13e^{-5}$ N.s^2; $I_x = 6.23e^{-3}$ kg.m^2; $I_y = 6.23e^{-3}$ kg.m^2; and $I_z = 1.121e^{-2}$ kg.m^2.

Handling of a load by a UAV produces three main effects: the total mass is increased, the inertia changes and so does the center of gravity. Assuming the center of mass is invariant, we have focused on the influence of the total mass increment and mostly on the variation of the inertia.

The rotational inertia of a set of objects is given by $I = \sum_i m_i r_i^2$, where m_i is the mass (Kg) of object i and r_i is the distance (m) between its center of mass and the turning axle [10]. We consider an object with mass M_L that is taken to a distance L_L from the center of the quadrotor in the Z axis.

If the object is infinitesimal and the mass is concentrated in one point, the Z axle rotational inertia is invariant. The object is attached to the UAV by a without-mass rigid structure. Thus, the total rotational inertia, Ix_T, Iy_T, Iz_T is:

$$Ix_T = Ix + M_L L_L^2, \qquad Iy_T = Iy + M_L L_L^2, \qquad Iz_T = Iz \qquad (6)$$

The terms $dist_m$ and $dist_l$ are introduced in (3–5) to represent the influence of the mass and rotational inertia variation in the UAV dynamics. These disturbances are modelled as a step function at the moment when the object is grabbed by the UAV, t_{dist}. After t_{dist} the total mass of the system is $m + M_L = K_m \cdot m$ and the rotational inertias are, respectively, $Ix + M_L L_L^2 = K_{Ix} \cdot Ix$ and $Iy + M_L L_L^2 = K_{Iy} \cdot Iy$, where K_m, K_{Ix} and K_{Iy} are multipliers gains, with the constraint $(K_{Ix} - 1) \cdot Ix = (K_{Iy} - 1) \cdot Iy$.

$$\ddot{\phi} = (\dot{\theta}\dot{\psi}(I_y + dist_{Iy} - I_z) + lb \cdot u_2)/(I_x + dist_{Ix}), \qquad (7)$$

$$\ddot{\theta} = (\dot{\phi}\dot{\psi}(I_z - I_x - dist_{Ix}) + lb \cdot u_3)/(I_y + dist_{Iy}) \qquad (8)$$

$$\ddot{\psi} = \dot{\phi}\dot{\theta}(I_x + dist_{Ix} - I_y - dist_{Iy})/I_z + (d/I_z)u_4 \qquad (9)$$

$$\ddot{X} = -(sin\theta\,cos\phi)(b/(m_Q + dist_m))u_1 \qquad (10)$$

$$\ddot{Y} = (sin\phi)(b/(m_Q + dist_m))u_1 \qquad (11)$$

$$\ddot{Z} = -g + (cos\theta\,cos\phi)(b/(m_Q + dist_m))u_1 \qquad (12)$$

$$dist_m = M_L \cdot step(t - t_{dist}), dist_{Ix} = dist_{Iy} = M_L \cdot L_L^2 \cdot step(t - t_{dist}) \qquad (13)$$

3   Description of the Neural Controller

3.1   Description of the Architecture of the Control System

Figure 1 shows the control system based on neuro-estimators architecture. It is made up of a position controller, an attitude controller and two neuro-estimators for the mass and the inertia.

The position controller has as input the desired trajectory $(Xref, Xref, Zref)$ and three output: u_1, Φ_{ref} and θ_{ref}. It is based on feedback linearization and three PIDs, one for each coordinate axis, to stabilize the trajectory. It requires the mass value that is changing so it is estimated with a neural network. The Z coordinate is directly controlled by u_1 while the control of X and Y coordinates is carried out by first obtaining the roll and the pitch references $(\Phi_{ref}, \theta_{ref})$. These are the inputs of the attitude controller, plus an external reference of the yaw (Ψ_{ref}). The attitude controller has the same structure as the position one. The rotational inertia directly affects the angular dynamics and indirectly the translational dynamics. For this reason, the rotational inertia values (Ix_{est}, Iy_{est}) are real-time estimated by the inertia neuro-estimator and introduced as inputs of the controller. Assuming that the inertia in the Z axis does not suffer variations during the operation, the attitude controller generates three outputs: u_2, u_3 and u_4. The roll angle is controlled by u_2, the pitch angle by u_3 and the yaw angle by u_4.

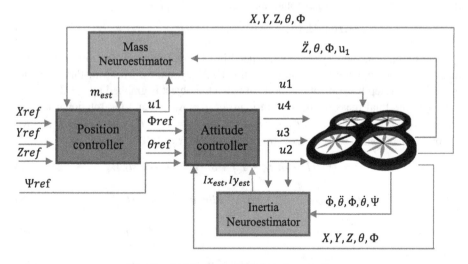

Fig. 1.   Architecture of the neural controller.

The internal equations of the position controller are (14–19):

$$rZ(t_i) = K_{PZ}\big(Z_{ref}(t_i) - Z(t_i)\big) + K_{DZ}\big(\dot{Z}_{ref}(t_i) - \dot{Z}(t_i)\big) \tag{14}$$

$$u_1(t_i) = \begin{cases} (rZ(t_i) + g)\frac{m_{est}}{b} & cos\theta_{i-1}cos\phi_{i-1} = 0 \\ (rZ(t_i) + g)\left(\frac{m_{est}}{b \cdot cos\theta_{i-1}cos\phi_{i-1}}\right) & cos\theta_{i-1}cos\phi_{i-1} \neq 0 \end{cases} \quad (15)$$

$$rY(t_i) = K_{PY}\left(Y_{ref}(t_i) - Y(t_i)\right) + K_{DY}\left(\dot{Y}_{ref}(t_i) - \dot{Y}(t_i)\right) \quad (16)$$

$$\mathbf{\Phi}_{ref}(t_i) = \begin{cases} \Phi_{ref}(t_{i-1}) & u_1(t_i) = 0 \\ -a\sin\left(rY(t_i)\left(\frac{m_{est}}{b \cdot u_1(t_i)}\right)\right) & u_1(t_i) \neq 0 \end{cases} \quad (17)$$

$$rX(t_i) = K_{PX}\left(X_{ref}(t_i) - X(t_i)\right) + K_{DX}\left(\dot{X}_{ref}(t_i) - \dot{X}(t_i)\right) \quad (18)$$

$$\mathbf{\theta}_{ref}(t_i) = \begin{cases} \theta_{ref}(t_{i-1}) & u_1(t_i)\cos(\Phi(t_i)) = 0 \\ asin\left(rX(t_i)\left(\frac{m_{est}}{b \cdot u_1(t_i)\cos(\Phi(t_i))}\right)\right) & u_1(t_i)\cos(\Phi(t_i)) \neq 0 \end{cases} \quad (19)$$

Where m_{est} is the estimated mass (Kg) and $(K_{PX}, K_{DX}, K_{PY}, K_{DY}, K_{PZ}, K_{DZ}) \in \mathbb{R}^6$ are the gains of the internal PIDs. If the neuro-estimator is not used, the value m_{est} is substituted by the mass of the quadrotor m_Q.

The equations of the attitude controller are obtained in a similar way, but in this case the angular dynamic is considered in the linearization (20–25):

$$r\Phi(t_i) = K_{P\Phi}\left(\Phi_{ref}(t_i) - \Phi(t_i)\right) + K_{D\Phi}\left(\dot{\Phi}_{ref}(t_i) - \dot{\Phi}(t_i)\right) \quad (20)$$

$$u_2(t_i) = \left(Ix_{est} \cdot r\Phi(t_i) - (Iy_{est} - Iz) \cdot \dot{\theta}(t_i)\dot{\Psi}(t_i)\right)/(l \cdot b) \quad (21)$$

$$r\theta(t_i) = K_{P\theta}\left(\theta_{ref}(t_i) - \theta(t_i)\right) + K_{D\theta}\left(\dot{\theta}_{ref}(t_i) - \dot{\theta}(t_i)\right) \quad (22)$$

$$u_3(t_i) = \left(Iy_{est} \cdot r\theta(t_i) - (Iz - Ix_{est}) \cdot \dot{\Phi}(t_i)\dot{\Psi}(t_i)\right)/(l \cdot b) \quad (23)$$

$$r\Psi(t_i) = K_{P\Psi}\left(\Psi_{ref}(t_i) - \Psi(t_i)\right) + K_{D\Psi}\left(\dot{\Psi}_{ref}(t_i) - \dot{\Psi}(t_i)\right) \quad (24)$$

$$u_4(t_i) = \left(Iz \cdot r\Psi(t_i) - (Ix_{est} - Iy_{est}) \cdot \dot{\Phi}(t_i)\dot{\theta}(t_i)\right)/(d) \quad (25)$$

Where Ix_{est} and Iy_{est} (Kg.m^2) are the rotational inertias in the X and Y axis, respectively, estimated by the inertia neuro-estimator and $(K_{P\Phi}, K_{D\Phi}, K_{P\theta}, K_{D\theta}, K_{P\Psi}, K_{D\Psi}) \in \mathbb{R}^6$ are the gains of the internal PIDs. The values Ix_{est} and Iy_{est} are substituted by Ix and Iy when the neuro-estimator is not used.

3.2 Description of the Neuro-Estimators

The generic structure of the neuro-estimators is show in Fig. 2. They need an analytic model of the UAV and a neural network. The model is used to obtain a measurement of the value to be estimated in order to train the network. But even when the model does not find a right value due to any singularity, the neural network is still able to provide a valid output, generating new knowledge.

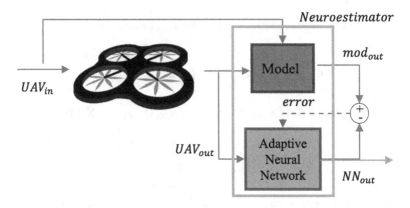

Fig. 2. Structure of the neuro-estimators.

First, the neural network is offline trained with the outputs of the model (mod_{out}) as targets and the outputs of the UAV (UAV_{out}) as inputs. Then, while the UAV system is working, the network is on-line learning from new inputs and targets. So it adapts to changes in the system parameters, and the controller can reject these disturbances.

The equations that represent the performance of the neuro-estimator are given by:

$$NN_{out}(t_i) = f_{NN}(w_{NN}(t_i), UAV_{out}(t_i)) \tag{26}$$

$$w_{NN}(t_i) = f_{adapt}(w_{NN}(t_{i-1}), UAV_{out}(t_{i-1}), mod_{out}(t_{i-1})) \tag{27}$$

$$mod_{out}(t_i) = f_{mod}(UAV_{in}(t_{i-1}), UAV_{out}(t_{i-1})) \tag{28}$$

Where f_{NN} represents the neural network during the simulation phase, w_{NN} is the set of internal parameters of the network, f_{adapt} is the adaption function, f_{mod} is the analytic model, UAV_{in} is the set of inputs from the UAV and UAV_{out} the set of UAV. For each estimator (mass or inertia), UAV_{in}, UAV_{out}, and the f_{mod} are different.

- Mass Adaptive Neuro-Estimator

$$f_{mod} = b/((Z+g)/(u_1 \cos \theta \cos \phi)), NN_{out}(t_i) = f_{NN}(w_{NN}, \ddot{Z}, \cos \theta, \cos \phi) \tag{29}$$

- Inertia Adaptive Neuro-Estimator

$$f_{mod} = \frac{\ddot{\phi} lbu_3 + \dot{\phi}\dot{\phi}\dot{\psi} Iz - \dot{\phi}\dot{\psi} lbu_2 + \dot{\phi}\dot{\psi}\dot{\theta}\dot{\psi} Iz}{\dot{\phi}\dot{\psi}\dot{\theta}\dot{\psi} + \dot{\phi}\dot{\theta}}, NN_{out}(t_i) = f_{NN}(w_{NN}, \ddot{\phi}, \ddot{\theta}). \tag{30}$$

4 Discussion of the Simulation Results

Simulation results have been obtained during 16 s. The tuple of multipliers (K_m, K_{Ix}, K_{Iy}) has been set to (2, 10, 10) and t_{dist} to 5 s. The neural controller is offline trained during the first 3 s; from them on the online learning is applied. In both neuroestimators the neural network is a MLP with one hidden layer of 30 neurons. The Levenberg-Marquardt algorithm has been used for the training with $\mu = 0.001$.

Figure 3 (left) shows the UAV tracking of a helical lemniscate trajectory (reference in blue and trajectory with neuro-estimators in red). Without neuro-estimators (blue line) the trajectory tracking was wrong (Fig. 3, right).

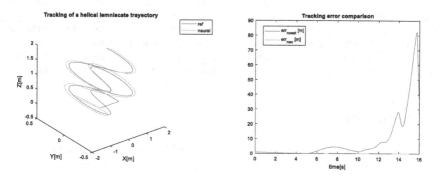

Fig. 3. Tracking of a helical lemniscate (left) and tracking error (right). (Color figure online)

Figure 4 shows the UAV trajectory tracking of each coordinate. The results with neuro-estimators (red line) and without them (blue line) regarding the reference (yellow line) proves the efficiency of the neuro-estimators and how the error grows and the system becomes unstable without them. The worst error is in the y-axis.

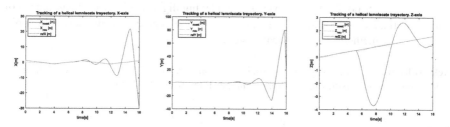

Fig. 4. UAV tracking of X, Y and Z coordinates, respectively, with mass and inertia variation. (Color figure online)

Results for different trajectories are shown in Table 1. The columns labelled "Neur" means results with the neuro-estimators vs "Nest", without them. Values below 0.0001 have been fixed to 0. The best results per column have been underlined. The evaluation criteria are the MSE, U1 (control effort), and PERF (performance of the controller). The latter is calculated multiplying the inverse of MSE_T, MAX_T, and U1, therefore bigger values indicate better results.

For every trajectory and evaluation criteria the best results are obtained with the neuro-estimators. In general, the best performance is obtained for the helical lemniscate trajectory that also shows the biggest difference depending on the application of the neuro-estimators; indeed, regarding the MAX_T criterium, this value is up to 80 times smaller with the neuro-estimators.

Table 1. Tracking error, control effort and performance comparison for different trajectories.

Trajectory	MSE_X		MSE_Y		MSE_Z		MSE_T		MAX_T		U1		PERF	
	Neur	Nest	Neur	Nest	Neur	Nest	Neur	Nest	Neur	Nest	Neur	Nest	Neur	Nest
Linear	**0,094**	0,098	**0,094**	0,098	**1,208**	2,362	**1,223**	2,377	**10,10**	10,10	**8,650**	8,681	**0,009**	0,004
Circular	**0,192**	0,226	**0,098**	0,172	**1,208**	2,362	**1,246**	2,418	**10,24**	10,24	**8,656**	8,736	**0,009**	0,004
Helical	**0,161**	0,350	**0,037**	0,452	**0,040**	1,120	**0,185**	1,557	**1,007**	4,642	**8,644**	9,993	**0,618**	0,013
Cyc. helical	**0,140**	0,192	**0,012**	0,151	**0,574**	1,335	**0,651**	1,476	**1,141**	5,850	**8,685**	8,832	**0,154**	0,013
Lemniscate	**0,098**	0,271	**0,013**	0,219	**1,208**	2,363	**1,220**	2,521	**10,05**	10,05	**8,652**	9,201	**0,009**	0,004
Hel. lemnisci	**0,156**	3,029	**0,037**	6,635	**0,025**	1,098	**0,180**	8,629	**1**	82,32	**8,684**	42,34	**0,638**	0
Step	**0,983**	1,018	**0,985**	1,012	**1,212**	2,367	**2,011**	3,173	**17,32**	17,32	**8,949**	8,993	**0,003**	0,002

5 Conclusions

In this work a new control strategy based on the hybridization of feedback linearization, PIDs and adaptive neural network estimators is proposed. This approach helps to deal with changes in the rotational inertia of UAV while flying.

The neuro-estimators combine an analytic model of the UAV and MLP neural networks. Online learning of the neural network allows the system to improve the estimation of the changing mass and inertia of the quadrotor.

Simulation results with different trajectories prove the validity of this control strategy with neuro-estimator to stabilize the UAV even under noticeable variations in the mass and in the rotational inertia of the quadrotor.

Among other possible future works, we may highlight the study of the influence of other disturbances such as the ones generated by the engines and the desirable application of this methodology to a real system.

Acknowledgment. This work was partially supported by the Spanish Ministry of Science, Innovation and Universities under Project number RTI2018-094902-B-C21.

References

1. Schäfer, B.E., Picchi, D., Engelhardt, T., Abel, D.: Multicopter unmanned aerial vehicle for automated inspection of wind turbines. In: 2016 24th Mediterranean Conference on Control and Automation (MED), pp. 244–249. IEEE, June 2016
2. Sierra, J.E., Santos, M.: Modelling engineering systems using analytical and neural techniques: hybridization. Neurocomputing **271**, 70–83 (2018)
3. San Juan, V., Santos, M., Andújar, J.M.: Intelligent UAV map generation and discrete path planning for search and rescue operations. Complexity (2018)

4. Santos, M., Lopez, V., Morata, F.: Intelligent fuzzy controller of a quadrotor. In: 2010 International Conference on Intelligent Systems and Knowledge Engineering (ISKE), pp. 141–146). IEEE, November 2010
5. Yanez-Badillo, H., Tapia-Olvera, R., Aguilar-Mejia, O., Beltran-Carbajal, F.: On line adaptive neurocontroller for regulating angular position and trajectory of quadrotor system. Revista Iberoamericana de Automática e Informática Industrial, **14**(2), 141–151 (2017). ISSN: 1697–7912, https://doi.org/10.1016/j.riai.2017.01.001
6. Min, B.-C., Hong, J.-H., Matson, E.T.: Adaptive robust control (ARC) for an altitude control of a quadrotor type UAV carrying an unknown payloads. In: 2011 11th International conference on control, automation and systems Korea, pp. 26–29 (2011)
7. Wang, C., Nahon, M., Trentini, M., Nahon, M., Trentini, M.: Controller development and validation for a small quadrotor with compensation for model variation. In: 2014 International Conference on Unmanned Aircraft Systems (ICUAS). IEEE (2014)
8. Wang, C., Song, B., Huang, P., Tang, C.: Trajectory tracking control for quadrotor robot subject to payload variation and wind gust disturbance. J. Intell. Robot. Syst. **83**(2), 315–333 (2016)
9. Nicol, C., Macnab, C.J.B., Ramirez-Serrano, A.: Robust neural network control of a quadrotor helicopter. In: 2008 Canadian Conference on Electrical and Computer Engineering, pp. 001233–001238. IEEE, May 2008
10. Serway, R.A., Jewett, J.W.: Physics for scientists and engineers with modern physics. Cengage learning (2018)

Special Session on Finance and Data Mining

A Metric Framework for Quantifying Data Concentration

Peter Mitic[1,2,3(✉)]

[1] Risk, Santander UK, London NW1 3AN, UK
peter.mitic@santandercib.co.uk
[2] Department of Computer Science, UCL, Gower Street, London WC1E 6BT, UK
[3] Laboratoire d'Excellence sur la Régulation Financière (LabEx ReFi), Paris, France

Abstract. Poor performance of artificial neural nets when applied to credit-related classification problems is investigated and contrasted with logistic regression classification. We propose that artificial neural nets are less successful because of the inherent structure of credit data rather than any particular aspect of the neural net structure. Three metrics are developed to rationalise the result with such data. The metrics exploit the distributional properties of the data to rationalise neural net results. They are used in conjunction with a variant of an established concentration measure that differentiates between class characteristics. The results are contrasted with those obtained using random data, and are compared with results obtained using logistic regression. We find, in general agreement with previous studies, that logistic regressions out-perform neural nets in the majority of cases. An approximate decision criterion is developed in order to explain adverse results.

Keywords: Copula · Hypersphere · Cluster · Herfindahl-Hirschman · HHI · Credit · Concentration · Decision criterion · Tensorflow · Neural Net

1 Introduction

Successful applications of artificial neural net (hereinafter *ANN*) methods, and also of other AI methods, are numerous, and particular successes are often reported in the press. A notable recent success in the field of cancer diagnosis is [1]. AI methods have been less successful for credit risk: some credit risk datasets are the 'wrong shape' (the term will be formalised in Sect. 5). This view is prompted by the following observations:

1. Insensitivity to *ANN* configuration or tuning
2. Low correlations of single explanatory variables with class

---

The opinions, ideas and approaches expressed or presented are those of the author and do not necessarily reflect Santander's position. The values presented are just illustrations and do not represent Santander data.

© Springer Nature Switzerland AG 2019
H. Yin et al. (Eds.): IDEAL 2019, LNCS 11872, pp. 181–190, 2019.
https://doi.org/10.1007/978-3-030-33617-2_20

3. Insensitivity to data transformations (e.g. reducing to principal components)
4. Insensitivity to attempts to redress the imbalance (e.g. SMOTE, gradient boosting, under-sampling or over-sampling).

Our underlying assumption is that distributional properties of the credit data inhibit prediction of a correct classification. The 'wrong shape' phenomenon is illustrated in Fig. 1 which shows two contrasting marginal distributions from two of the data sets considered in this study (see Sect. 4.1). Data set *LCAB* with class *credit not approved* on the left shows a loose scatter with no discernable trend or 'shape'. Data set *AUS* with class *credit approved* on the right shows a concentrated scatter with a trend and a triangular 'shape'. The former type is more typical of credit-related data.

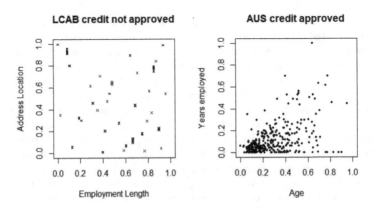

Fig. 1. Marginal Distribution Examples showing contrasting data concentrations

1.1   Economic Consequences of Credit Default

Credit default is very costly for the lender and is a social burden for the borrower and for society. A broad estimate of the amounts involved can be made from UK Regulator figures (https://www.fca.org.uk/data/mortgage-lending-statistics/commentary-june-2019). The outstanding value of all residential mortgage loans at Q1 2019 was £1451bn, of which 0.99% was in arrears. The 2018 capital disclosures from https://www.santander.co.uk/uk/about-santander-uk/investor-relations/santander-uk-group-holdings-plc show that approximately 88% (which is typical) of arrears can be recovered. Therefore the worst case net loss to lenders in the first 3 months of 2019 was $1451 \times 0.99 \times (1 - 0.88) = £1.724bn$, a very substantial sum!

1.2   Nomenclature and Implementation

In this paper the variable values to be predicted are referred to as *classes*. Typically in the context of credit risk, class determination is a binary decision. The two classes are usually expressed as categorical variables: 'approved' (alternatively 'pass' or 'good'), and 'not approved' (alternatively 'fail' or 'bad'). Explanatory variables are referred to as *features*. In credit-related data they usually include items such as income, age, address, mean account balance, prior credit history etc. There can be many hundreds of them. The term *tuple* will be used to refer to a single instance of a set of features. Each *tuple* is associated with a single class. The acronyms are: *LR* for *Logistic Regression* and *AUC* for *Area under Curve*.

The metric calculations were done using the *R* statistical language, and *TensorFlow* was used for neural net calculations. All computations were done using a 16GB RAM i7 Windows processor.

2 Review of Neural Net Applications in Credit Risk

Louzada [2] has an extensive review of the success rate of credit-related applications prior to 2016, using the German and Australian data sets (Sect. 4.1). The mean success rates of all 30 cases considered were: German: 77.7% and Australian: 88.1%. Those figures are consistently good compared to some we have encountered, but are not comparable to the worst result for the Yala's [1] medical application: 96.2%. More generally, Atiya's pre-2001 review [3] is similar: 81.4% and 85.50% success for two models. Bredart's bankruptcy [4] prediction result is marginally lower: 75.7%.

The results reported by West [5] indicate a general failure of *ANN* methods to improve on results obtained using regressions for the German and Australian data. We used the same data, as well as our own, in Sect. 4.1. We concur with the conclusion that LRs often perform better than AI-based methods: 11.8% greater error rate for *ANN*s. Lessmann [6] gives a lower margin of about 3.2%, using 8 data sets.

There are some better results post-2016. Kvamme et al. [7] reports high accuracy (given as optimal AUC 0.915) using credit data from the Danmarks Nationalbank with a convolutional *ANN*. Addo et al. [8], used corporate loan data, and report AUC = 0.975 for their best deep learning model, and 0.841 for their worst. These results are surprisingly good, and we suspect that either the data set used contains some behavioural indicator of default, or that loans in the dataset are only for 'select' customers who have a high probability of non-default. The LC and LCAB data.Sect. 4.1 have some behavioural indicators (such as amount owing on default, added later), and they are omitted in our analysis. More recently, Munkhdalai et al. [9] reports more relative LR successes: 5.2% better error rate than an *ANN* using a two-stage filter feature selection algorithm, and 7.5% better using a random forest-based feature selection algorithm.

Yampolskiy [10] gives a similar general explanations of AI failure which is particular applicable in the context of credit risk. If a new or unusual situation is encountered in an AI learning process, it will be interpreted, wrongly, as a 'fail' within the context of that process. We suspect that, in the context of assessing credit-worthiness, those new or unusual situations are future events that can only be anticipated with some degree of probability (such as illness, loss of income, mental incapacity).

3   The Concentration Metric Framework

We propose a framwork to measure data concentration, which we think is responsible for the 'wrong shape' phenomenon for credit data. The proposed framework comprises three metrics, each used within a concentration component where the values of the metrics for each class are combined. The idea of a 'framework' is one of extensibility: further metrics can be incorporated in a simple way (see the end of Sect. 3.1).

3.1   Inter-class Concentration Measure

The illustrations in Fig. 1 show one instance of a high class concentration and another of low concentration. In order to quantify them, we develop inter-class concentration metrics. Data are partitioned by class, and a concentration metric is calculated for each. They are combined using a variant of an established concentration measure, the Herfindahl-Hirschman Index (HHI - see for example [11]). The HHI is usually used in economic analysis to measure concentration of production in terms of, for example, percentage of market share or of total sales. We define the index in terms of a metric M_i for class i, associated with a weight w_i (the weight was not part of the original HHI formulation). Let M be the sum of the M_i for n classes: $M = \sum_{i=1}^{n} M_i$. Then the HHI for metric M is given by $\hat{H}$ in Eq. 1.

$$\hat{H} = \sum_{i=1}^{n} w_i \left(\frac{M_i}{M} \right)^2 \tag{1}$$

In the context of ANN classification problems, we use three different interpretations of the metric M_i: M_C, the $Copula$ metric M_S, the $Hypersphere$ metric, and M_N the k-$Neighbours$ metric. The first measures data correlation. The second measures data dispersion and the third measures clustering. For all metrics the weights used (Eq. 1) are the proportions of the number of tuples in each class in a training set. The metrics are combined to form the geometric mean concentration measure $\hat{H}$ in Eq. 2, which is a general expression for m metrics. The term $framework$ in this paper is used to refer to the applicability of the 'concentration measure + metrics' approach to any required value of m. The geometric mean is used because multiplying the metrics exaggerates the differentiation that each introduces.

$$\hat{H} = \left(\prod_{i}^{m} \hat{H}_i \right)^{\frac{1}{m}} \quad \in (0, 1) \tag{2}$$

In the case of three metrics, Eq. 2 reduces to $\hat{H} = (\hat{H}_C \hat{H}_S \hat{H}_N)^{\frac{1}{3}}$.

3.2   The Copula Metric, M_C

A copula is a mechanism for modelling the correlation structure of multivariate data, and thereby generating random samples of any desired distribution. An initial fit to some appropriate distribution is required. Of the common *Ellip-tic* copulas we choose the multivariate *t*-copula, as it can capture the effect of extreme values better than the multivariate normal equivalent is able to (see [12] and [13]). Extreme values are often observed in financial return data. It is not necessary to use *Archimedean* copulas, *Clayton*, *Gumbel* or *Frank*, that emphasise extremes even more.

The calculation of the *Copula* metric proceeds by first using a *Fit* function to fit, using maximum likelihood, normal distributions to each of n features data $\{x_i\}$, giving a set of normal parameter pairs $\{\mu_i, \sigma_i\}$. Then we define a t-copula $C_t(c, \nu)$, with $\nu = 3$ degrees of freedom using the covariance matrix c of all the data, and generate a random sample of $m \sim 100000$ U[0,1]-distributed random variables U_i from it using the R *copula* package random number generator, denoted here as $r(C_t)$. The inverse normal distribution function F^{-1} is then applied to the parameter pairs and the values derived from the copula, resulting in a matrix of normal distributions $\{N_i\}$. The row sums of that matrix are then summed to derive the required metric, M_C (Eq. 3).

$$\{\mu_i, \sigma_i\} = \{Fit(x_i)\}$$
$$\{U_i = r(C_t(\nu, c), m\}$$
$$\{N_i = F^{-1}(U_i, \mu_i, \sigma_i)\}$$
$$M_C = \Sigma(N_i(*, n)) \tag{3}$$

3.3 The Hypersphere Metric, M_S

The *Hypersphere* metric measures the deviation of each tuple that lies within a prescribed hypersphere centred on the centroid of all tuples. For a set of n tuples $t_i, i = 1..n$, denote their centroid by $\bar{t}$, and let the covariance matrix of the set of tuples be c. Then the deviation for tuple t_i is calculated from the Mahanalobis distance, D_i of t_i from $\bar{t}$. The hypersphere refers to the subset of D_i that is within 95% of the maximum of the D_i, and is denoted by $D_i^{(95)}$. The required metric is the sum of the elements of $D_i^{(95)}$ (Eq. 4).

$$\{D_i\} = \{\sqrt{(t_i - \bar{t})^T c \,(t_i - \bar{t})}\}$$
$$D_i^{(95)} = \{D_i : D_i \leq 0.95 \, max(D_i)\}$$
$$M_S = \Sigma_{i=1}^n D_i^{(95)} \tag{4}$$

In practice it makes very little difference if the 95% hypersphere is replaced by, for example, a 90% or a 99% hypersphere.

3.4 The k-Neighbours Metric, M_N

The k-*Neighbours* metric uses a core k-*Nearest Neighbours* calculation. Empir-ically, we have found that maximal differentiation between classes is achieved by considering the more distant neighbours. Therefore we use the farthest 20%

neighbours, not the nearest. The calculation proceeds, for each class, by calculating the Euclidean distances D_i of all the tuples $t_i, i = 1..n$ in each class to the centroid, $\bar{t}$, of that class. The set of distances in excess of the 80th quantile, $Q_{80}(D_i)$ is extracted and summed. We have found that with large datasets, calculating the Mahanalobis distance in place of the Euclidean distance is not always possible due to singularity problems with some covariance matrices. The details are in Eq. 5

$$\{D_i\} = \{\sqrt{\Sigma(t_i - \bar{t})^2}\}$$
$$D_{i,80} = \{D_i : D_i > q_{80}(D_i)\}$$
$$M_N = \Sigma(D_{i,80}) \tag{5}$$

3.5 Theoetical Metric Minimum Value

The metric formulations in Eqs. 1 and 2 admit a theoretical minimum result when using random data with a binary decision. The value of each metric with weights w_i should be $w_i(\frac{1}{2})^2 + (1 - w_i)(\frac{1}{2})^2 = \frac{1}{4}$ (from Eq. 1 with $H_1 = H_2$) since random data should yield no useful predictive information. Then, for m metrics, Eq. 2 gives the theoretical minimum concentration measure $\hat{H}_{min}$, independent of m in Eq. 6

$$\hat{H}_{min} = \left(\left(\frac{1}{4}\right)^m\right)^{\frac{1}{m}} = \frac{1}{4} \tag{6}$$

4   Results

The *ANN* configuration used was: 2 hidden layers with sufficient neurons (always ≤ 100) in each to optimise AUC; typically 100 epochs; ReLU activation in the hidden layers, Sigmoid in the input layer, Softmax in the output layer; categorical cross entropy loss, 66.67% of data used for training.

4.1   Data

Details of the data used in this study are in Table 1. L-Club is the Lending Club (https://www.lendingclub.com/info/download-data.action). UCI is the University of California Irvine Machine Learning database [14]. SBA is the U.S. Small Business Administration. [15]. BVD is Bureau Van Dijk, the Belfirst database (https://www.bvdinfo.com). RAN-P is a randomly generated predictive dataset with two classes, and two highly correlated features. It represents a near minimal concentration with a high predictive element. RAN-NP is similar but is designed to have no predictive element. In all cases, all features are normalised to range [0,1], and there are no missing entries. Where relevant, categorical variables have been replaced by numeric.

Table 1. Data sources

Data	Source	Notes
INT	Internal	Retail short-term loans
LC	L-Club	All credit grades: LoanStats3b
LCAB	L-Club	Best credit grades A and B only: LoanStats3b
GERMAN	UCI	Statlog German Credit Data
CARD	UCI	Default of credit card clients [16]
AUS	UCI	Statlog Australian Credit Approval
JP	UCI	Japanese Credit Screening
IND	UCI	Qualitative Bankruptcy India
POL5	UCI	Polish Companies Bankruptcy (5-year) [17]
POL1	UCI	Polish Companies Bankruptcy (1-year) [17]
SBA	SBA	'SBA Case' dataset
BVD	BVD	filtered on W. Eur. + Manufacturing Financials
RAN-P	Random	Randomly generated predictive
RAN-NP	Random	Randomly generated non-predictive

4.2   Metric and Concentration Results

Table 2 shows the values obtained for the three concentration metrics and the concentration measure (Eqs. 3, 4, 5 and 2 respectively). The error rates (*Err* columns) are given as proportions, rather than as percentages.

Table 2. Distributional Indicators: metrics, $\hat{H}$ and *ANN* results, in $\hat{H}$ order.

Name	$\hat{H}_C$	$\hat{H}_S$	$\hat{H}_N$	$\hat{H}$	ANN Err	ANN AUC	LR Err	LR AUC
RAN-NP	0.289	0.917	0.885	0.617	0.083	0.540	0.083	0.560
POL1	0.250	0.923	0.911	0.595	0.032	0.590	0.219	0.630
POL5	0.251	0.862	0.877	0.575	0.033	0.62	0.122	0.705
LCAB	0.246	0.769	0.765	0.526	0.330	0.618	0.108	0.635
LC	0.244	0.778	0.606	0.486	0.345	0.679	0.160	0.682
SBA	0.256	0.724	0.521	0.459	0.551	0.680	0.370	0.775
CARD	0.241	0.685	0.470	0.427	0.181	0.775	0.728	0.720
BVD	0.250	0.464	0.426	0.367	0.533	0.870	0.063	0.995
IND	0.374	0.343	0.271	0.326	0.428	0.885	0.012	0.985
GER	0.259	0.350	0.373	0.323	0.245	0.770	0.299	0.820
INT	0.243	0.299	0.349	0.294	0.341	0.815	0.280	0.760
JP	0.253	0.309	0.259	0.273	0.140	0.930	0.252	0.945
AUS	0.249	0.304	0.260	0.270	0.168	0.930	0.342	0.930
RAN-P	0.250	0.262	0.252	0.255	0.305	0.928	0.496	0.680
RAN	0.250	0.250	0.250	0.250	0.501	0.507	0.501	0.506

It is noticeable from the results in Table 2 that a low $\hat{H}$ value is associated with datasets which work well with ANN processing. Conversely, a high $\hat{H}$ value indicates that ANN processing may not be successful in class determination. LC, LCAB, POL1 and POL5 are the worst cases. The $\hat{H}$ values are more aligned with the AUC values. Figure 2 shows the $\hat{H}$-AUC scatter with a linear trend line ($AUC \sim 1.2 - \hat{H}$, $R^2 = 0.88$), and the $\hat{H}$-Error Rate scatter for comparison. We note that error rate variation with $\hat{H}$ is more volatile than the variation with AUC. Ordinates for the randomly-generated datasets RAN-P and RAN-NP are shown separately. RAN-P represents a borderline *wrong/right shape* boundary and RAN-NP represents a 'worst case' with a minimal predictive element. A further result, not in Table 2 is for randomly generated features with randomly allocated classes (50% in each class). Consistent with Eq. 6, we obtained $\hat{H}_C = \hat{H}_S = \hat{H}_N = \hat{H} = 0.25$, with AUC and % success values for ANN and LR all marginally greater than 0.5. Therefore, even 'badly-shaped' datasets are not random!

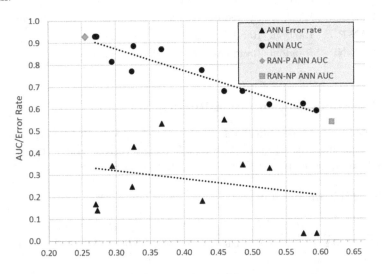

Fig. 2. AUC- and Error rate-Concentration trends.

4.3   Significance Tests

Table 3 shows the results of significance tests for the correlation coefficients for the covariates used to calculate of the two fitted lines in Fig. 2 (random data is excluded). The table shows the values of the sum of measured correlation coefficients, r, the calculated t-values and their corresponding p-values. For a theoretical correlation coefficient ρ, with Null hypothesis is $\rho = 0$ and Alternative hypothesis $\rho \neq 0$, the 95% critical t-value is $t_c = 2.228$. The result for the covariate pair $\{ANNAUC/\hat{H}\}$ falls just short of the 95% critical value (at significance level 5.9%).

Table 3. Paired $\hat{H}$ t-test

Covariates	r	t	p
ANN AUC-$\hat{H}$	−0.559	2.13	0.059
ANN Error-$\hat{H}$	−0.347	1.17	0.134

A Sign test on the difference of the *ANN* and LR AUC results (columns *ANN AUC* and *ANN AUC* in Table 2) gives a probability that LR will produce a higher AUC than the *ANN* AUC of 0.0537 (9 cases out of 12): again, just short of a 5% significance level.

5 Discussion

The empirical results in Table 2 give an indication of how the concentration measure $\hat{H}$ can be used to explain any poor results obtained in a *ANN* analysis. Given the result for RAN-P in particular, a decision boundary, $\hat{H}_B$ set at 0.3 is a useful guide. Therefore, a calculated a value of $\hat{H}$, $\hat{H} > \hat{H}_B$ implies that *ANN*-treatment might be unsuccessful or marginally successful (the data are 'wrong'-shaped). Few datasets are successful: {JP and AUS}, and INT is borderline. Dataset RAN-P has been configured specifically to produce a good separation of features so that class can be determined with a high degree of success.

Some characteristics of 'badly-shaped' datasets can be isolated from the metric calculations. A large *Copula* (H_C) metric is often associated with imbalanced data and almost coincident tuples in two or more classes. For example, RAN-NP tuples in class 0 are a random perturbation of its class 1 tuples, corresponding to the {POL1, POL5, LC, LCAB} group. The *Hypersphere* (H_S) metric measures the effect of outliers: either many of them or a smaller number of extremes, or both. Coincident clustering in more than one class is indicated by a high value of the *k-Neighbours* metric M_N.

The value of the concentration metric, $\hat{H}$, should only be seen either as a guide or as an explanatory element of the *ANN* analysis. A high value $\hat{H}$ implies that either the data are too noisy or that they provide insufficient predictive information. When trying to predict credit-worthiness, cases that appear to be high risk sometimes turn out not to be, and vice versa. These cases look like 'noise' in the data, but they are significant because they provide alternative paths to 'success'. It is better to be able to predict a higher proportion of potential credit failures going to deny credit to borrowers who are apparently low risk. Therefore the within-class error rates (i.e. type I and II errors) are also important.

References

1. Yala, A., Lehman, C., Schuster, T., Portnoi, T., Barzilay, R.: A Deep Learning Mammography-based Model for Improved Breast Cancer Risk Prediction (2019). Radiology https://doi.org/10.1148/radiol.2019182716. 07 May 2019

2. Louzada, F., Ara, A., Fernandes, G.B.: Classification methods applied to credit scoring. Surv. Oper. Res. Manage. Sci. **21**(2), 117–134 (2016). https://doi.org/10.1016/j.sorms.2016.10.001

3. Atiya, A.F.: Bankruptcy prediction for credit risk using neural networks. IEEE Trans. Neural Networks **12**(4), 929–935 (2001)

4. Bredart, X.: Bankruptcy prediction model using neural networks. Account. Finance Res. **3**(2), 124–128 (2014)

5. West, D.: Credit scoring models. Comput. Oper. Res. **27**(11), 1131–1152 (2000). https://doi.org/10.1016/S0305-0548(99)00149-5

6. Lessmann, S., Baesens, B., Seow, H., Thomas, L.C.: Benchmarking state-of-the-art classification algorithms for credit scoring, Eur. J. Oper. Res. (2015). https://doi.org/10.1016/j.ejor.2015.05.030

7. Kvamme, H., Sellereite, N., Aas, K., Sjursen, S.: Predicting mortgage default using convolutional works. Expert Syst. Appl. **102**, 207–217 (2018). https://doi.org/10.1016/j.eswa.2018.02.029

8. Addo, P.M., Guegan, D., Hassani, B.: Credit risk analysis using machine and deep learning models. Risks **6**(38) (2018). https://doi.org/10.3390/risks6020038

9. Munkhdalai, L., Munkhdalai, T., Namsrai, O., Lee, J.Y., Ryu, K.H.: An empirical comparison of machine-learning methods on bank client credit assessments. Sustainability **11**(699) (2019). https://doi.org/10.3390/su11030699

10. Yampolskiy, R.V.: Predicting future AI failures from historic examples. Foresight (2018). https://doi.org/10.1108/FS-04-2018-0034

11. Bikker, J.A., Haaf, K.: Measures of competition and concentration in the banking industry. Econ. Finan. Model. **9**(2), 53–98 (2002)

12. Demarta, S., McNeil, A.J.: The t-copula and related copulas. Int. Stat. Rev. **73**(1), 111–129 (2005)

13. Rodriguez, C.: Measuring financial contagion: a copula approach. J. Empirical Financ. **14**(3), 401–423 (2007). https://doi.org/10.1016/j.jempfin.2006.07.002

14. Dua, D. and Graff, C.: UCI Machine Learning Repository Irvine CA (2019). http://archive.ics.uci.edu/ml

15. Li, M., Mickel, A., Taylor, S.: Should this loan be approved or denied? J. Stat. Educ. **26**(1), 55–66 (2018). https://doi.org/10.1080/10691898.2018.1434342

16. Yeh, I.C., Lien, C.H.: The comparisons of data mining techniques for the predictive accuracy of probability of default of credit card clients. Expert Syst. Appl. **36**(2), 2473–2480 (2009)

17. Zieba, M., Tomczak, S.K., Tomczak, J.M.: Ensemble boosted trees with synthetic features generation in application to bankruptcy prediction. Expert Syst. Appl. **58**(1), 93–101 (2016)

Adaptive Machine Learning-Based Stock Prediction Using Financial Time Series Technical Indicators

Ahmed K. Taha[✉], Mohamed H. Kholief, and Walid AbdelMoez

College of Computing and Information Technology, Arab Academy for Science,
Technology and Maritime Transport, Alexandria, Egypt
ahmedengu@student.aast.edu, {kholief,
walid.abdelmoez}@aast.edu

Abstract. Stock market prediction is a hard task even with the help of advanced machine learning algorithms and computational power. Although much research has been conducted in the field, the results often are not reproducible. That is the reason why the proposed workflow is publicly available on GitHub [1] as a continuous effort to help improve the research in the field. This study explores in detail the importance of financial time series technical indicators. Exploring new approaches and technical indicators, targets, feature selection techniques, and machine learning algorithms. Using data from multiple assets and periods, the proposed model adapts to market patterns to predict the future and using multiple supervised learning algorithms to ensure the adoption of different markets. The lack of research focusing on feature importance and the premise that technical indicators can improve prediction accuracy directed this research. The proposed approach highest accuracy reaches 75% with an area under the curve (AUC) of 0.82, using historical data up to 2019 to ensure the applicability for today's market, with more than a hundred experiments on a diverse set of assets publicly available.

Keywords: Stock price prediction · Technical indicators feature importance · Adaptive stock prediction · Machine learning · Feature selection

1 Introduction

Despite the abundance and the quality of stock market data, it suffers from a high noise to signal ratio that makes the task of predicting the stock market direction an extremely challenging task. It has been established by the random walk hypothesis [2] and the efficient market hypothesis [3] that no one can beat the markets, however, the extraordinary results of quant funds may indicate otherwise. All of that is due to the uniqueness of the stock market unlike any other problem; the financial markets require careful handling of the data to prevent look-ahead bias and overfitting that results in false discoveries from experienced and inexperienced data scientists alike.

Following a strict research process is the key to achieve high accuracy market forecasts, as suggested by López de Prado [4]. De Prado also places emphasis on the importance of avoiding overfitting at all costs by ensuring test and training data

© Springer Nature Switzerland AG 2019
H. Yin et al. (Eds.): IDEAL 2019, LNCS 11872, pp. 191–199, 2019.
https://doi.org/10.1007/978-3-030-33617-2_21

separation; reviewing data preprocessing results to make sure there is no data leakage; and always start with a hypothesis to avoid fitting to the noise, which can happen due to the prolonged exposure to the same data.

Japanese Candlestick Charting Techniques is considered to be a good indication of the timeframe it is covering, as it resamples the trades into Open, High, Low, Close, and Volume as a snapshot of a timeframe. However, a bar chart over a long time can be considered to be very complicated to use alone to understand the market. That is why technical analyst [5] ordinarily makes use of technical indicators [5], which are mathematical formulas that try to reduce the market noise to help identify opportunities.

Technical analysis [5] is a method that uses statistical analysis to evaluate market assets and involves looking at asset price movements over the years. Trying to identify recurring patterns and price action for entering and exiting the market, this can be conducted manually by a technical analyst who stares at the market charts for hours constructing support and resistance lines to identify the market trend, which is not scalable.

The proposed work aims at designing an adaptive machine learning workflow, with the following contributions: (1) Inclusion of one hundred fifty-four technical indicators plus seventeen time-based features, (2) Employing multiple machine learning algorithms and feature selection techniques to generate a scoring leaderboard, which isused to achieve consistent performance over multiple timeframes and assets, (3) Exploring multiple unrelated markets, timeframes, and targets to illustrate the benefits of the proposed approach; with publicly accessible codebase on GitHub [1] to allow results validation, improvements, and future modifications, (4) Exploring rarely researched machine learning algorithms and feature selection techniques such as generalized linear model and feature importance extraction from deep learning models using the Gedeon method, alongside a diverse set of feature selection and machine learning algorithms, (5) Introducing the ability to improve prediction using alternative targets such as the High price which is explored below, using the High price as a label allows capturing better signals from the market that can then get used within a trading strategy.

This paper is structured as follows: Sect. 2 presents the field literature. Section 3 describes the proposed model layers in detail from the dataset, data preprocessing to model training. Section 4 highlights and analyzes the results of the experiments, and finally, Sect. 5 concludes the paper.

2   Related Work

Ding et al. [6], introduced a long term and short term stock price prediction based on a convolutional neural network with input processed using an event-embedding layer. Ding proposed approach improved the accuracy by 6% compared to reviewed papers, thus achieving individual stock prediction accuracy of 65.48%. The paper's event embedding layer is a CNN that trained using input datasets of events from Reuter's financial news and Bloomberg financial news using Open IE. The prediction model training makes use of the previous layer output divided into long-term, medium-term, and short-term events.

Si et al. [7] introduced a topic-based twitter sentiment technique. Si proposes one of the first attempts to make use of a non-parametric continuous Dirichlet process model (cDPM) based on twitter sentiments for stock prediction in an autoregressive framework experimented on real-life Standard & Poor's 100 (S&P 100) market index. The author experiments involved three phases. First, the author makes use of cDPM to identify the daily topic set, which helps solve the broad topic diversity of Twitter tweets. Next, the proposed technique makes use of an opinion lexicon to determine every tweet sentiment to build a sentiment time series. Finally, Si presents a vector autoregressive layer that uses sentiment time series and stock index to predict the stock index. The best prediction the paper was able to achieve was with using a window size of (21, 22) days of data with an average accuracy of 68.0%.

In [8], Jung and Aggarwal introduce a feature generation layer that makes use of 24 technical indicators and generate a total of 24 numerical features and 126 binary features as the input dataset to the prediction algorithms. Jung makes use of a Bayesian Naive Classifier and a Support Vector Machine (SVM) as prediction algorithms. The author used six years of S&P 500 daily stock prices as the raw dataset, and the experiment revealed a performance accuracy of 73.9% for Naive Bayes and 74.3% for SVM. Moreover, the Naive Bayes was able to finish within 10 s while the SVM training time was 600 s.

In Zhang et al. [9], a comparative study has been conducted based on 13 years' worth of daily data from the Chinese stock market, to observe the features effect on the model used. The author proposed a causal feature selection (CFS) algorithm to select the most relevant set of features to improve the model predictability. Moreover, Zhang compares it with principal component analysis (PCA) [10], decision trees (DT) [11], and the least absolute shrinkage and selection operator [12]. The author chooses to work with 50 fundamental and technical features from various categories, such as Valuation factors, Profitability factors, and Growth factors. Zhang trains the model on a sliding window of the dataset one year at a time to prevent overfitting; the CFS approach on average selected the highest number of features achieving higher prediction accuracy over PCA, DT, and LASSO, with a reported accuracy of 54.70%.

3 Proposed Approach

The proposed model workflow in Fig. 1 constructs an adaptive stock prediction model that it is usable for any market or timeframe. The model consists of four different layers. The first layer is the data-gathering layer that is configurable to collect data from multiple sources such as Yahoo Finance, Dukascopy, and Kaggle. The second layer is the data-preprocessing layer that generates the targets and 154 technical indicators from the OHLCV data, alongside the Exploratory Data Analysis (EDA) of the input dataset and the generated features to help understand and evaluate the results, then finally scaling and splitting. The third layer is where the model training, hyperparameter optimization, and feature elimination takes place; Model training happens first on all the features to define a performance benchmark, then once more after the feature elimination process as a two-way process with adaptive feedback. The fourth layer evaluates and scores the training performance then sorts based on AUC as it considers the randomness and imbalance of the testing dataset.

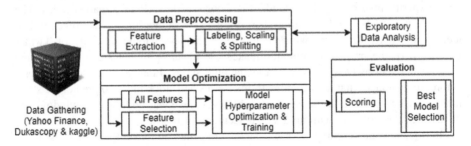

Fig. 1. Illustration of the proposed machine learning workflow for adaptive stock prediction using technical indicators.

The published workflow on GitHub [1] can be configured and reused to test any other asset available on Yahoo Finance or Dukascopy, with no code modifications nor local environment, as the code can be executed from the browser using Google Colab.

3.1   Datasets

The dataset consists of time series based Open, High, Low, Close, and Volume bars of the asset price to represent a snapshot of the trades that happened in the bar period. The reported results derived from multiple datasets and timeframes, as shown in Table 1. To show how the model performs under different markets and timeframes.

Table 1. A list of the used tickers and data sources.

Asset ticker	Asset name	Daily		Hourly	
		Range	Source	Range	Source
AAPL	Apple US	2009–2019	Yahoo Finance	2018–2019	Dukascopy
EURUSD	EUR/USD	2009–2019	Yahoo Finance	2018–2019	Dukascopy
BTCUSD	Bitcoin USD	2012–2019	Bitstamp	2018–2019	Bitstamp

The conducted experiment uses multiple data sources as shown in Table 1; that is due to Dukascopy one year constraint on historical data; thus Yahoo Finance data is used as the source of the ten-year daily timeframe and Dukascopy for the hourly data.

3.2   Data Preprocessing

The target used for this supervised ML problem is a binary label. The label equals (1) when the next bar is higher than the current bar and equals (0) when it is lower. The label calculated for the Close price and the High price within two separate experiments to compare the information gain for each of them.

An extensive list of technical indicators hasbeen used, with all of the datasets in Table 1; at the data-preprocessing layer before training the model. One hundred fifty-four technical indicators [13] was used for this research to allow the evaluation of all technical indicators to capture weak market signals and improving accuracy. Not only that, those indicators have been incorporated with multiple parameters and time-based feature extraction that generates seventeen new features from the timestamp to capture time-related market behavior such as holidays, which increases the number of columns to one thousand twenty-four features.

Then min-max scaling gets applied per row on the entire dataset to ensure no look forward bias and to help the model train. Finally, before training the dataset to get divided into training, validation, and testing sets with the following ratios 70%, 20%, 10% respectively.

3.3 Feature Selection Algorithms

Multiple feature selection techniques tested within the research process, such as principal component analysis, high correlation filters, and low variance filters. However, based on the results of the experiments, this list contains the best performing algorithms: (1)All of the generated features (ALL), (2) Deep Learning Feature (DLF) importance, (3) XGB Feature (XGBF) importance, (4)Recursive Feature Elimination (REF), (5) Family-Wise Error (FWE) rate.

Where ALL represents all of the technical indicators and time-based features generated, the DLF feature's importance is extracted from the trained deep learning model using the Gedeon method that uses the first two-layer of the neural network to assign weights to the input features. XGBF calculates the variable importance from the trained XGBoost model to represent the new input set for the training of all the models. RFE makes use of an extremely randomized tree classifier as the RFE estimator that calculates the feature importance for every step. FWE is a univariate selection technique that filters the features ANOVA f-value based on the p-value set for the experiment.

3.4 Machine Learning Algorithms

This research goal is building a versatile predictive workflow, with many machine-learning algorithms implemented. This experiment used only the three best-performing algorithms listed below: (1) Multi-layer Feedforward Neural Network Deep Learning (DL) [14], (2) eXtreme Gradient Boosting (XGBoost) [15], (3) Generalized Linear Model (GLM) [16].

All of the machine learning algorithms listed above go through the hyperparameter optimization layer, which trains up to 10 models until it reaches the optimal set of parameters that achieve the highest accuracy.

Model training initially took place on the original training dataset as a performance benchmark, then trained with the features and data calculated from the dimensionality reduction algorithms stated before.

4   Experiment Results and Analysis

Table 2 shows the results of the conducted experiments on APPL, Bitcoin, and EURUSD. Moreover, an emphasis on the advantage of the proposed workflow, despite the weakness of some models alone, the entire workflow reports better final results; it can adapt to different markets, timeframes, targets, and regimes. For more than 50 additional experiments, detailed results, exploratory data analysis, and different metrics review the most recent implementation on GitHub [1] and run the experiments in the browser using Google Colab for free.

Table 2. The results of the various experiments in this study, reported in the accuracy metric, with the best accuracy in bold for each asset, target pairs.

Timeframe		Daily						Hourly					
Target		Close			High			Close			High		
Model		DL	GLM	XGB	DL	GLM	XGB	DL	GLM	XGB	DL	GLM	XGB
Asset	Features												
APPL	ALL	0.51	0.51	0.563	0.494	0.494	0.721	0.541	0.553	0.553	0.659	0.688	0.671
	DLF	0.51	0.51	0.538	0.494	0.494	0.725	0.547	0.571	0.582	0.647	0.671	0.694
	FWE	0.51	0.51	0.563	0.692	0.725	**0.729**	0.571	0.553	0.553	0.676	0.659	0.671
	RFE	0.534	**0.563**	0.547	0.7	0.725	0.721	0.559	0.553	0.547	0.641	0.653	0.641
	XGBF	0.51	0.51	0.538	0.494	0.494	0.725	0.571	0.571	**0.582**	0.659	0.671	**0.694**
Bitcoin	ALL	0.546	0.538	0.55	0.602	0.594	0.697	0.587	0.576	0.576	0.651	0.665	**0.692**
	DLF	0.538	0.53	0.55	0.622	0.61	0.709	0.576	0.572	0.579	0.668	0.662	0.69
	FWE	0.582	0.562	0.57	0.622	0.598	**0.717**	0.582	0.584	0.58	0.656	0.673	0.688
	RFE	0.57	0.538	**0.59**	0.653	0.602	0.709	**0.588**	0.583	0.586	0.665	0.68	0.683
	XGBF	0.558	0.53	0.55	0.602	0.61	0.709	0.576	0.572	0.579	0.66	0.662	0.69
EURUSD	ALL	0.719	0.738	**0.754**	0.57	0.551	**0.648**	0.556	0.543	0.562	0.724	0.72	0.733
	DLF	0.742	0.738	0.738	0.555	0.547	0.613	0.562	0.536	0.551	0.726	0.733	**0.734**
	FWE	0.711	0.711	0.703	0.57	0.559	0.57	0.548	0.543	0.562	0.712	0.726	0.721
	RFE	0.746	0.73	0.742	0.586	0.582	0.621	0.543	0.553	**0.562**	0.728	0.715	0.721
	XGBF	0.746	0.738	0.738	0.613	0.547	0.613	0.546	0.536	0.551	0.721	0.733	0.734

In Table 2, Highlights the best results in bold; however, in the case of multiple feature selection approaches achieving the same best accuracy, the approach with the highest average accuracy is highlighted.

Apple achieved 56% on the Daily-Close with GLM-RFE model. However, it reported AUC of 0.53, which may indicate a slight data imbalance; however, with the XGB-ALL approach, it reports the same accuracy with a higher AUC of 0.54 which emphasizes the advantage of using AUC for scoring and sorting, instead of accuracy.

EURUSD reported AUC of 0.82 for the Daily-Close price, 0.66 for the Daily-High price, 0.55 for the Hourly-Close price, and 80% for the Hourly-High price.

Despite Bitcoin volatility, the model achieved 59% accuracy predicting Daily-Close and 71% for the Daily-High, with AUC of 0.56 and 0.74 respectively.

From Table 2 results, it appears that the Close-price sometime has a weaker signal in comparison with the High, Low, and Mid-price targets, that is why it is essential to use an ensemble of multiple algorithms and strategies in production, not just one machine learning model with an emphasis on sentiment-based models that can act as a circuit breaker. In the case of a black swan event, the system can shut down the system immediately due to instability.

On every run, the proposed architecture captures a different set of essential features that help maximize the predictability of the market direction from technical factors, as shown in Fig. 2. Technical indicators based features such as Balance of Power, Stochastic Fast, and True Range, and time-based features such as Day of the Week are some of the top features that increased helped increase the accuracy of predicting EURUSD direction.

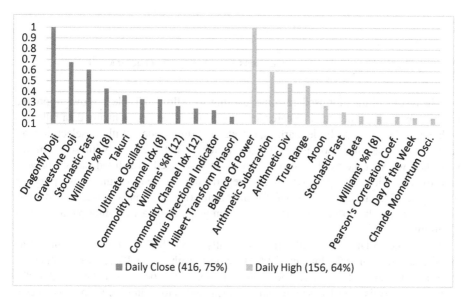

Fig. 2. A comparison of EURUSD top 10 selected features.

EURUSD Daily-Close reports an accuracy of 71% with all the features as an input using the deep learning model, however, after using XGBF the accuracy increases up to 74% with the best accuracy of 75% with the XGB model, and the number of features drops to 416 as in Fig. 2. The drop in the number of features helps improve the model accuracy, as it captures a specific market characteristic of the data used, which leads to adaptive accuracy and less computation requirement for the extra-unrelated features to get predictions on time during back-testing and production.

Figure 3 visualizes the model's prediction, which reports good accuracy. However, using it directly on the market can result in red profit and loss (PnL) despite having high accuracy. It is best to use the model within an algorithmic trading strategy or combined with another model to minimize the downside and protect against black swans.

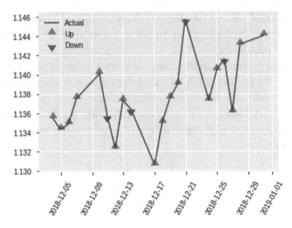

Fig. 3. Euro Daily-Close predictions from XGB with an accuracy of 75%.

The results of the proposed machine learning workflow in Table 2 shows improved results over the current actual research state that does not suffer from any data fallacies, with applicability for multiple markets, targets, and timeframes. With no selection bias, survivorship bias, nor look-forward bias; which is not the case with some of the published work in the field, that only works on a selected over-fitted dataset or with look-forward bias. The proposed model reports the accuracy of the Close price on worst-case 53% and up to 75% and for the High price 58% on worst-case, and up to 74%—those results of the experiments over ten years of data with multiple assets and timeframes.

Compared to Ding et al. [6] Reported results of 65.48%, Si et al. [7] Achieved an average accuracy of 68%, Jung et al. [8] Have best results of 74.3% and Zhang et al. [9] Report best results of 54.70%. The proposed approach reports better results than the sentiment analysis based research and technical indicators literature, on more recent data, on different markets and timeframes.

5   Conclusion

The stock market prediction problem is not easy. Even with the discovery of a trading edge, it will not work for long due to alpha decay. This paper focuses on experimenting with all the financial time series technical indicators available and explores their importance over multiple models, markets, and parameters with different feature elimination algorithms. To ensure the selection of the most valuable features to help with the high noise to signal ratio characteristics of the stock market and capture hidden patterns and asset unique characteristics. The reported results may seem weak individually; however, considering the entire workflow accuracy achieve the level of 72% predicting apple stock, 71% with bitcoin, and 75% EURUSD.

The public repository presents an opportunity to improve the state of research in the field with more than a hundred experiments results available to be evaluated and improved in the future.

Future directions would involve optimizing the implementation performance to allow the experimentations with more datasets and feature variations, which going to undergo general code optimizations and the development for clustered machines.

References

1. Taha, A.: feature importance for ml stock prediction. https://github.com/ahmedengu/feature_importance
2. Fama, E.F.: Random walks in stock market prices. Financ. Anal. J. **51**, 75–80 (1995). https://doi.org/10.2469/faj.v51.n1.1861
3. Fama, E.F.: Efficient capital markets: a review of theory and empirical work. J. Finance. **25**, 383 (1970). https://doi.org/10.2307/2325486
4. López de Prado, M.M.: Advances in financial machine learning (2018)
5. Murphy, J.J., Murphy, J.J.: Technical Analysis of the Financial Markets: A Comprehensive Guide to Trading Methods and Applications. Institute of Finance, New York (1999)
6. Ding, X., Zhang, Y., Liu, T., Duan, J.: Deep Learning for Event-Driven Stock Prediction. aaai.org
7. Si, J., Mukherjee, A., Liu, B., Li, Q., Li, H., Deng, X.: Exploiting topic based Twitter sentiment for stock prediction. In: Proceedings of the 51st Annual Meeting of the Association for Computational Linguistics (Volume 2 Short Pap.), pp. 24–29 (2013)
8. Jung, H.J., Aggarwal, J.K.: A binary stock event model for stock trends forecasting: Forecasting stock trends via a simple and accurate approach with machine learning. In: International Conference on Intelligent Systems Design and Applications, ISDA, pp. 714–719. IEEE (2011). https://doi.org/10.1109/ISDA.2011.6121740
9. Zhang, X., Hu, Y., Xie, K., Wang, S., Ngai, E.W.T., Liu, M.: A causal feature selection algorithm for stock prediction modeling. Neurocomputing **142**, 48–59 (2014). https://doi.org/10.1016/J.NEUCOM.2014.01.057
10. Jolliffe, I.T.: Principal Component Analysis. Springer, New York (2002). https://doi.org/10.1007/b98835
11. Quinlan, J.R.: Learning efficient classification procedures and their application to chess end games. In: Michalski, R.S., Carbonell, J.G., Mitchell, T.M. (eds.) Machine Learning, pp. 463–482. Springer, Heidelberg (1983). https://doi.org/10.1007/978-3-662-12405-5_15
12. Tibshirani, R.: Regression shrinkage and selection via the lasso. J. R. Stat. Soc. Ser. B. **58**, 267–288 (1996). https://doi.org/10.1111/j.2517-6161.1996.tb02080.x
13. TA-Lib : Technical Analysis Library. http://ta-lib.org/
14. LeCun, Y., Bengio, Y., Hinton, G.: Deep learning. Nature **521**, 436–444 (2015). https://doi.org/10.1038/nature14539
15. Chen, T., Guestrin, C.: XGBoost. In: Proceedings of the 22nd ACM SIGKDD International Conference on Knowledge Discovery and Data Mining - KDD 2016, pp. 785–794. ACM Press, New York (2016). https://doi.org/10.1145/2939672.2939785
16. McCue, T., Carruthers, E., Dawe, J., Liu, S.: Evaluation of generalized linear model assumptions using randomization (2008). http://www.mun.ca/biology/dschneider/b7932/B7932Final10Dec2008.pdf

Special Session on Knowledge Discovery from Data

Exploiting Online Newspaper Articles Metadata for Profiling City Areas

Livio Cascone, Pietro Ducange(✉)(iD), and Francesco Marcelloni(iD)

Department of Information Engineering, University of Pisa,
Largo L. Lazzarino 1, 56122 Pisa, Italy
{pietro.ducange,francesco.marcelloni}@unpi.it

Abstract. News websites are among the most popular sources from which internet users read news articles. Such articles are often freely available and updated very frequently. Apart from the description of the specific news, these articles often contain metadata that can be automatically extracted and analyzed using data mining and machine learning techniques. In this work, we discuss how online news articles can be integrated as a further source of information in a framework for profiling city areas. We present some preliminary results considering online news articles related to the city of Rome. We characterize the different areas of Rome in terms of criminality, events, services, urban problems, decay and accidents. Profiles are identified using the k-means clustering algorithm. In order to offer better services to citizens and visitors, the profiles of the city areas may be a useful support for the decision making process of local administrations.

Keywords: Information retrieval · Smart cities · Data mining · Machine learning

1 Introduction

In the last years, a considerable amount of research activities and projects have been developed under the general umbrella of *smart cities* [10]. In the literature, several definitions of smart cities have been proposed, but essentially smart cities are those cities that have the greatest quality of life and economic well-being for their citizens. Indeed, in 2009 about 50% of the world population was living in towns and cities. In 2050, this value is expected to reach almost 70%[1].

---

[1] https://www.un.org/development/desa/en/news/population/2018-revision-of-world-urbanization-prospects.html.

This work was partially supported by Tuscany Region, in the context of the projects Talent and Sibilla in the framework of regional program "FESR 2014–2020", and by the Italian Ministry of Education and Research (MIUR), in the framework of the CrossLab project (Departments of Excellence).

H. Yin et al. (Eds.): IDEAL 2019, LNCS 11872, pp. 203–215, 2019.
https://doi.org/10.1007/978-3-030-33617-2_22

Thus, smart services for citizens, city administrations, security officers and so on, will become increasingly important.

Characterizing and profiling city areas and neighborhoods, in terms of activities, behaviors, events and issues, may be strategic for designing appropriate smart services for citizens. An area in a city can be featured in terms of (i) social, commercial, fun and touristic activities, (ii) people and traffic flows, (iii) living and property costs, (iv) crime hot spots, etc. A number of heterogeneous data stream sources, such as smartphones, cameras, weather stations, traffic management stations, and web sites/services, may be used for profiling the different areas of a city (i.e a district, a neighborhood, a main square). In this context, recently, some contributions have been proposed. As an example, in [12], authors discuss CrimeTelescope, a web platform that continuously collects open crime data (not always available, especially in Europe), Foursquare[2] point of interests (POIs) and Tweets, for predicting crime hot spots in the city of New York. In [3], the authors present CityPulse, a platform which exploits online social data (geo-located tweets, post, check-ins, photos) to support decision making of citizens, city visitors, and city administration. In 2018, we have proposed an original framework to profile city areas [2]. The framework allows us to collect available geo-located data (about POIs, posts, traffic information, etc.) from online web services and sites related to the considered city. Each city area is described in terms of a set of meaningful features, extracted from the collected data. In order to identify similarities and differences among the areas of the city the framework exploits data mining techniques.

In recent decades, almost all traditional newspaper companies have provided digital versions of news articles on their websites. Moreover, the readers can find several additional sources of news on the web, including also digital-only news sites, television websites, and search engine providing specific news aggregation sections, such as Google News[3] and Yahoo News[4]. Finally, several editorial companies, which offer free high quality news articles thanks to advertising revenues, have recently appeared on the web. Some of them have also local versions of the news portals offering articles regarding specific cities and news are updated almost in real-time.

Since we are experiencing the *Internet of Everything and Everywhere era* [4], online news is currently the most popular channel for Internet users to read news. The huge amount of news available online represents a very attractive source of valuable information to be mined and the term *Web News Mining* has been recently coined [5]. Indeed, information retrieval strategies along with machine learning and data mining algorithms have been successfully applied for approaching several tasks, such as news recommending systems [6], news categorization [1,11], article source detection and geo-localization [9], and crime mapping [8].

---

[2] http://www.foursquare.com/.

[3] https://news.google.com/.

[4] https://news.yahoo.com/.

Online news are usually characterized by a semi-structured data organization: they may include, apart the text and the date, images, videos, tags describing the categories of the news, comments written by the readers, information regarding the location of the news, and other metadata. With the aim of mining useful information for profiling city areas, in this work, we exploit tags and location metadata extracted for online newspaper articles. We integrate these new sources of information into the aforementioned framework, introduced in [2]. Specifically, after a geo-localization process, the news are first associated with the different areas of a specific city. Then, each city area is described by using a set of features obtained counting, for each news category, the total number of geo-located articles. In this way, we are able to provide maps describing the different city areas also in terms of specific events and issues, such as criminality, urban decay, traffic and accidents, and immigration problems.

2 On Line Newspapers Articles Metadata

In our previous work discussed in [2], we highlighted that online web sites and services may be used as sources of data for profiling city areas. We can extract from such sources both raw data and aggregated information about points of interest disseminated in the city, costs of living, traffic, etc. For the sake of simplicity, in the following we will refer to all the possible elements characterizing a city area with the general term *city feature*. The majority of web data sources is general and provides information about any city or geographical region, whereas other data sources are specific to the city, i.e., the open data website about the city local government, and city-related news/events websites.

Among the different web data sources, online newspapers have recently attracted a lot of attention from researchers. Indeed, in a recent contribution in [9], authors introduce and discuss an interesting dataset containing 100,000 news articles, published from January 1st to December 31th 2014, enriched with a set of metadata. The articles have been extracted from different online newspapers websites, including BBC News, The Telegraph, The Guardian, The Washington Post. All the articles include at least one image and possibly tags, user comments extracted from social networks, geographic coordinates, information regarding the number of shares. Moreover, all the original articles have been enriched with additional pictures extracted from Google Images, using the full title as search query. As shown in [7–9], metadata extracted from on-line newspapers articles may be useful to enrich the set of information to be used for solving different kinds of tasks related to article analysis, such as news categorization, geolocation prediction, identification of crime hotspots, image classification.

In this work, we exploit two types of metadata extracted from online newspapers articles: the tags, which specify the categories of the news, and the place, where the events described in the news are occurred. As source of information, we use the different versions of Today[5] websites, owned by the CityNews company. These websites offer free online news updated in almost real-time, covering

---

[5] http://www.today.it/.

48 Italian cities, with more than 250 journalists collaborating with the different editorial offices. Each day, CityNews publishes more than 1,300 news and its websites count more than 720,000 registered users. In January 2019 CityNews was ranked as the first online editorial company in terms of number of visits for its websites[6].

In Table 1 we summarize the data extracted from the news articles. The table also shows sample values extracted from a recent news article and its translation in English.

Table 1. Raw Data Extracted from news articles

Type of raw data	Sample value	Translation in English
Link	http://parioli.romatoday.it/salario/via-di-priscilla-chiusa-bus-deviati.html	
Time stamp	09:04:2019-09:47	
Title	Roma sprofonda, chiusa via di Priscilla: quartiere in tilt	Rome collapses, closed via di Priscilla: neighborhood in chaos
List of Tags	Deviazioni Bus, Strade Chiuse	Bus deviations, closed roads
Location	Roma, Via di Priscilla	
Summary	La strada del Salario interdetta al traffico tra piazza Vescovio e via Monte delle Gioie: accertamenti tecnici in corso sull'avvallamento	The Salario road is closed to traffic between piazza Vescovio and via Monte delle Gioie: expert inspections in progress on the dip
Text of the news	Via di Priscilla chiusa a causa di un avvallamento. La strada del quartiere Salario é interdetta al traffico tra piazza Vescovio e via Monte delle Gioie...	Via di Priscilla closed to the traffic due to a dip. The street of the Salario district is closed to traffic between Piazza Vescovio and Via Monte delle Gioie...

Since the categories defined by the tags are too specific, we grouped these categories into macro-categories as shown in Table 2. In the following section, we discuss how we extract the data from online newspapers articles and how we pre-process and analyze them, integrating this new source of information in our framework for profiling city areas.

3   The Framework for Smart Profiling of City Areas

As stated in Section Introduction, in [2] we proposed the architecture of a framework for profiling city areas, based on web data. We show the architecture

---

[6] https://www.primaonline.it/2019/02/01/284200/classifica-informazione-online-a-dicembre-su-dati-comscore/.

of the framework in Fig. 1: it is composed by four main functional modules, namely Data Retrieval, Data Preparation, Data Mining Analysis and Result Visualization.

Table 2. Correpondence between macro-categories and tag categories.

Macro-category	List of corresponding original tag categories
Criminality	Thieves, Theft, Robbery, Sexual Violence, Wife Beaten, Stalking, Harassment, Stabbings, Brawl, Assault, Vandalism, Murders, Rapes, Drugs, Drug dealing, Fraud, Cocaine, Pursuit, Pick pocketing, Arrest, Seizure, Mafia, Camorra, 'Ndrangheta, Racket, Extortion
Accidents	Accident, Rear-end collisions, Investment
Politics	Elections, Politics, Referendum, Strike, City Council, Regional Council, Political Parties
Events and Services (E&S)	Entertainment, Theater, Concert, Show, Expo, Presentation, Game, Music, Events, Event, Evening, Feast, Services, Library, Wifi, Subway, Hospitals, Car parks, Stations, Bus
Urban Problems and Decay (UP&D)	Rom, Nomads, Foreigners, Mosque, Homelessness, Muslim, Islam, Refugees, Antisemitism, Counterfeits, Racism, Migrants, Illegals, Fire, Crash, Blackout, Bus deviations, Lighting, Fire, Flood, Inconvenience, Holes, Work in Progress, Closed roads, Construction, Public Works, Redevelopment, Degradation, Social housing, Graffiti, Mural, evictions, Prostitution, Writer

The module *Data Retrieval* is devoted to collect raw data, about cities or geographical regions, from different meaningful web data sources and store them in a database. Usually, data are extracted in batch mode, stored in the database and eventually processed. The data about a city can be occasionally updated *e.g.*, by extracting the most recent online news, or by collecting data from a new source. The module requires to select the city to be analyzed and the web data sources. Whenever the web pages and services do not provide developers with API for retrieving data, web crawlers and scrapers may be used. In this paper, we present preliminary results achieved considering the data collected from the different online newspaper portals of Today.it. In order to extract raw data from these portals, we adopted a web scraper that executes the following actions: first, it downloads all the links published in a specified time interval; then, it accesses and downloads the HTML page of each article and finally it extracts and stores all the metadata in the data base. We wish to point out that whenever the terms of service on data usage have some retrieving and/or storage limitations, we avoid to download data in batch mode and/or we delete the data after the elaboration.

The module *Data Preparation* prepares the data for the subsequent analysis, *i.e.*, the raw data are filtered, preprocessed and aggregated, according to the definition of the areas. First of all, we need to define the boundaries of the city areas. To this aim, we build a uniform virtual grid of squared cells on the city. Hence, primarily, the areas to be profiled correspond to the cells of the grid. The size of the cells, and consequently the number of cells of the grid, is a parameter of the framework and can be adjusted as desired. Then, once the grid is created, for each cell, representative descriptive features are extracted from the raw data collected for the specific city. The features describing the areas can contain data from a single source or heterogeneous data from different sources. The description of the cells changes accordingly to the data sources. More precisely, the module allows carrying out the following operations:

(i) *grid definition:* the city grid is set up by choosing the size of the cells;

(ii) *data pre-processing:* the data collected from the web sources and stored in the database are filtered so as to remove useless or redundant information (*e.g.*, repeated information collected from different data sources). Data normalization may be also applied.

(iii) *data projection in the cells:* each cell is associated with the list of data which can be projected into the cell itself on the basis of the corresponding geographic coordinates. In the experiments, the cell is associated with the list of news articles whose coordinates of the corresponding place fall into the cell. In this work, since the geographic coordinates are not directly available as metadata of the articles, we extract them using the Geocoding API[7] provided by Google Maps. The Geocoding API, accessible using an HTTP interface, given a specific address, returns the geographic coordinates.

(iv) *feature extraction:* features characterizing each cell are extracted by aggregating the raw data, and are projected into the specific cell, in order to obtain its description as a multi-dimensional vector. In the experiments, we describe each cell by counting the number of news associated with a specific macro-category, as shown in Table 2. We recall that a single news can be associated with more than one macro-category.

In Table 3 we show a simple example of cell description where we consider just 4 cells and 4 types of news article macro-categories for each cell, namely Criminality, Accidents, Events and Services (E&S), and Urban Problems and Decay (UP&D). The output of the Data Preparation module is a dataset prepared in the format required by the data mining algorithms, *i.e.*, being F the number of significant features to take into account, the cells of the grid are represented in terms of an F-dimensional feature vector.

The *Data Mining Analysis* module allows extracting useful information from the available data. Indeed, different machine learning algorithms (classification, regression, clustering, etc.) can be applied on the dataset prepared by the Data

---

[7] https://developers.google.com/maps/documentation/geocoding/start.

Table 3. An example of cells description

Cell id	GPS coordinates of the cell center	UP&D	Accidents	Criminality	E&S
1	Lat_1, Lon_1	1	2	3	2
2	Lat_2, Lon_2	10	0	4	0
3	Lat_3, Lon_3	2	10	2	0
4	Lat_4, Lon_4	5	0	2	1

Preparation module. Currently, the framework includes only clustering algorithms for profiling city areas. However, we envision to extend the framework by adding predictive models so as to add functionalities such as event detection and forecasting, and estimation of missing data. A clustering analysis can be performed with the aim of grouping similar areas, *i.e.*, cells of the grid. In the experiments, the groups of similar areas are described in terms of the predominant news macro-categories: the set of these macro-categories defines the profile of the group. In order to identify the profiles, we employed the k-means partitioning clustering algorithm. The algorithm groups the objects (in our case, cells), represented in the F-dimensional feature space, into k clusters (k is chosen by the user), with the aim of minimizing a certain convergence criterion, *i.e.*, the intra-cluster variance. Each cluster is described by its centroid corresponding to a prototype for the cluster. The results of the run of a clustering algorithm are stored in the database. The framework allows exporting as a simple text file such results and visualizing them by means of the "Result Visualization" module.

Finally, the module *Result Visualization* is in charge of supporting the interpretation of the collected data and of the results obtained after a data mining analysis. Indeed, it allows visualizing data and results in terms of graphs, charts, heat maps, and statistics. For our analysis, we adopted discrete heat maps built directly on the virtual grid.

We designed a general and easy-to-extend framework: it can be adapted to different web data sources, cities, and area definitions. The framework was developed mainly using the Python language. We employed the Flask microframework[8] as web server, and the scikit-learn[9] and Pandas[10] libraries for the data processing and the data mining elaborations. We employed the NoSQL database MongoDB[11] with the aim of avoiding constraints on the representation of data and on the operations to be performed.

---

[8] http://flask.pocoo.org.
[9] https://scikit-learn.org/.
[10] https://pandas.pydata.org/.
[11] https://www.mongodb.com.

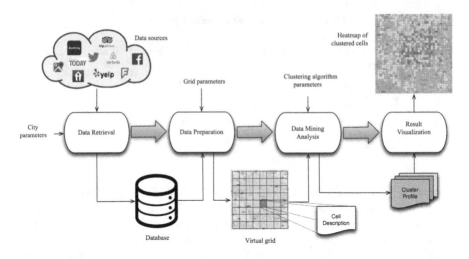

Fig. 1. The general scheme of the framework for profiling city areas.

4   Experimental Analysis

In the following, we discuss the experimentation of our framework considering the integration of the online newspapers articles as a new data source. In our experiments, we extracted news articles from RomaToday[12], which provides news regarding the Metropolitan City of Rome. However, in our analysis, we concentrate our attention on the main urban area of Rome, i.e. inside and near the Ring Road (Grande Raccordo Anulare) and considered just the news extracted in 2018. In total, we extracted 6951 news. Among them, 6537 news have at least an associated tag and 4001 news have been geolocated. Finally, in our experiments we selected just news that belong to the following macro-categories: E&S, Accidents, Criminality, and UP&D (1716 news in total).

Figure 2 shows a heat map describing the distributions of the selected news on the city grid. On the basis of our experience, as discussed in [2], we adopted a grid composed by 1936 squared cells with a size of 500 m. The most dense cells are concentrated in the city centre, which represents the most crowded area, especially of tourists and visitors. As shown in Fig. 2, some cells result to be empty, i.e., they do not contain any news. In order to improve both the accuracy and the speed up of the clustering algorithms, these cells are inserted by the framework into a dummy cluster and filtered out before applying any clustering algorithm. In order to profile the city areas, we applied the k-means clustering algorithm, considering the macro-categories of the selected news as features. Table 4 numerically describes these features.

We experimented different values of k, but we chose to show the results achieved with $k = 4$, since it produces easily interpretable clusters.

---

[12] www.romatoday.it.

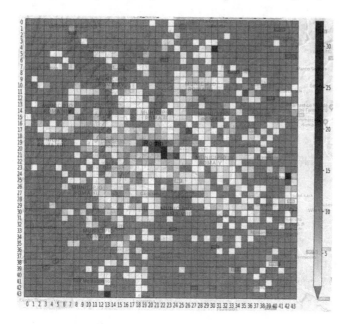

Fig. 2. Distribution of the Selected News on the City Map (2018)

Table 4. Distribution of the selected news in the macro-categories

Macro-category	Total extracted	Max in cell
Criminality	1093	24
UP&D	373	10
Accidents	318	26
E&S	31	2

The result of the clustering of the city areas is summarized in Fig. 3, where we superimposed a heat map of the clusters, on the map of Rome. In addition to the k clusters generated by the clustering algorithm, we also have to add the dummy cluster. Thus, the total number of clusters is 5 and their centroids are shown in Table 5.

In Table 6, we describe the profile of each cluster in terms of the following four characteristics [2]:

(i) the n $(n \leq F)$ dominant macro-categories of news in the cells of the cluster ($DomCat$). For each macro-category, a value is computed as the number of news of that macro-category over the total number of news in the cluster. Then, the values obtained are ranked and the first n ones are selected;

(ii) the m $(m \leq F)$ distinctive macro-categories of news in the cells of the cluster ($DistCat$). For each macro-category, a value is computed as the number of news of that macro-category over the total number of news of that category

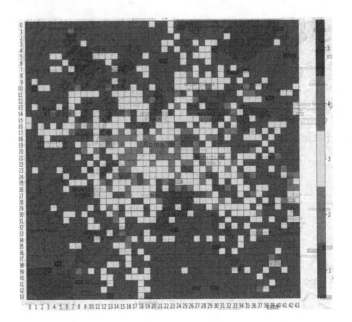

Fig. 3. The clustering result (heatmap with 5 clusters, source RomaToday) of 1936 cells of size 500 m.

Table 5. Description of the cluster centroids

Cluster id (color)	Centroids			
	Criminality	*E&S*	*Accidents*	*UP&D*
#1 (dark blue)	-	-	-	-
#2 (sky blue)	4.28	0.068	0.81	0.91
#3 (green)	0.88	0.04	0.39	0.51
#4 (orange)	14.53	0.23	0.38	1.84
#5 (red)	2	0	19	5

in all the clusters (*i.e.*, the whole city). Then, the values obtained are ranked and the first m ones are selected;

(iii) the density of news per cell in the cluster (*NewsDen*), computed as the number of news in the cluster over the number of cells in the cluster;

(iv) the overall number of cells (*ClNum*) in the cluster.

As regards *DomCat* and *DistCat*, in the following, we show only the first two ranked macro-categories of news.

Notice that Cluster #1 is the biggest cluster containing all the empty cells, mainly located in peripheral areas with low population density and very few tourist and commercial activities. Cluster #2 and Cluster #4 are characterized by the same dominant and distinctive macro-categories, namely Criminality and

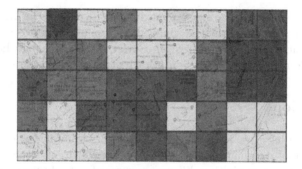

Fig. 4. A Map Zoom on the City Centre, including Termini and Tiburtina Railway Stations

UP&D, and Criminality and E&S, respectively. While Cluster #2 includes a high number of city areas mainly located in the west zone of the city, Cluster #4 mainly includes the most central areas. As expected, these areas are characterized by a very high density of news, mainly regarding Criminality and E&S. Indeed, as we can see from Fig. 4, these areas includes the two most important railways stations of Rome (Roma Termini and Roma Tiburtina) and some famous monuments such as Fontana di Trevi, Palazzo del Quirinale, Basilica di Santa Maria Maggiore.

As regards Cluster #3, it is the second biggest cluster, in terms of number of cells (454 in total), and mainly describes central zones characterized by events, services, urban problems and decay. If we compare this cluster with Clusters #2 and #4, we realize that UP&D appears as distinctive macro-category. Even though Cluster #3 includes several very central city areas, if we analyse Fig. 4, we realize that most of these central areas are just adjacent to the famous monuments. We may suppose that, perhaps, the city administration takes less care of them than the areas of Clusters #2 and #4. Finally, Cluster #5 is the one which includes the areas mainly characterized by news regarding the Accidents macro-category. This cluster includes just two areas in the North-east and South-west of the city: the first one, which includes most of the extracted news, correspond

Table 6. The profiles of the five clusters

| Cluster id (color) | Interpretability measure | | | |
	DomCat	DistCat	NewsDen	ClNum
#1 (dark blue)	-	-	0	1351
#2 (sky blue)	Criminality, UP&D	Criminality, E&S	6.09	116
#3 (green)	Criminality, UP&D	E&S, UP&D	1.84	454
#4 (orange)	Criminality, UP&D	Criminality, E&S	17	13
#5 (red)	Accidents, UP&D	Accidents, UP&D	26	2

to the cell into which all the news regarding accidents on the Ring Road are geolocated. The second area is the zone called Spinaceto (between Via Pontina and Via Cristofolo Colombo), where also most of the accidents happen while people leave Rome going towards South for reaching the seaside.

5   Conclusions

In this work, we have presented the integration of online newspaper articles as a further source of information to our framework, which has been developed for profiling city areas by exploiting different sources of data. In particular, we have defined a virtual grid on the map of a specific city. Then, we have employed the metadata, which specify the location of a news, for geolocating it in a cell of the grid. Each cell is characterized in terms of numbers of news belonging to a set of macro-categories. The macro-categories have been defined exploiting other metadata, namely the tags associated with the news.

We have shown some preliminary results related to the city of Rome, Italy. We have extracted online news articles from RomaToday, a web portal which offers frequently updated news regarding the entire province of Rome. In this work, we have focused our attention on the areas concentrated inside and near the Ring Road of Rome. The application of the k-means clustering algorithm allows us to easily describe different areas of Rome, especially of the city centre. Indeed, we realized that the areas characterized by the presence of the monuments are more interested by criminality but also by events and services. On the other hand, the central areas which are only adjacent to the monuments are also characterized by urban problems and decay. Finally, the Ring Road of Rome and the roads located in the south of Rome, are mostly characterized by news regarding accidents.

References

1. Abdullah, M.S., Zainal, A., Maarof, M.A., Kassim, M.N.: Cyber-attack features for detecting cyber threat incidents from online news. In: 2018 Cyber Resilience Conference, pp. 1–4. IEEE (2018)
2. D'Andrea, E., Ducange, P., Loffreno, D., Marcelloni, F., Zaccone, T.: Smart profiling of city areas based on web data. In: 2018 IEEE International Conference on Smart Computing, pp. 226–233. IEEE (2018)
3. Giatsoglou, M., Chatzakou, D., Gkatziaki, V., Vakali, A., Anthopoulos, L.: City-pulse:a platform prototype for smart city social data mining. J. Knowl. Econ. **7**, 344–372 (2016)
4. Hussain, F.: Internet of Everything. In: Hussain, F. (ed.) Internet of Things, pp. 1–11. Springer, Cham (2017). https://doi.org/10.1007/978-3-319-55405-1_1
5. Iglesias, J.A., Tiemblo, A., Ledezma, A., Sanchis, A.: Web news mining in an evolving framework. Inf. Fusion **28**, 90–98 (2016)
6. Karimi, M., Jannach, D., Jugovac, M.: News recommender systems-survey and roads ahead. Inf. Process. Manag. **54**(6), 1203–1227 (2018)

7. Lin, A.Y., Ford, J., Adar, E., Hecht, B.: VizByWiki: mining data visualizations from the web to enrich news articles. In: Proceedings of the 2018 World Wide Web Conference on World Wide Web, pp. 873–882. International World Wide Web Conference Committee (2018)

8. Po, L., Rollo, F.: Building an urban theft map by analyzing newspaper crime reports. In: 2018 13th International Workshop on Semantic and Social Media Adaptation and Personalization, pp. 13–18. IEEE (2018)

9. Ramisa, A., Yan, F., Moreno-Noguer, F., Mikolajczyk, K.: BreakingNews: article annotation by image and text processing. IEEE Trans. Pattern Anal. Mach. Intell. **40**(5), 1072–1085 (2018)

10. Silva, B.N., Khan, M., Han, K.: Towards sustainable smart cities: a review of trends, architectures, components, and open challenges in smart cities. Sustain. Cities Soc. **38**, 697–713 (2018)

11. Teske, A., Falcon, R., Abielmona, R., Petriu, E.: Automatic identification of maritime incidents from unstructured articles. In: 2018 IEEE Conference on Cognitive and Computational Aspects of Situation Management, pp. 42–48. IEEE (2018)

12. Yang, D., Heaney, T., Tonon, A., Wang, L., Cudré-Mauroux, P.: Crimetelescope: crime hotspot prediction based on urban and social media data fusion. World Wide Web **21**, 1323–1347 (2018)

Modelling the Social Interactions in Ant Colony Optimization

Nishant Gurrapadi[1] , Lydia Taw[2] , Mariana Macedo[3](✉) ,
Marcos Oliveira[4] , Diego Pinheiro[5] , Carmelo Bastos-Filho[6] ,
and Ronaldo Menezes[3]

[1] Department of Computer Science, University of Texas at Dallas, Richardson, USA
[2] Department of Computer Science, George Fox University, Newberg, USA
[3] BioComplex Lab, Department of Computer Science,
University of Exeter, Exeter, UK
mmacedo@biocomplexlab.org
[4] Computational Social Science,
GESIS–Leibniz Institute for the Social Sciences, Mannheim, Germany
[5] Department of Internal Medicine, University of California, Davis, USA
[6] Polytechnic School of Pernambuco, University of Pernambuco, Recife, Brazil

Abstract. Ant Colony Optimization (ACO) is a swarm-based algorithm
inspired by the foraging behavior of ants. Despite its success, the efficiency of ACO has depended on the appropriate choice of parameters,
requiring deep knowledge of the algorithm. A true understanding of ACO
is linked to the (social) interactions between the agents given that it is
through the interactions that the ants are able to explore-exploit the
search space. We propose to study the social interactions that take place
as artificial agents explore the search space and communicate using stigmergy. We argue that this study bring insights to the way ACO works.
The interaction network that we model out of the social interactions
reveals nuances of the algorithm that are otherwise hard to notice. Examples include the ability to see whether certain agents are more influential
than others, the structure of communication, to name a few. We argue
that our interaction-network approach may lead to a unified way of seeing swarm systems and in the case of ACO, remove part of the reliance
on experts for parameter choice.

Keywords: Swarm intelligence · Swarm-based algorithms · Ant
colony optimization · Interaction network · Social interactions

1 Introduction

Swarm intelligence algorithms have been successfully applied to solve a wide
range of optimization problems due to the simultaneous use of multiple artificial
agents on high dimensional search problems [4,5]. Even though they are effective,
the usability of swarm-based algorithms is limited by the lack of knowledge on
why the interaction of simple reactive agents lead to such a complex system.

© Springer Nature Switzerland AG 2019
H. Yin et al. (Eds.): IDEAL 2019, LNCS 11872, pp. 216–224, 2019.
https://doi.org/10.1007/978-3-030-33617-2_23

Another challenge is the diversity of algorithms inspired by different animals such as ants, bees, fish, wolves and birds. Knowing what is the best swarm-based algorithm and its initialization to each type of problem requires deep expertise.

Bratton and Blackwell [1] proposed a simplified version of the Particle Swarm Optimization (PSO) algorithm [3] by removing the randomizing factors from the equations of the algorithm. Their simplification gives insight into the swarm behavior and helps us understand what makes PSO an effective algorithm. However, this still could not quite capture why the social behaviour emerges from the simple rules.

In our previous works, we showed the value of using an interaction network to analyze the social interactions occurring during executions of swarm-based algorithm [7–10]. In this paper, we show that applying the same principles of analyzing social interactions to Ant Colony Optimization can provide a better way to initialize the algorithm for higher usability and understanding. This paper contributes to the literature in ACO and Interaction Networks because it is the first work that tracks social interaction even though these interactions are indirect via stigmergy.

2   ACO and TSP

Ant Colony Optimization (ACO) is a meta-heuristic technique inspired by the behaviour of ants [2]. In nature, ants solve complex problems by indirectly communicating via the environment (stigmergy). While an ant travels to a food source, it drops an amount of pheromone along the path that other ants can follow. Then, the ants that come next choose a path probabilistically; the paths already taken by previous ants are more likely to be selected because the pheromone amount is longer. Given that ACO was extensively used in TSP, we also chose the TSP to model the interaction network in ACO.

Artificial ants move on a fully connected graph where the vertices represent cities and the edges represent the paths to go from one city to another. The goal for each ant is to find the shortest path that visits each city and returns back to the origin city. During the tour of the cities, an ant is presented with many choices on which city to visit next. As the ants travel through the cities, they drop pheromone on the path allowing them to use it and make decisions on which city to go to next. The pheromone τ_{ij} associated with the edge joining cities i and j, is updated as follows:

$$\tau_{ij}(t+1) = \rho \cdot \tau_{ij}(t) + \sum_{k=1}^{m} \Delta\tau_{ij}^{k}(t), \tag{1}$$

where ρ is the evaporation rate, m is the number of ants, and $\Delta\tau_{ij}^{k}(t)$ is the amount of pheromone dropped by ant k on the edge (i, j) at iteration t calculated based on a heuristic that represents the goodness of the path taken.

Ants drop more pheromone on edges which lead to good solutions, and the pheromone from sub-optimal solutions is evaporated over time. Thus, an ant k going from city i chooses city j using the following probability equation:

$$p_{ij}^k(t) = \frac{\tau_{ij}^\alpha(t) \cdot \eta_{ij}^\beta(t)}{\sum_{u \in \mathcal{N}_k(t)} \tau_{iu}^\alpha(t) \cdot \eta_{iu}^\beta(t)}, \qquad \text{if } j \in \mathcal{N}_k(t), \qquad (2)$$

where $\mathcal{N}_k$ is the list of cities that the ant k has not yet visited, and $\eta_{ij} = 1/d_{ij}$ is the visibility, where d_{ij} is the distance between cities i and j. The parameters α and β control, respectively, the importance of pheromone and edge visibility.

3   Interaction Network

Swarm-based algorithms, like ACO, consist of artificial agents interacting with each other and with the environment [2]. These algorithms are useful tools to solve real-world problems [4,5]. Despite their effectiveness, they lack *explainability*; we are still unable to explain the dynamics that make these techniques useful [10]. Notably, the pivotal feature in these algorithms is the social interactions enabling the system to solve problems. Indeed, each algorithm has its own rules defining the way agents interact with others and determining how they change themselves as result of interaction. This interplay among agents leads to the emergence of a network of interaction which provides a mezzo-level perspective of swarms [10].

Such network-based viewpoint was first introduced by Oliveira et al. in the context of Particle Swarm Optimizers [7,9] and then extended to swarm-based algorithms in general [10]. The concept of *interaction network* enables the analysis of social influence among the agents in the swarm [10]. The approach unveils the dynamics of swarm systems via tracking the evolution of social interactions.

In the interaction network $\mathbf{I}(t)$ nodes represent the agents and links (or edges) between the nodes represent the influence between the agents. This network allows us to capture the social behavior of the swarm at different points in time during an execution of the algorithm. Formally, the network $\mathbf{I}(t)$ at iteration t is represented by an adjacency matrix where each element of the matrix can be defined by the presence, 1, or the absence, 0, of influence between the artificial agents i and j. However, this definition only tells us whether or not two artificial agents in a swarm interacted with each other over time. Oliveira et al. [7] developed an expanded version of the interaction network that keeps track of the history of information exchanges, by creating a separate interaction network for each iteration and summing all of them up at the end to get an accurate picture of all the interactions that took place among the artificial agents in the swarm. The matrix resulting from this sum is a weighted interaction network $\mathbf{I}_t^w$ as shown in Eq. 3:

$$\mathbf{I}_t^w = \sum_{t'=1}^{t} \mathbf{I}(t') \qquad (3)$$

The weighted interaction network $\mathbf{I}_t^w$ allows us to analyze the history of interactions of each agent in a swarm, and determine if there are any particular agents that had a major influence on the interactions of other agents. Moreover, we can analyze different time windows to identify the peculiarities of swarm-based techniques.

To capture the structure of the information flow within the swarm, Oliveira et al. also proposed a metric called *Interaction Diversity* (ID) to measure how quickly the interaction network can be destroyed by removing the edges from the network. The precise definition of ID can be found in [8]. If the graph becomes completely disconnected with the removal of only a few edges, it indicates that the swarm lacked diversity in its interactions. On the other hand, if the graph remains well connected even after the removal of edges, it indicates higher diversity in the interactions that occurred among the agents.

4   Interaction Network in ACO

In the Ant System (AS-ACO) [2], the decision-making of each ant (Eq. 2) considers the aggregated pheromone deposited by all ants. Given that ants indirectly communicate with one another, the interaction network for ACO algorithms attempts to estimate the interaction between ants based on the pheromone they deposit on the environment and how this pheromone is used by other ants.

Kromer et al. [6] previously defined an interaction network for the Ant System. According to their definition, the interaction between two ants depends on the similarity of their current tours. In this sense, it ranges from 0 (i.e., disjoint tours) to 1 (i.e., identical tours). This definition, however, neglects the fact that pheromone deposited in previous iterations still enables ants to influence each other in subsequent iterations regardless of the similarity of their current tours. Technically, having similar tours at a specific iteration does not necessarily imply ants exerted influence on each other towards the decision-making that led to their final tours. It implies that at the end of such a tour, ants will similarly deposit pheromone on the constituting edges of that tour, which in turn will influence ants in the subsequent interactions.

Conversely, our definition of the interaction network for the Ant System accounts for the effective interaction between ants as measured by the pheromone they leave on the environment. It captures how the decision-making of an individual ant is effectively influenced by other ants in terms of their deposited pheromone. This definition considers that the pheromone existing between cities i and j is actually an aggregation of pheromones deposited by all ants. If not completely evaporated, pheromone deposited at previous interactions can still influence the decision-making of ants at current iteration.

The interaction I_{kl} between ants k and l is formally defined as

$$I_{kl}(t) = \sum_{(i,j)\in \mathcal{T}_k(t)} \tau_{ij}^l(t), \tag{4}$$

where $\mathcal{T}_k(t)$ is the set of edges visited by ant k at iteration t, and $\tau_{ij}^l(t)$ is the cumulative pheromone deposited by ant l between cities i and j at iteration t after evaporation. Although it accumulates past pheromone, the interaction network does not necessarily become denser with more iterations because the cumulative pheromone of less visited edges tends to decrease (due to ρ in Eq. 1).

5   Analysis of ACO's Interaction Network

The AS algorithm was applied to the Symmetric Traveling Salesman Problem. As the TSP has been widely replicated in optimization problems, we argue that this work may refrain from the analyses of its fitness performance. We evaluated our implementation against four different instances of the TSP (Groetschel): 17 cities, 21 cities, 24 cities and 48 cities[1]. The considered initial parameters for the simulation of AS-ACO was 2,000 iterations, Q (used in the calculation of $\Delta\tau_{ij}^{k}(t)$ of Eq. 1) is 1.0, $\alpha = \beta = 0.85$ and five different values for $\rho : \{0.0, 0.3, 0.5, 0.7, 1.0\}$. The ρ value reflects the amount of memory each ant carries, so the five different values for ρ were chosen in order to test how different levels of memory affect the finding of the solution. The weighted degree analyzed in this section was normalized to ensure that the values are scaled to the same range for all the four problems.

Figure 1 compares the evolution of shortest paths found in the simulations for several problems, and several evaporation rates for the problem with 48 cities. Figure 1(A) shows that in each problem, the shortest path is found within the first hundred iterations. In Fig. 1(B), it can be noted that the excess of evaporation is not positive for the system because it removes the memory of the system.

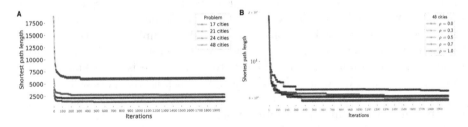

Fig. 1. (A) Fitness evolution of all the problems using the evaporation rate equal to 0.5. **(B)** Fitness evolution of the 48 cities problem varying the evaporation rate as 0.0, 0.3, 0.5, 0.7 and 1.0.

Figure 2 depicts the Empirical Distribution Function (EDF) from the final interaction network in each problem simulated using $\rho = 0.5$, and the final interaction network in the problem of 48 cities simulated using ρ equals to 0.0, 0.3, 0.5, 0.7 and 1.0. All the distributions are a Gaussian distribution which means that the majority of ants displays similar behaviour but a few of the ants can be seen as hubs or in the periphery. In Fig. 2(A), we can observe that the values of weighted degree change based on the magnitude of the paths. In Fig. 2(B), the excess of pheromone evaporation changes the distribution of weighted degrees, as the memory of system is removed as defined in Eq. 1.

---

[1] https://github.com/pdrozdowski/TSPLib.Net.

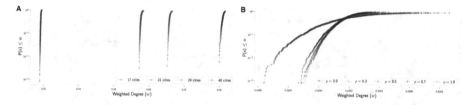

Fig. 2. (A) Empirical Distribution Function (EDF) of the weighted degree from the final interaction network on the problems: 17, 21, 24 and 48 cities using the evaporation rate as 0.5. **(B)** Empirical Distribution Function (EDF) of the weighted degree from the final interaction network on the problem of 48 cities. The important issue here is that the interaction network allows us to get a deeper understanding of the execution of the algorithm. Note that by studying such behaviour, one can make better decisions about the parameters in Eqs. 1 and 2.

In Fig. 3, both cumulative interaction networks from 24 and 48 cities are displayed at 200 and 2000 iterations. The red, orange, yellow, green and blue are the representations from the highest to the smallest amount of interaction. Each row indicates the strength of influence of each ant, and each column indicates how much an ant was influenced by another ant. The presence of homogeneous lines demonstrates that the ants influence more equally than are influenced. We observe that the lack of pheromone history has a different impact on both problems. On 24 cities, accounting a percentage of memory decreases the ants interaction. However, on 48 cities, the opposite is identified, the ants strength increases. As the presence of memory is positive for the system, the heatmaps for ρ equal to 0.5 are better for the problem. In this way, the high presence of more hubs seems to be benefit only for the 48 cities problem. As the majority of ants start to follow the same path over time, the contribution of pheromone could be more similar between them. However, some ants get lost in the process and they usually are less influenced by the other ants, which are perceived as blue dots/lines on the heatmap. In the beginning, some ants are more likely to contribute more because some ants might find first a good path than the others.

In order to understand the exploration-exploitation balance of the system, we analyzed the Interaction Diversity of the interaction networks in Fig. 4. In the beginning of the process, the interaction diversity goes down quickly. This decrease might be caused by the rules of the system because even when we remove the memory of previous iterations, we still notice such behaviour. In Fig. 4(A), we observe that the total evaporation of pheromone makes the convergence of the swarm more difficult by adding on the swarm a constant growth of exploration. When the evaporation rate gets bigger than 0.0, as in Fig. 4(B), the exploitation is maintained which helps on the convergence of the swarm. The 48 cities problem can be highlighted by its dynamic behaviour of increasing and decreasing the ID. If we increase the evaporation rate for bigger than 0.1, we identified that the dynamic behaviour gets smaller. Moreover, in 24 cities, we could notice that the dynamicity is higher than the other problems. In summary, we can observe

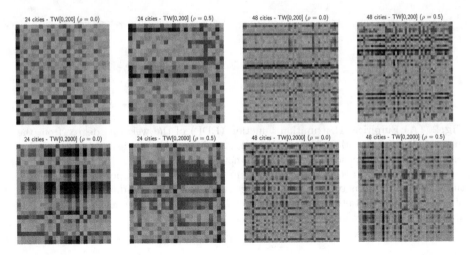

Fig. 3. Cumulative interaction network from different time windows on the problems 24 and 48 cities using the evaporation rate (ρ) set to 0.0 and 0.5. (Color figure online)

that the memory of the system is important for the convergence of the swarm, and that different problem sizes may show more deviation on the interaction diversity.

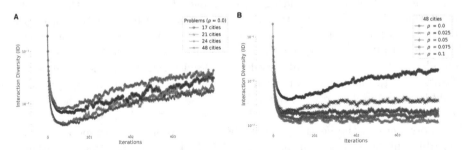

Fig. 4. Dynamics of Interaction Diversity (ID). (**A**) Problems with 17, 21, 24, and 48 cities, using the evaporation rate (ρ) as .0. (**B**) Problem with 48 cities using ρ equals to .0, .025, .05, .075, and .1. The ID is calculated over the weighted degree interaction network with time windows equal to [10, 20, 30, 40, 50, 100, 150, 200].

6   Conclusion

Swarm-based algorithms are computational models containing multiple agents simultaneously searching for optimal solutions while sharing with each other the solutions they find. Though swarm-based algorithms are effective, they depend on the appropriate adjustment of their parameters, and such adjustment requires significant knowledge about the algorithm. Previous research has demonstrated that the dynamics of these algorithms can be characterized and better understood by the social interactions among individuals and that, by changing the parameters of a swarm-based algorithm at the micro level, we actually create different conditions for social interactions to occur at the mezzo level which, in turn, can ultimately improve the overall performance of the algorithm at the macro level. In this work, we sought to answer the question of whether it would be possible to extract the social interactions of the Ant System—a well-known Swarm Intelligence algorithm—even though the communication in this system is based on stigmergy (indirect communication). We show that indeed the use of an Interaction Network framework can help us understand how to properly adjust its parameters without deep knowledge of the algorithm.

In order to show the effectiveness of the Interaction Network approach, we analyzed the social interactions occurring among ants of an Ant System while they solve four different instances of the Traveling Salesman Problem (TSP). Then, we examined the Ant System with different rates of pheromone evaporation. For future works, we argue that further experiments should be performed to understand the impact of other parameters to the interaction network. Also, we want to further explore why the current influence strength as measured by the weighted degrees in the network seems to be homogeneously distributed around a well-defined typical value. Such type of distribution implies that ants tend to exert similar levels of influence on each other and might prohibit the algorithm to appropriately balance the extent of exploration and exploitation.

Acknowledgment. The authors acknowledge support from National Science Foundation (NSF) grant No. 1560345 (http://www.nsf.gov/awardsearch/showAward? AWD_ID=1560345). Any opinions, findings, and conclusions or recommendations expressed in this material are those of the authors and do not necessarily reflect the views of the NSF. This work also used the Extreme Science and Engineering Discovery Environment (XSEDE) Bridges at the Pittsburgh Supercomputing Center through allocation TG-IRI180008, which is supported by National Science Foundation grant number ACI-1548562 [11].

References

1. Bratton, D., Blackwell, T.: Understanding particle swarms through simplification: a study of recombinant PSO. In: Proceedings of the 9th Annual Conference Companion on Genetic and Evolutionary Computation, pp. 2621–2628. ACM (2007)
2. Dorigo, M., Maniezzo, V., Colorni, A.: Ant system: optimization by a colony of cooperating agents. IEEE Trans. Syst. Man Cybern. Part B (Cybern.) **26**(1), 29–41 (1996)
3. Eberhart, R., Kennedy, J.: A new optimizer using particle swarm theory. In: Proceedings of the Sixth International Symposium on Micro Machine and Human Science, MHS 1995, pp. 39–43. IEEE (1995)
4. Engelbrecht, A.P.: Fundamentals of Computational Swarm Intelligence. Wiley, Chichester (2006)
5. Kennedy, J., Eberhart, R.C.: Swarm Intelligence, 1st edn. Morgan Kaufmann Publishers Inc., San Francisco (2001)

6. Krömer, P., Gajdo, P., Zelinka, I.: Towards a network interpretation of agent inter-action in ant colony optimization. In: 2015 IEEE Symposium Series on Computational Intelligence, pp. 1126–1132. IEEE (2015)
7. Oliveira, M., Bastos-Filho, C.J., Menezes, R.: Towards a network-based approach to analyze particle swarm optimizers. In: 2014 IEEE Symposium on Swarm Intelligence (SIS), pp. 1–8. IEEE (2014)
8. Oliveira, M., Pinheiro, D., Andrade, B., Bastos-Filho, C., Menezes, R.: Communication diversity in particle swarm optimizers. In: Dorigo, M., Birattari, M., Li, X., López-Ibáñez, M., Ohkura, K., Pinciroli, C., Stützle, T. (eds.) ANTS 2016. LNCS, vol. 9882, pp. 77–88. Springer, Cham (2016). https://doi.org/10.1007/978-3-319-44427-7_7
9. Oliveira, M., Pinheiro, D., Macedo, M., Bastos-Filho, C., Menezes, R.: Better exploration-exploitation pace, better swarm: examining the social interactions. In: 2017 IEEE Latin American Conference on Computational Intelligence (LA-CCI), pp. 1–6. IEEE (2017)
10. Oliveira, M., Pinheiro, D., Macedo, M., Bastos-Filho, C., Menezes, R.: Unveiling swarm intelligence with network science-the metaphor explained. arXiv preprint arXiv:1811.03539 (2018)
11. Towns, J., et al.: XSEDE: accelerating scientific discovery. Comput. Sci. Eng. 16(5), 62–74 (2014). https://doi.org/10.1109/MCSE.2014.80

An Innovative Deep-Learning Algorithm for Supporting the Approximate Classification of Workloads in Big Data Environments

Alfredo Cuzzocrea[1]($\boxtimes$), Enzo Mumolo[2], Carson K. Leung[3],
and Giorgio Mario Grasso[4]

[1] Universitá della Calabria, Rende, Italy
cuzzocrea@si.dimes.unical.it
[2] Universitá di Trieste, Trieste, Italy
mumolo@units.it
[3] University of Manitoba, Winnipeg, MB, Canada
kleung@cs.umanitoba.ca
[4] Universitá di Messina, Messina, Italy
gmgrasso@unime.it

Abstract. In this paper, we describe *AppxDL*, an algorithm for approximate classification of workloads of running processes in big data environments via deep learning (deep neural networks). The Deep Neural Network is trained with some workloads which belong to known categories (e.g., compiler, file compressor, etc...). Its purpose is to extract the type of workload from the executions of reference programs, so that a Neural Model of the workloads can be learned. When the learning phase is completed, the Deep Neural Network is available as Neural Model of the known workloads. We describe the *AppxDL* algorithm and we report and discuss some significant results we have achieved with it.

Keywords: Workload classification · Virtualized environment · Deep learning

1 Introduction

Virtualization technology has become fundamental in modern computing environments such as cloud computing [6,14–16] and server farms [1,7]. By running multiple virtual machines on the same hardware, virtualization allows us to achieve high utilization of available hardware resources. Furthermore, virtualization brings several advantages in terms of security, reliability, scalability and resource management. Resource management in the virtualization context can be performed by classifying the workload of virtualized applications. Consequently, the characterization of virtualized workloads has been extensively studied during the past years. More recently, significant amount of work has been done to characterize the workload in data center environments. On the other hand, modeling

© Springer Nature Switzerland AG 2019
H. Yin et al. (Eds.): IDEAL 2019, LNCS 11872, pp. 225–237, 2019.
https://doi.org/10.1007/978-3-030-33617-2_24

and classification of workload in virtualization environments has been addressed, while virtualized workload balancing based on the found workload has not been addressed here and will be the topic of future research.

We use a Deep Neural Network with three hidden layers, 32 inputs and 12 outputs. The basic idea is that by training a Deep Neural Network with some executions of a program with a certain workload category, the Network should learn the workload category itself. Using a Deep Neural Network for all the workloads categories, after learning we have a Neural Model tuned on the known workloads. Therefore, by giving as input to the Neural Model the execution of a program with unknown workload, then the output of the Neural Model corresponding to the unknown workload, or the most similar to, will be the highest.

From a methodological point of view, the workload classification is hence performed by collecting the measurements of quantities that characterize the execution of reference applications. By measuring the same quantities from the execution of an application whose workload is not known, the most likely class of the unknown workload is determined.

At a basic level, the workload can be classified as CPU intensive or I/O intensive. At a finer level, the workload can be classified as memory intensive, disk read/write intensive or Network in/out intensive. The classification of workloads is performed by extracting information on memory accesses. The workload Model is characterized by this sequence of addresses. In this paper our reference applications are that contained in the SPEC CINT2006 Dataset [19], which contains several programs belonging to the workloads listed in Table 1. The programs of the reference dataset are executed in a virtualized environment. From these executions, some features are extracted using the Virtual Machine Monitor APIs [20]. Our analysis approach is to consider memory references as one-dimensional signals, which is analyzed using spectral approaches. Collecting the spectral characteristics and using the collected data for learning a Deep Neural Network, we obtain a workload Model of the reference programs. The Deep Neural Network is then used as classifier. Unknown programs are executed in the same environment and their spectral parameters are extracted. Using the Neural Workload Model, we get to which workload Model the features extracted from the unknown program whose workload is unknown can be associated. The proposed approach is not only significant in cloud computing, but it is also relevant for modern collaboration and cooperation scenarios, as the problem of categorizing the workload can be tackled collaboratively and cooperatively on top of cloud environments. It is important to note that the input data of the reference programs as well as of the unknown program are different for each execution. Therefore the execution sequences (processes) are always different.

In data mining, the detection of anomalies (malware detection) [17], that is, the identification of elements, events or observations that do not conform to an expected Model or to other elements present in a data set, can be an activity very complex. Typically abnormal objects lead to some kind of problem, such as bank fraud, a structural defect, medical problems or errors in a text.

Table 1. Dataset SPEC CINT2006

Program	Workload
h264	Video Coding
gcc	C Compiler
perlbench	Interpreter
bzip2	File Compression
gobmk	Go Game
mcf	Vehicle Scheduling
hmmer	Hidden Markov Model
sjeng	Chess Game
libquantum	Quantum Computer Simulation
omnetpp	Discrete Events Simulation
astar	Path Finding
xalancbmk	XML to HTML

The anomalies can also be novelties, noise, deviations and exceptions. Using the Neural classification we can manage to face this problem of anomaly detection. For example, if the benchmarks are chosen appropriately, it can be determined which are the workloads of the processes running in the virtual machine, in order to be sure they are what they should be. Otherwise an alarm will be triggered. Another possibility could be to know which processes a given client typically performs. In this regard, the running processes can be monitored to see if their workload is the same or if it changes over time. Experimental evaluation and analysis clearly confirm the benefits of our workload classifier.

As far as classifiers are concerned, classifiers of the Hidden Markov Model (HMM) type [5] have been used previously, while here we use Deep Neural Networks with three hidden layers. Using learning algorithms, a Model is then obtained for all the workloads. Unknown workloads are then classified using the workload Model. The classification among the application workload running in virtualization offers interesting potential applications.

2   Related Work

Islam et al. [2] evaluated the performance of the Neural Networks for forecasting the resources' usage. In particular, they presented a workload prediction algorithm by combining the artificial Neural Network and the linear regression for virtual machines. Following this track, Akindele and Samuel described in [4] an algorithm for predicting the CPU utilization and the parameters of service level agreements. Their prediction results demonstrate that by combining Neural Network, linear regression and the support vector machine good CPU utilization results are obtained. Garg et al. [3] described a Neural Network-based algorithm for resource allocation such as CPU utilization in a cloud data center. Kousiouris et al. [8] proposed an approach to predict the application workload using time series analysis and artificial Neural Networks. A recurrent Neural Network is used by Chang et al. [11] to predict the workload of cloud servers. They consider the workload as the number of processes assigned to servers. In this paper, Chang et al. show that this method performs better than the regression scheme in case of rapid changes of the workload. Chen et al. presented a resource demand prediction Model based on fuzzy Neural Network [12]. Also Ramezani and Naderpour [13] used Neural Networks and the fuzzy logic system to predict the CPU usage patterns of virtual machines available in cloud. An algorithm to predit resources usage in cloud environments by Neural Networks and reinforcement learning is described by Amiri et al. in [9].

3   Approximate Classification of Workloads in Big Data Environments: Principles and Case Studies

The experimental virtualized environment used in this paper is VirtualBox [21]. Virtualbox is an open source multi-platform virtualization program that provides a rich set of APIs that allow you to collect different metrics of virtualized programs. The SDK (Software Development Kit), equipped with Virtual Box, allows third parties to develop applications that interact with Virtual Box. Virtual Box is designed in levels. Below we find the VMM (hypervisor). It is the heart of the virtualization engine, monitoring the performance of virtual machines and providing security and the absence of conflicts between virtual machines and hosts. Above the hypervisor (virtual machine monitor) there are modules that provide additional features, such as the RDP server (Remote Desktop Protocol). The API level is implemented above these functional blocks. VirtualBox is equipped with a web service that, once executed, acts as an HTTP server, accepts SOAP (Simple Object Access Protocol) connections [22] and proceeds them. The interface is described in a Web service description language file (WSDL) [23]. In this way it is possible to write client programs in any programming language that has provided the tools to process WSDL files, such as Java, C++, PHP NET, Python, Perl. In addition to Java and Python, the SDK contains libraries ready for use. Inside, the API is implemented using the Component Object Model (COM) as a reference Model. In Windows it is natural to use Microsoft COM. In other hosts, where COM is not present, XPCOM which is a free COM application, can be used. In this environment we developed the acquisition system which is described in the block diagram shown in Fig. 1. The acquisition of the performance characteristics is driven by the host; all the commands and data acquired use the COM (Component Object Model) interface. There is also a web interface VMM (Virtual Machine Manager) but it is much slower. The measured quantities used for the characterization of workloads are of two types, namely memory references and resource requests. Both quantities can be collected using the VirtualBox VMM API classes. The acquired data is collected in a $<timestamp>$ $<address>$ format. Using the PerformanceCollector API we can also collect the functionality of the resource request generated by the virtualized process, such as the CPU used in user mode, the CPU used in system mode and memory fragmentation (free memory/total memory).

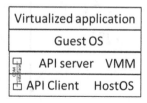

Fig. 1. Acquisition system

Memory references are the instruction addresses generated by the virtualized process. The VMM's MachineDebugger API can collect the value of the program counter related to the instructions every 0.5 ms. The memory references are translated into a virtual page reference. Page references are produced with the clock frequency of the CPU instruction. This speed is too high to make reasonable evaluations of the workload and, as a result, the number of page references is too large. Therefore, it is necessary to perform some operations to eliminate the redundant information and reduce the data speed. Hence, data processing is required. According to the idea of considering the sequence of page references as a 1D signal, we compute a spectral description of the page reference sequences. Spectral description can well describe the basic behaviour of the execution, such as sequential or looping instructions. For example, loops introduce spectral peaks with values related to the looping widths while sequential address sequences produces a high DC component.

4   *AppxDL*: Approximate Classification of Workloads in Big Data Environments via Deep Learning

4.1   Overview of the *AppxDL* Algorithm

In Fig. 2 we report a block diagram representing the training approach of the Deep Neural Network. Each of the 12 programs of SPEC 2006 represents a workload, as described in Table 1. While a single program represents a single workload, if each single program is executed several times with different input values, many different processes are obtained, each for every set of input data. In Fig. 2 we show that there are M different programs, and to each program a set of N elements of different data is given as input data. Therefore we have $M * N$ different processes, i.e. N processes for each program. Each of these processes represents in a hidden way the workload of each program. We expect that by Deep learning our Neural Model, the Model would learn the hidded workloads. The Deep Neural Network has 12 output lines, one for each workload.

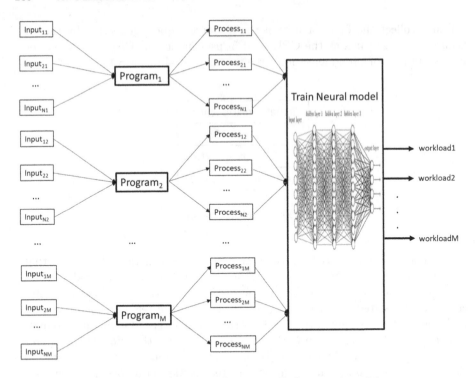

Fig. 2. Workload classification with the neural model

The supervised learning of the neural model is performed by supplying the pair (nth workload – 'high value') to the input layer and to the output line corresponding to the nth workload, respectively. Then, the weight optimization is performed. In this work we set 'high value' = 1. When the Neural Model is used for classification, its input are the parameters obtained from the execution of an unknown new program. Then, the output of the Neural Model corresponding to the workload more similar to that of the new program, should be the highest. This is shown in Fig. 3.

4.2   Features

The sequence of pages of virtual memory is time varying, and therefore is divided into short sections (frames). Each frame is analyzed by mean of the Discrete Fourier Transform, implemented through the Discrete Cosine Transform (DCT) computation. The amplitude of the computed spectra is log-transformed. Let us call $x(n)$ the input page sequence and

$$X(\omega) = \sum_{-\infty}^{+\infty} x(n)e^{j\omega n} \tag{1}$$

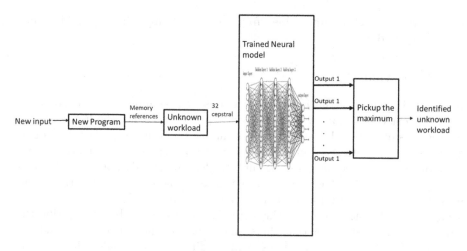

Fig. 3. Training of the neural model

its Fourier spectrum, where ω is the frequency. Then the *cepstrum* of the input sequence $c(n)$ is obtained through the inverse-Fourier transform of the logarithm of the spectrum amplitude. In other words,

$$\log(X(\omega)) = \sum_{-\infty}^{+\infty} c_n e^{j\omega n}. \tag{2}$$

A fundamental issue concerns the comparison between the log-spectral data of two frames [13] called also log-spectral distance. To show how to define the log-spectral distance, we start with the definition of the Euclidean distance $e(\omega)$ between the log spectra of the two frames $x(n)$ and $y(n)$:

$$\begin{aligned} e(\omega) &= \log|X(\omega)|^2 - \log|Y(\omega)|^2 = 2(\log|X(\omega)| - log|Y(\omega)|) \\ &= 2Re\{\log|X(\omega)| - log|Y(\omega)|\} \end{aligned} \tag{3}$$

Hence, turning to the distance $e(\omega)$ and calling c_n^x and c_n^y the Cepstrum of the sequences $x(n)$ and $y(n)$ respectively we can write

$$e(\omega) = 2Re\{\log|X(\omega)| - log|Y(\omega)|\} = 2Re\{\sum_{-\infty}^{+\infty}(c_n^x - c_n^y)e^{j\omega n}\} = \{\sum_{-\infty}^{+\infty}(c_n^x - c_n^y)e^{j\omega n}\} \tag{4}$$

because the sequences c_n are symmetrical, since the sequence of the input reference page is real. Finally, the spectral distance between two sequences $x(n)$ and $y(n)$ is:

$$d(X, Y) = \int_{-\infty}^{+\infty} e(\omega)^2 d(\omega) = \sum_{-\infty}^{+\infty}(c_n^x - c_n^y)^2 \tag{5}$$

Only few initial Cepstrum values from the DCT sequence are used, which removes redundancy.

4.3 Data Analysis

It is well known that the initial instructions of a running code generally do not represent the stationary behavior of the program. In fact the first blocks of instructions of each code usually perform file I/O and memory allocation operations, since the data structures are set before arriving at the calculation of the function that the program must execute. In this work we do not use techniques to discover programs phases as described for example in [18] to find the beginning of the steady state phase of the programs. Much more simply we by-pass a first block of instructions (in this work 10^9 instructions) before starting the data analysis. We consider that the sequence of references in memory and the sequence of the request of resources as mono-dimensional signals. Therefore, as reported in the previous Section, similar to what happens in signal processing, we use a parametric description of the sequences. We observe that the events of the process, such as cycles or sequential instructions, produce important events in the signal spectrum. Furthermore, the sequences dynamically change their properties. For these reasons, we have used a short-time spectral description of the memory reference sequences. Thus, the sequences are divided into overlapping frames of analysis with a given size. The frames are further divided into blocks. Likewise, the sequence of resource requests is divided into overlapped frames and blocks.

In our case, to simplify the analysis we have extracted, from every memory reference sequence, blocks of 510^6 references. Each block is then segmented into frames of size 1000 which is the final size of the analysis frames.

The next step is to perform a spectral analysis of the frames. As mentioned before, we use the DCT representation. DCT [10] is a well-known signal processing operation with important properties. For example, it is useful for reducing signal redundancy because it puts as much energy as possible into a few coefficients (energy compaction). The first DCT coefficients are further processed. So, for each frame we get 1000 spectral coefficients from which we compute 32 Cepstrum coefficients. Moreover, the first coefficient of the Cepstrum sequence represents the energy of the frame and for this reason it is eliminated.

Summarizing what previously described, each frame of 1000 referenced is analyzed by Cepstral analysis. The first 32 Cepstral values are then selected as input to our Deep Neural Network, omitting the first coefficient which is a gain parameter. It is worth noting that since the basic point of our approach is to make the Network learn the workload of the reference programs, the training data is built by running, for each workload, the same program with different input data set. Therefore, various instances of the same workload are generated. In this way the Deep Neural Network should learn the inherent identity of the workload itself. Supervised learning is performed by setting to one the output corresponding to each workload when that workload is given as input.

In output we therefore have a vector of 12 elements that correspond to the 12 workload Models. This process is repeated for all the 12 workloads obtained before, so we constructed a Deep Neural Network Model of the SPEC 2006 workloads.

4.4   Deep Neural Network

The Deep Neural Model has three hidden layers with 32 inputs (the Cepstrum coefficients) and 12 outputs (the workloads).

4.5   Network Training

Training is performed using stochastic gradient descent with momentum $= 0.9$, which is the algorithm to optimize the network weights. To reduce over-fitting we use L2 regularization with $\lambda = 1.0e - 6$. The Neural Model is trained using a number of executions of reference programs with different inputs. Each execution will add knowledge to the network. Therefore, training of the Deep Neural Network is performed according to an incremental approach, where each execution corresponds to a time epoch t_k, where a new Training Data $TD(t_k)$ is available. The next execution, performed at t_{k+1}, shall update the previous weights of the Neural Model $w(t_k)$ which become $w(t_{k+1})$ using the weights optimization algorithm of the Deep Neural Network.

Several different inputs are given to each of the 12 programs in order to obtain several different address sequences (processes) for each program. With these sequences the Deep Neural Network with 12 outputs is trained using the approach described above. The last Training is further analyzed. The Network obtained with the last training was analyzed in terms of learning convergence or progress report during optimization. In Fig. 4 we report the Progress Report during convergence of stochastic gradient descent. The curve is the average of 12 network trainings. The number of neurons per hidden layer is 80.

5   Experimental Assessment and Analysis

Summarizing what previously described, each frame of 1000 references is analyzed by Cepstral analysis. The first 32 Cepstral values obtained are then selected, omitting the first coefficient. The 32 Cepstral values are set as input to our Deep Neural Network. We note that here, the Deep Neural Network once trained succeeds in extracting from this input a behavior related to the category of workload from which they come. In output we therefore have a vector of 12 elements that correspond to the 12 workload Models. This process is repeated for all the 12 workloads obtained before, so we constructed a Deep Neural Network Model of the SPEC 2006 workloads.

Trained Deep Neural Networks have been used to classify sequences of addresses from unknown workloads. Our experiments, reported in this paper, we assumed that the unknown workloads are extracted from the 12 programs of

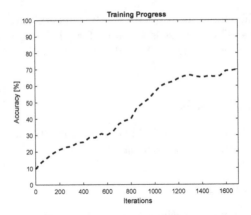

Fig. 4. Progress report (average data)

Spec2006. In other words, as described above the Deep Neural Network Model is trained with a train subset of the twelve SPEC2006 data set. For testing, we extracted a test subset from the same data set. Each of the test data set program is executed with input data totally different from all the previously used input data. For testing purposes we therefore know the workload of every testing program. Accuracy is computed by counting the total of correctly classified workloads. In other words, we do not use any program different from SPEC2006, just for accuracy evauation. The results justify our Two tests were done, one with 80 neurons in the hidden layer and one with 96 neurons. The result is that the average accuracy with 80 neurons is about 68%. The average accuracy with 80 neurons raise to 71% with 96 neurons per hidden layer (Figs. 5 and 6).

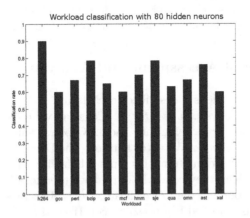

Fig. 5. Deep neural network with 80 neurons in each hidden layer

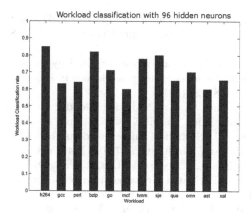

Fig. 6. Deep neural network with 96 neurons in each hidden Layer

6   Conclusions and Future Work

In this paper, we focused on classifying application workloads in a virtualized environment. Then, starting from the known programs (benchmarks), a workload Model is built for each of them by training a Deep Neural Network using execution sequences of each instruction in the program. This process actually allows to characterize the program to which its workload Model is associated. With these workload Models collected by each program, we implemented in Matlab a 32-input, three hidden layer with 96 and 80 hidden neurons and 12 outputs per layer Deep Neural Network, trying to capture the information of each workload, that is, making the Network learn these "features" so that the Network is capable of roughly identifying a workload. We thus build a Deep Neural Network Model of SPEC 2006 workloads and testing is performed with a test subset of the same data set. Preliminary results indicate that with 80 and 96 neurons per hidden layer the accuracyis around 68% and 71% respectively, indicating that the Deep Neural Model is able to satisfyingly learn the workloads of SPEC2006. Future work will be on one hand directed towards an extension of the testing conditions in terms of amount of workloads identified, by using other available data sets. On the other hand a different number of hidden layer and amount of neurons will be tested. Another possible future work will be to use *AppxDL* for workload balancing. Finally, we plan to incorporate specific *big data features* (e.g., [24–26]) in our framework.

Acknowledgments. This project is partially supported by NSERC (Canada) and University of Manitoba.

References

1. Van Do, T.: Comparison of allocation schemes for virtual machines in energy-aware server farms. Comput. J. **54**(11), 1790–1797 (2011)

2. Islam, S., Keung, J., Lee, K., Liu, A.: Empirical prediction models for adaptive resource provisioning in the cloud. Future Generation Comp. Syst. **28**(1), 155–162 (2012)
3. Garg, S.K., Toosi, A.N., Gopalaiyengar, S.K., Buyya, R.: SLA-based virtual machine management for heterogeneous workloads in a cloud datacenter. J. Netw. Comput. Appl. **45**, 108–120 (2014)
4. Akindele, A.B., Samuel, A.A.: Predicting cloud resource provisioning using machine learning techniques. In: IEEE CCECE, pp. 1–4 (2013)
5. Baum, L.E., Petrie, T.: Statistical inference for probabilistic functions of finite state Markov chains. Ann. Math. Stat. **37**(6), 1554–1563 (1966)
6. Deng, Y., Shen, S., Huang, Z., Iosup, A., Lau, R.W.H.: Dynamic resource management in cloud-based distributed virtual environments. In: ACM Multimedia, pp. 1209–1212 (2014)
7. DiFranzo, D., Graves, A.: A farm in every window: a study into the incentives for participation in the windowfarm virtual community. In: WebSci 2011, p. 14 (2011)
8. Kousiouris, G., Menychtas, A., Kyriazis, D., Gogouvitis, S., Varvarigou, T.: Dynamic, behavioral-based estimation of resource provisioning based on high-level application terms in cloud platforms. Future Gener. Comput. Syst. **32**, 27–40 (2014)
9. Amiri, M., Feizi-Derakhshi, M.R., Mohammad-Khanli, L.: IDS fitted Q improvement using fuzzy approach for resource provisioning in cloud. J. Intell. Fuzzy Syst. **32**(1), 229–240 (2017)
10. Hou, H.S., Tretter, D.R., Vogel, M.J.: Interesting properties of thediscrete cosine transform. J. Vis. Commun. Image Represent. **3**(1), 73–83 (1992)
11. Chang, Y., Chang, R., Chuang, F.: A predictive method for workload forecasting in the cloud environment. In: Proceedings of Advanced Technologies, Embedded and Multimedia for Human-Centric Computing, pp. 577–585 (2014)
12. Chen, Z., Zhu, Y., Di, Y., Feng, S.: Self-Adaptive prediction of cloud resource demands using ensemble model and subtractive-fuzzy clustering based fuzzy neural network. Comput. Intell. Neurosci. **2015**, 17:1–17:14 (2015)
13. Ramezani, F., Naderpour, M.: A fuzzy virtual machine workload prediction method for cloud environments. In: FUZZ-IEEE, pp. 1–6 (2017)
14. G. Bruder, F. Steinicke, A. Nchter, Poster: Immersive point cloud virtual environments. In: Proceedings of 3DUI 2014, pp. 161–162. Publisher, Location (2010)
15. Chuang, I.-H., Tsai, Y.-T., Horng, M.-F., Kuo, Y.-H., Hsu, J.-P.: A GA-based approach for resource consolidation of virtual machines in clouds. In: Nguyen, N.T., Attachoo, B., Trawiński, B., Somboonviwat, K. (eds.) ACIIDS 2014. LNCS (LNAI), vol. 8397, pp. 342–351. Springer, Cham (2014). https://doi.org/10.1007/978-3-319-05476-6_35
16. Bleikertz, S., Vogel, C., Gro, T.: Cloud radar: near real-time detection of security failures in dynamic virtualized infrastructures. In: ACSAC, pp. 26–35 (2014)
17. Hsiao, S.-W., Chen, Y.-N., Sun, Y.S., Chen, M.C.: Combining dynamic passive analysis and active fingerprinting for effective bot malware detection in virtualized environments. In: Lopez, J., Huang, X., Sandhu, R. (eds.) NSS 2013. LNCS, vol. 7873, pp. 699–706. Springer, Heidelberg (2013). https://doi.org/10.1007/978-3-642-38631-2_59
18. Vandromme, N., et al.: Life cycle assessment of videoconferencing with call management servers relying on virtualization, In: ICT4S 2014 (2014)
19. Standard Performance Evaluation Corporation. http://www.spec.org/cpu2006/CINT2006/

20. Virtualbox API. http://www.virtualbox.org/sdkref/index.html
21. Virtualbox. http://www.virtualbox.org/manual
22. SOAP. http://www.w3.org/tr/soap
23. WSDL. http://www.w3.org/tr/wsdl20
24. Braun, P., Cameron, J.J., Cuzzocrea, A., Jiang, F., Leung, C.K.: Effectively and efficiently mining frequent patterns from dense graph streams on disk. Procedia Comput. Sci. **35**, 338–347 (2014)
25. Wu, Z., Yin, W., Cao, J., Xu, G., Cuzzocrea, A.: Community detection in multi-relational social networks. In: International Conference on Web Information Systems Engineering, pp. 43–56 (2013)
26. Cuzzocrea, A., Bertino, E.: Privacy preserving OLAP over distributed XML data: A theoretically-sound secure-multiparty-computation approach. J. Comput. Syst. Sci. **77**(6), 965–987 (2011)

Control-Flow Business Process Summarization via Activity Contraction

Valeria Fionda$^{(\boxtimes)}$ and Gianluigi Greco

DeMaCS, University of Calabria, via Pietro Bucci 30B, 87036 Rende, CS, Italy
{fionda,greco}@mat.unical.it

Abstract. Organizations collect and store considerable amounts of process data in event logs that are subsequently mined to obtain process models. When the business process involves hundreds of activities, executed according to complex execution patterns, the process model can become too large and complex to identify relevant information by manual and visual inspection only. Summarization techniques can help, by providing concise and meaningful representations of the underling process. This paper describes a business process summarization algorithm based on the hierarchical grouping of activities. In the proposed approach, activity grouping is guided by the existence of some relations, between pairs of activities, mined from the associated event log.

Keywords: Business process · Summarization · Process Mining

1 Introduction

A business process is a collection of structured activities that in a specific sequence produces a service or product thus serving a particular business goal. In general, the sequence that allow to obtain the business goal is not univocal and the sequences that happen in practice are registered as traces of a log. Such log can be used to extract a control-flow model useful to analyze and improve the underlying process. The most basic control-flow modeling approach transforms each business process into a directed graph, such that nodes are in one-to-one correspondence with activities and edges connect two activities that have happened to be executed one after the other in at least one trace of the log. The edges can be weighted by the number of traces providing evidence for the sequentiality of activities.

If the process involves hundreds of activities implemented according to highly variable execution sequences, the above graph representation becomes too large and complex to grasp significant information just by manual and visual inspection. Thus, it becomes necessary to design abstraction and summarization techniques suitable to provide concise and meaningful representations of the process.

Contributions. In this paper, we discuss the major findings of our investigation into summarization techniques for process graphs involving control-flow information. We present a flexible framework that allows to build summaries of business

© Springer Nature Switzerland AG 2019
H. Yin et al. (Eds.): IDEAL 2019, LNCS 11872, pp. 238–248, 2019.
https://doi.org/10.1007/978-3-030-33617-2_25

processes with different levels of abstraction. We also discuss an implementation of the proposed approach and the results of an experimental evaluation.

Related Work. Graph summarization touches several research fields such as graph mining, data management and graph visualization. The techniques most similar to ours are the hierarchical node grouping methods returning supergraphs[1], where a supergraph consists of collections of original nodes and superedges between them. Some node grouping approaches employ existing clustering techniques to find clusters that then map to supernodes of the summary (e.g., [5]). Even if node grouping and clustering are related, since they both result in collections of nodes, they have different goals. In the context of graph summarization, node grouping is based on application-dependent optimization functions while clustering usually targets (a variant of) the minimization of cross-cluster edges and its end goal is not to produce a summary. Furthermore, graph summarization techniques find groups of nodes that are linked to the rest of the graph in a similar way, while clustering results in densely connected groups of nodes. Several node-grouping methods have been proposed with the objective of minimizing some version of the approximation or reconstruction error [3,8,10] or to maintain specific properties of the original graph, such as diffusive properties related to the spectrum of the graph, and specifically its first eigenvalue [6]. In addition, also grouping-based methods that aggregate nodes into supernodes based on both structural properties and node attributes have been proposed [7,9]. As for approaches in the visualization domain, several summarization techniques exist that simplify graphs in order to enhance graph visualization. For example, Dunne and Shneiderman [1] introduce motif simplification that replaces common links and common subgraphs with compact glyphs. The work presented in the paper differs from related approaches since none of them has been proposed to specifically target graphs encoding business processes.

2   Preliminaries

Logs. Let $\mathcal{A}$ be a finite set of symbols, univocally identifying all process activities. An event a is the occurrence of an activity $a \in \mathcal{A}$. A trace $\pi = \pi[1]...\pi[n]$ with $\pi[i] \in \mathcal{A}$, for each $i \in \{1,...,n\}$, is a finite sequence of n events, where n is the size of the trace indicated by $|\pi|$. A log $L = \{\pi_1,...,\pi_m\}$ is a finite non-empty set of traces, where m is the number of traces it contains, indicated by $|L|$. The size of a log L is the number of events in it, indicated by $||L|| = \sum_{\pi \in L} |\pi|$. The set of distinct activities occurring in L is indicated as $\mathcal{A}_L \subseteq \mathcal{A}$.

Example 1. The trace $\pi = a_1 a_2 a_1$ has size $|\pi| = 3$ and denotes that the first activity executed is a_1, then a_2 and finally again a_1. The log $L = \{a_1 a_2 a_1, a_1 a_2 a_3\}$ consists of two traces $a_1 a_2 a_1$ and $a_1 a_2 a_3$, i.e., $|L| = 2$, and has size $||L|| = 6$.

---

[1] The reader is pointed to [4] for a comprehensive survey on graph summarization techniques including also other categories.

Process Graph. A process graph is a graph encoding a log, in which the vertices represent activities and edges connect pairs of activities that are executed subsequently in at least one of the traces of the log. More formally, given a log L, $G_L = (A_L, E_L, \lambda)$ is a directed labeled graph such that: *(i)* A_L is the set of nodes which are in one to one correspondence with activities appearing in L (i.e., $\mathcal{A}_L$); *(ii)* $E_L \subseteq A_L \times A_L$ is the set of directed edges; *(iii)* $\lambda : E_L \to \mathbb{Z}$ is a function assigning to each edge (a_i, a_j) a natural number indicating how many times the activity a_j is executed immediately after a_i in L.

Example 2. Figure 1(a) shows an example of log, reporting the execution traces of the evaluation process of loan applications, and the graph modeling it. The process always starts with the check for the completeness of the application form (CAFC), then, if the form is not complete it is returned back to the applicant (RABA) and then a updated form is received (RUA). Otherwise, if the form is complete the activity of property appraising (AP) is executed simultaneously to the checking of the applicant credit history (CCH) that is followed by the Assessment of loan risk (ALR). After such activities have been completed eligibility for the loan is assessed (AE) and the loan is either approved, by performing in sequence the activities of approving the application (AA) and send the loan approved confirmation (LAA), or rejected, by performing in sequence the activities of rejecting the application (RA) and send the loan rejected confirmation (LAR).

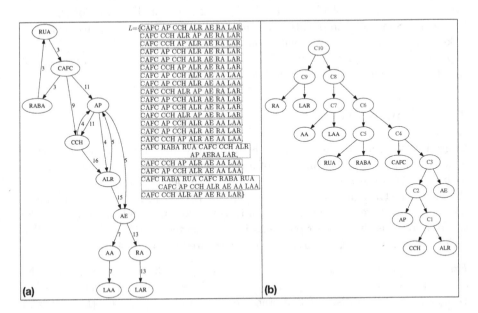

Fig. 1. (a) A process graph and its associated log; (b) the process tree according to the log-based activity node similarity.

3   Contraction-Based Framework

This section discusses the framework at the bases of summary construction. In the following, let L be a log and $G_L = (A_L, E_L, \lambda)$ its associated process graph.

Contraction Operator. The contraction operator $\Gamma(G_L, a_i, a_j, \hat{a})$, where $\{a_i, a_j\} \subset A_L$, returns a process graph $\hat{G}_L = (\hat{A}_L, \hat{E}_L, \hat{\lambda})$, such that

- $\hat{A}_L = (A_L \cup \{\hat{a}\}) \setminus \{a_i, a_j\}$;
- $\hat{E}_L = \{(a_l, a_k) \in E_L \mid a_l, a_k \in A_L \setminus \{a_i, a_j\}\} \cup \{(a_l, \hat{a}) \mid a_l \in \hat{A}_L \setminus \hat{a}$ and $(a_l, a_i) \in E_L \vee (a_l, a_j) \in E_L\} \cup \{(\hat{a}, a_l) \mid a_l \in \hat{A}_L \setminus \hat{a}$ and $(a_i, a_l) \in E_L \vee (a_j, a_l) \in E_L\}$;
- $\hat{\lambda}((a_l, a_k)) = \lambda((a_l, a_k))$ for all $(a_l, a_k) \in \hat{E}_L$ such that $a_l, a_k \in A_L \setminus \{a_i, a_j\}$;
- $\hat{\lambda}((a_l, \hat{a})) = \lambda((a_l, a_i)) + \lambda((a_l, a_j))$ (respectively, $\hat{\lambda}((\hat{a}, a_l)) = \lambda((a_i, a_l)) + \lambda((a_j, a_l))$.

Activity Node Log-Based Similarity. The similarity between activity nodes of G_L is computed by analyzing the log. In particular, from the log we try to identify some relations between activities. We exploit three basic relations that are commonly used to mine business process models (see, e.g. [2]);

- *Sequentiality ($\rightarrow$):* $\rightarrow (a_1, a_2)$ means that from L it can be derived the sequential composition of a_1 and a_2, i.e., a_2 is frequently executed immediately after a_1;
- *Parallelisms ($\|$):* $\|(a_1, a_2)$ means that from L it can be derived the parallel composition of a_1 and a_2, i.e., a_1 and a_2 represent parallel (interleaved) activities;
- *Alternatives ($\times$):* $\times(a_1, a_2)$ means that from L it can be derived the exclusive choice of a_1 and a_2, i.e., a_1 and a_2 are frequently executed as alternative activities.

To each mined relation $\otimes(a_i, a_j)$, with $\otimes \in \{\rightarrow, \times, \|\}$, is associated a confidence value $s(\otimes(a_i, a_j)) \in [0, 1]$, that measures how likely is the relation to hold. Let $n(a_i)$ be the number of occurrences of the activity a_i, $n(a_i, a_j)$ the number of occurrences of the subtrace $a_i a_j$ and $p(a_i, a_j)$ the number of occurrences of non-overlapping subtraces $a_i a_j$ or $a_j a_i$ in L. Intuitively,

$$s(\rightarrow (a_i, a_j)) = 2 \cdot n(a_i, a_j) / (n(a_i) + n(a_j))$$

that is the fraction of occurrences of a_i and a_j in which a_j is executed immediately after a_i. Similarly,

$$s(\| (a_i, a_j)) = p(a_i, a_j) / (n(a_i) + n(a_j))$$

that is the fraction of occurrences of a_i and a_j in which either a_j is executed immediately after a_i or a_i is executed immediately after a_j with the additional constraint that $s(\rightarrow(a_i, a_j)) \approx s(\rightarrow(a_j, a_i))$. Finally, $\times(a_i, a_j)$ holds with confidence $s(\times(a_i, a_j)) = \alpha$ if in $\alpha \cdot (n(a_i) + n(a_j))$ occurrences of a_i and a_j it appears either the subtrace $a_k a_i a_l$ or the subtrace $a_k a_j a_l$, for some $a_k, a_l \in \mathcal{A} \setminus \{a_i, a_j\}$, with the additional constraint that $s(\rightarrow(a_k, a_i, a_l)) \approx s(\rightarrow(a_k, a_j, a_l))$ (where sequentiality is extended to three activities).

Then, the similarity between two activity nodes a_i and a_j is defined as $sim(a_i, a_j) = \max\{s(\rightarrow(a_i, a_j)), s(\rightarrow(a_j, a_i)), s(\times(a_i, a_j)), s(||(a_i, a_j))\}$. Given a process graph G_L and its associated log L, the cost of computing the similarity between each pair of activities is $\mathcal{O}(|A_L|^2 + ||L||)$.

Hierarchical Contraction of Activities. Our framework is based on a hierarchical contraction algorithm that starts with a process graph where each activity node is a singleton node and, then, contracts pairs of nodes (via the contracting operator) until only one node is obtained. At each contraction step the algorithm *(i)* contracts the pair of activity nodes having the maximum similarity, *(ii)* replaces the id of the new contracted node in the log wherever the two contracted activities appear and, *(iii)* computes the similarity values between the contracted node and all the other nodes of the process graph. The contraction sequence is stored in a rooted binary tree $\tau(G_L) = (V_{\tau(G_L)}, E_{\tau(G_L)}, \gamma)$, called contraction tree. The leaves of the tree are the activity nodes of the initial process graph and a node a_k, having a_i and a_j as children nodes, encodes the fact that a_i and a_j have been contracted into a_k. Nodes are labeled by the function $\gamma : V_{\tau(G_L)} \rightarrow \mathbb{Z}$, where $\gamma(a_k)$ indicates the number of edges that will be subsumed by considering the (contracted) node a_k in place of its children nodes a_i and a_j, that is the number of edges between a_i and a_j plus the number of common neighbors of a_i and a_j in the process graph before contraction. The contraction tree of the process graph in Fig. 1(a), computed according to the log-based similarity measure, is reported in Fig. 1(b).

The contraction algorithm can be implemented by using a priority queue P such that $P[i]$ stores, in decreasing order of similarity $sim(a_i, a_j)$, the nodes a_j. At the beginning the similarity between pairs of nodes of G_L is computed and stored in P. Overall, this step costs $\mathcal{O}(|A_L|^2 \cdot \log(|A_L|) + ||L||)$ for an implementation of priority queues that supports insertions in $\mathcal{O}(\log(|A_L|))$. Then, the algorithm performs $|A_L| - 1$ contraction steps, until it obtains a graph with one node only. In each contraction step, the two nodes having the maximum similarity are contracted and each contraction operation has cost $\mathcal{O}(|A_L| + ||L||)$ since G_L is a simple graph and the log must be updated with the is of the new contracted activity. Furthermore, if at the generic step the algorithm is contracting the nodes a_i and a_j into the node a_k, then a_k is added to $V_{\tau(G_L)}$ and the edges (a_k, a_i) and (a_k, a_j) are added to $E_{\tau(G_L)}$. All the entries of P involving a_i and a_j are deleted and, for all the nodes a_s in the contracted graph, the similarity value between a_s and a_k is computed and inserted into P. This last step has cost $\mathcal{O}(|A_L| \cdot \log(|A_L|) + ||L||)$ for an implementation of priority queues that also supports deletions in $\mathcal{O}(\log(|A_L|))$. Therefore, the overall complexity of the algorithm is $\mathcal{O}(|A_L|^2 \cdot \log(|A_L|) + |A_L| \cdot ||L||)$.

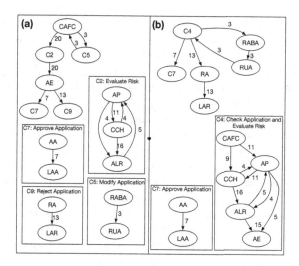

Fig. 2. Two examples of summaries built from the process graph and the contraction tree in Fig. 1.

4   Building Business Process Summaries

In this section we formalize the problem of computing business process summaries and discuss an algorithm to build them.

Process Graph Summaries. Given a process graph G_L and a contraction tree $\tau(G_L) = (V_{\tau(G_L)}, E_{\tau(G_L)}, \gamma)$, a summary $\alpha(G_L) = (V_{\alpha(G_L)}, E_{\alpha(G_L)}, \delta)$ is a supergraph of G_L such that:

1. $V_{\alpha(G_L)} \subset V_{\tau(G_L)}$ and for each $n \in V_{\alpha(G_L)}$ it holds that $n' \notin V_{\alpha(G_L)}$ for all n' that are in the subtree of $\tau(G_L)$ rooted at n.
2. For each pair of activity nodes a_i and a_j in $V_{\alpha(G_L)}$, we have that $(a_i, a_j) \in E_{\alpha(G_L)}$ holds if, and only if, there exist two activity nodes a_k and a_l that are, respectively, in the subtrees of a_i and a_j and such that the edge (a_k, a_l) is in G_L.
3. For each $(a_i, a_j) \in E_{\alpha(G_L)}$ we have that $\delta((a_i, a_j)) = \sum_{a_k \in st(\tau(G_L), a_i), a_l \in st(\tau(G_L), a_j)} \gamma((a_k, a_l))$, where $st(x, y)$ indicates the subtree of x rooted at y.

Example 3. Figure 2 shows two summaries of six nodes of the process graph in Fig. 1(a) built on the contraction tree in Fig. 1(b).

Building Summaries. We focus on the following problem.

MAXIMUM CONTRACTION SUMMARY:
Input: A process graph G_L, a contraction tree $\tau(G_L)$, an integer m
Output: A summary $\alpha(G_L)$ of G_L, built on $\tau(G_L)$, with m nodes that maximize the number of edges subsumed by contracted nodes.

Solving the MAXIMUM CONTRACTION SUMMARY problem aims at computing a summary of G_L having m nodes that allows to hide the maximum number of edges within the contracted nodes. It can be solved in polynomial time by using the dynamic programming algorithm reported in Fig. 3. The idea is to associate to each node n of $\tau(G_L)$ an array a_n whose size is equals to the number of leaves in the subtree rooted at n. The array a_n in position i stores: *(i)* the maximum number of edges subsumed by a partial summary of i nodes built by considering the subgraph induced by the leaves in the subtree rooted at n (i.e., $a_n[i].e$); and *(ii)* the partial summary that allows to obtain the value $a_n.e$ (i.e., $a_n[i].\alpha$). Values are propagated from the leaves to the root (Function MCS line 1) by performing in the procedure Propagate a postorder traversal of the tree, by visiting the left subtree first (line 5), then the right subtree (line 6) and finally the parent node (lines 7–13). In particular, let $lchild$ and $rchild$ be the two children of the node n, then the position k of a_n is filled by combining the values of $a_{lchild}[i]$ and $a_{rchild}[j]$, such that $i+j=k$ (lines 12–13), that allow to subsume the highest number of edges (line 11). Eventually, the best summary having m nodes is the one stored in a_{root} at position m (Function MCS, line 2).

The correctness of the algorithm can be proven by structural induction on $\tau(G_L)$, where the base case are the node n that are leaves and for which the algorithm is clearly correct since a_n has size 1. Then, assume that it is correct for all nodes at depth $\geq i$ and let n be a node at depth $i-1$ with $lchild$ and $rchild$ be its left and right child, respectively. First, note that a summary of the subtree rooted at n can be built either by considering n alone (Propagate, lines 7–8) or by combining two partial summaries covering, respectively, the left and right subtree. By induction hypothesis the best summaries for $lchild$ and $rchild$ are those stored in a_{lchild} and a_{rchild}, respectively, and, thus, the best possible summaries for n can be computed by combining them (Propagate, lines 9–13).

The running time of MCS is $\mathcal{O}(|A_L|^2)$ in the worst case, that is when $\tau(G_L)$ is not balanced. Indeed, in this case the height of the tree is $|A_L|-1$ and the cost of Propagate for the node at height i is $|A_L|-1-i$. In the case of balanced contraction trees the running time of MCS is $\mathcal{O}(|A_L|)$.

5   Experimental Evaluation

This section reports the results of an experimental evaluation of: *(i)* the performance of the framework in terms of running time; *(ii)* the compression ratio that can be obtained by applying the proposed method; and, *(iii)* the effectiveness of summaries as compact representations of the original process.

Dataset. We used a public benchmark dataset consisting of logs describing executions of realistic processes, collected under the placement of the IEEE task-force (www.win.tue.nl/ieeetfpm) on Process Mining. The benchmarks are publicly available within the 4TU.Datacentrum (http://data.4tu.nl), a longterm archive for storing scientific research data. The dataset we used consists of 9 benchmarks (Loan application, Review Example, Artificial, isbpm2013, book, Business Process Drift, Digital Photo, bpm2013 benchmark, Bank Transaction)

Function MCS($G_L, \tau(G_L), m$)

Input: G_L: process graph, $\tau(G_L)$: contraction tree, m: integer **Output:** a contraction having m nodes

1: **Propagate**($root$) — $root$ is the root of $\tau(G_L)$
2: **return** $a_{root}[m].\alpha$

Procedure Propagate(n)

Input: n: a node of $\tau(G_L)$

```
1:  if n is a leaf then
2:      a_n[1].e = 0
3:      a_n[1].α = {n}
4:  else
5:      Propagate(lchild) — lchild is the left child of n
6:      Propagate(rchild) — rchild is the right child of n
7:      a_n[1].e = a_lchild[1].e + a_rchild[1].e + γ(n)
8:      a_n[1].α = {n}
9:      for i:1 to a_lchild.size do
10:         for j:1 to a_rchild.size do
11:             if a_n[i + j].e < a_lchild[i].e + a_rchild[j].e then
12:                 a_n[i + j].e = a_lchild[i].e + a_rchild[j].e
13:                 a_n[i + j].α = a_lchild[i].α ∪ a_rchild[j].α
```

Fig. 3. Dynamic-programming algorithm for the MAXIMUM CONTRACTION SUMMARY problem.

Table 1. Characteristics of the benchmark dataset.

Dataset	#logs	min # traces	max # traces	min # distinct event	max # distinct event
Artificial	1	100	100	30	30
Digital Photo	1	100	100	68	68
Review Example	1	10000	10000	14	14
Bank Transaction-BkT	2	2000	2000	113	113
bpm2013 benchmark	3	500	1200	311	363
Loan Application	4	70	200	5	8
isbpm2013	17	2000	2000	30	110
book	26	300	300	6	42
Business Process Drift	72	2500	10000	19	31
	127				

and its salient features are summarized in Table 1. In the experiments, for each log: *(i)* we first computed the contraction tree by running the hierarchical contraction algorithm (discussed in Sect. 3) and, then, *(ii)* we built three process graph summaries by considering a number of node equals to the 25%, 50% and 75% of the original number of activities. All the tests were conducted on a Mac-Book Pro 3300MHz/16GB Intel Core i7 and, in order to reduce statistical bias, running times were computed as the average of 3 runs.

Performances. For each log we measured the time required by: *(i)* the hierarchical contraction algorithm, *(ii)* the MCS function and *(iii)* the procedure to

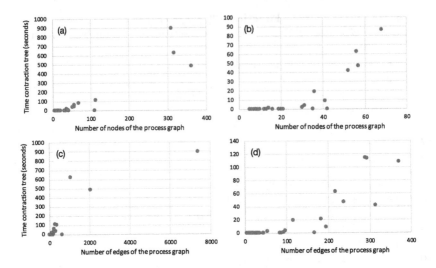

Fig. 4. Contraction tree construction time with respect to: (a) the number of distinct activities of each log; (b) zoom-out of (a) among the range from 0 to 80 activities; (c) the number of edges of the process graph obtained from each log; (d) zoom-out of (c) among the range from 0 to 400 edges.

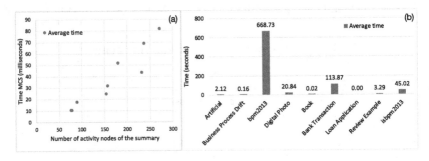

Fig. 5. Computation time (average) of: (a) the MCS algorithm on the logs of the bpm2013 dataset; (b) the contraction tree construction over the different datasets.

reconstruct the summary process graph starting from the result of the MCS function. The last step has running times that varies from 1 to 5 ms, and, thus, is negligible w.r.t. the two previous steps. The running time of MCS is, in most of the cases, of few millisecond; it was tangible only for the bpm2013 dataset, that is the one having the largest number of activities (see Table 1), on which the running time was of few tens of milliseconds. Results are reported in Fig. 5 (a) where time is reported in milliseconds on the y-axis and the number of activity nodes of the summary is reported on the x-axis.

The slowest step corresponds to the construction of the contraction tree. The performances are depicted in Fig. 5(b), which reports the average time (in seconds) per benchmark required to run the hierarchical contraction algorithm.

Table 2. Compression ratio obtained on the benchmark dataset.

Dataset	avg # nodes	avg # edges	25% nodes		50% nodes		75% nodes	
			avg # edges	σ	avg # edges	σ	avg # edges	σ
Artificial	30	640	51	8%	200	31%	390	61%
Digital Photo	68	182	21	12%	45	25%	82	45%
Review Example	14	54	7	13%	14	26%	30	56%
Bank Transaction-BkT	113	292	33	11%	69	24%	132	45%
bpm2013 benchmark	330	3510	170	7%	460	16%	1206	37%
Loan Application	7	9	1	15%	3	43%	6	73%
isbpm2013	54	216	19	10%	39	20%	78	38%
book	14	22	3	15%	8	34%	13	60%
Business Process Drift	20	30	4	15%	10	35%	17	59%
	35	169	12	14%	30	32%	66	56%

In order to shed further lights on the scalability of our method, we compared the computation time with respect to: *(i)* the number of distinct activities per log (that corresponds to the number of activity nodes of the process graph) and, *(ii)* to the number of edges of the process graph of each log (that is strictly related to the size $||L||$ of the log). Results are reported in Fig. 1(a) and (c), respectively (where time is reported in seconds on the y-axis) and they evidence that, in general, larger numbers of activity nodes (edges, respectively) require more time for the computation of the contraction tree (as one would expect after the analysis reported in Sect. 3). The intuition is confirmed by Fig. 1(b) and (d) that explode the results by focusing on the cases where the number of activity nodes is limited to 70 and the number of edges is limited to 400, respectively. However, note that the contraction tree building and the MCS algorithm must be run only once and their results can be used offline to build summaries at different level of abstraction. Thus, the cost of the contraction tree building, that is the most onerous one, must be paid only once per process log.

Compression Ratio. The second line of investigation of our experimental evaluation regarded the compression ratio, in terms of number of edges of the process graph summary, obtained in the benchmark dataset. We define the compression ratio σ as the fraction of edges that are kept in the summary, computed as the ration between the number of edge of the summary and the number of edges in the original process graph. More formally, if $G_L = (A_L, E_L, \lambda)$ is the process graph and $\alpha(G_L) = (V_{\alpha(G_L)}, E_{\alpha(G_L)}, \delta)$ is the summary then $\sigma = \frac{|E_{\alpha(G_L)}|}{|E_L|}$. Table 2 shows for each benchmark dataset the average compression rate obtained on its process logs by specifying a number of nodes for the summary construction equals to the 25%, 50% and 75% of the number of nodes of the original process graph. It is evident from the experiments that the highest the difference between the number of nodes and edges of the process graph, the highest ithe obtained compression ratio.

Effectiveness. In our experimental validation we also analyzed how effective the summaries are to compactly represent the original process graph. To this hand we focused on the Business Process Drift dataset for which the baseline process models are available and we compared them to the obtained contraction tree. In particular, we checked for each log in the dataset the percentage of the activity relation in the model that are correctly identified in the contraction tree. This is actually a measure of the suitability of the log-based similarity measure discussed in Sect. 3. The average accuracy obtained is 79.14%, whith a variance of 0.74 (with a minimum value equals to 61% and a maximum value equals to 94%.

References

1. Dunne, C., Shneiderman, B.: Motif simplification: improving network visualization readability with fan, connector, and clique glyphs. In: Proceedings of CHI, pp. 3247–3256 (2013)
2. Kopp, O., Martin, D., Wutke, D., Leymann, F.: The difference between graph-based and block-structured business process modelling languages. EMISA **4**(1), 3–13 (2009)
3. LeFevre, K., Terzi, E.: GraSS: graph structure summarization. In: Proceedings of SDM, pp. 454–465 (2010)
4. Liu, Y., Safavi, T., Dighe, A., Koutra, D.: Graph summarization methods and applications: a survey. ACM Comput. Surv. **51**(3), 62 (2018)
5. Newman, M.E.J., Girvan, M.: Finding and evaluating community structure in networks. Phys. Rev. E **69**(2), 026113 (2004)
6. Purohit, M., Prakash, B.A., Kang, C., Zhang, Y., Subrahmanian, V.S.: Fast influence-based coarsening for large networks. In: Proceedings of SIGKDD, pp. 1296–1305 (2014)
7. Raghavan, S., Garcia-Molina, H.: Representing web graphs. In: Proceedings of ICDE, pp. 405–416 (2003)
8. Riondato, M., García-Soriano, D., Bonchi, F.: Graph summarization with quality guarantees. Data Min. Knowl. Discov. **31**(2), 314–349 (2017)
9. Song, Q., Wu, Y., Lin, P., Dong, L., Sun, H.: Mining summaries for knowledge graph search. IEEE Trans. Knowl. Data Eng. **30**, 1887–1900 (2018)
10. Toivonen, H., Zhou, F., Hartikainen, A., Hinkka, A.: Compression of weighted graphs. In: Proceedings of SIGKDD, pp. 965–973 (2011)

Classifying Flies Based on Reconstructed Audio Signals

Michael Flynn$^{(\boxtimes)}$ and Anthony Bagnall

The University of East Anglia, Norwich, UK
`Michael.Flynn@UEA.ac.uk`

Abstract. Advancements in sensor technology and processing power have made it possible to create recording equipment that can reconstruct the audio signal of insects passing through a directed infrared beam. The widespread deployment of such devices would allow for a range of applications previously not practical. A sensor net of detectors could be used to help model population dynamics, assess the efficiency of interventions and serve as an early warning system. At the core of any such system is a classification problem: given a segment of audio collected as something passes through a sensor, can we classify it? We examine the case of detecting the presence of fly species, with a particular focus on mosquitoes. This gives rise to a range of problems such as: can we discriminate between species of fly? Can we detect different species of mosquito? Can we detect the sex of the insect? Automated classification would significantly improve the effectiveness and efficiency of vector monitoring using these sensor nets. We assess a range of time series classification (TSC) algorithms on data from two projects working in this area. We assess our prior belief that spectral features are most effective, and we remark on all approaches with respect to whether they can be considered "real-time".

Keywords: Insect classification · Time series classification · Spectral features

1 Introduction

Over the last century there have been many attempts at solving insect classification problems. Increased interest in classifying insects has been fuelled by a number of factors. Insects are responsible for the pollination of the majority of crop species, but are also vectors for disease and responsible for a massive number of fatalities. Monitoring the presence and abundance of mosquitoes is crucial in understanding the population dynamics and effectiveness of interventions. Quantifying the abundance of an insect specie in a natural setting is challenging. Typically, expert entomologists are required to manually identify species using morphological differences. This can result in a lengthy delay in detection and quantification and the amount of data that can be collected is limited.

© Springer Nature Switzerland AG 2019
H. Yin et al. (Eds.): IDEAL 2019, LNCS 11872, pp. 249–258, 2019.
https://doi.org/10.1007/978-3-030-33617-2_26

However, recent advances in sensor technology has made the collection of large datasets more feasible [2,14,16,19]. These approaches, described in more detail in Sect. 3, record data as an object passes through a target area. This results in a data segment that can be interpreted as audio and used to classify the object. As with most audio problems, standard classification approaches to this problem use features in the frequency domain. We assess a range of standard techniques in addition to recently proposed algorithms for general time series classification (TSC) problems. The most successful algorithms for TSC (the classification of real valued, ordered series such as audio) are based on transformation and ensembling. The effectiveness of these techniques is explored in the Hierarchical Vote Collective of Transformation-based Ensembles (HIVE-COTE) [7]. We perform a thorough experimental evaluation of a range of classifiers, three of which are used in HIVE-COTE, and conclude that in this case TSF is most effective.

The rest of this paper is structured as follows. In Sect. 2 we discuss what motivates insect classification, and describe current techniques used for this problem. In Sect. 3 we describe two data sets collected by research groups in USA and Germany which we use in the experimental evaluation. In Sect. 4 we outline the methods used in our evaluation and in Sect. 5 we present our results. We conclude in Sect. 6.

2   Background

Traditionally, insects are classified by their morphological differences. This becomes more difficult as you move down through taxonomic ranks. In some cases, it is not possible to classify species without gene sequencing [4]. Early investigations into the classification of flying insects were focused on wingbeat frequency [1,15] which was collected manually via the use of a stroboscope. The effectiveness of this data as a class predictor was then quantified using standard statistical modelling. These studies also quantified the effects of air temperature and location on wingbeat frequency. However, findings were often inconclusive and lacked robustness due to small sample sizes. As technology advanced, recording insect wingbeats became feasible. This allowed additional spectral information to be utilised, such as harmonics [10,18]. These early studies concluded that wingbeat frequency alone is not an adequate predictor of class. It follows a normal distribution and exhibits significant intra class variability. This results in substantial overlap between wingbeat frequencies of different classes, a problem that is only made worse as the number of classes increase.

The development of robust phototransitor recording techniques, capable of operating for extended periods of time, provided the first medium size datasets to work with [9]. The advent of artificial neural networks also provided a novel approach to classification. Many studies went on to report increases in accuracy when including or using entirely spectral features [6,11]. It was also noted that classification could be confounded by the insect behaviour, such as the relative insect trajectory, the fact that some species tendency to buzz their wings before takeoff and the effect of the circadian rhythm on behaviour.

Researchers at the University of California, Riverside (UCR) have developed a low cost recording system and used it to produce the first large high dimension multiclass insect wingbeat problem [2]. They also went on to establish a benchmark accuracy using a relatively simple Bayesian based approach, augmented with temporal information. A similar performance has been achieved via a combination of power spectral density features and Gaussian mixture models [12] and, via the convolutional neural network AlexNet, which was used to extract features from spectrograms before they were classified, via an support vector machine [20].

3 Datasets

We use two publicly available datasets for insect classification, summarised in Table 1. The first, InsectWingbeat, comes from the ongoing project at the University California Riverside (UCR) [2] and is part of the UCR TSC archive[1]. The second, MosquitoWingbeat, was produced during the development of a low cost insect sensor at TEII [14], and was recently used in a Kaggle competition[2].

In the case of both datasets, perspex boxes were used to confine flies of each class for recording. In the case of the InsectWingbeat dataset, the four mosquito species were also separated by sex. Both systems use a combination of infrared LEDs and photodiodes to record fly behaviour. Recordings are made as the wings and body occlude the signal from the LEDs during flight. The signal produced can be interpreted as audio and captures data similar to that of conventional audio recording devices [13].

The InsectWingbeat dataset contains 50,000 one second audio segments recorded with a sample rate of 16 kHz. There are ten equally distributed classes, comprised of four mosquito species (differentiated into male and female) from two genera and two other fly species from different genera (not differentiated by sex). These are Ae. aegypti, Cx. stigmatosoma, Cx. tarsalis, Cx. quinquefasciatus, Mu. domestica and Dr. simulans.

The MosquitoWingbeat dataset is comprised of six mosquito species from three genera. These are Ae. aegypti, Ae. albopictus, An. arabiensis, An. gambiae, Cu. pipiens, Cu.quinquefasciatus. There is no differentiation between sexes. It consists of 279,566 instances of 0.625 s of audio segments samples at 8 kHz. For the purpose of this paper, the number of instances per class has been reduced to 5000, reducing runtime and creating equal class distribution.

Table 1. Summary of attributes for datasets.

Dataset	Instances	Classes	Attributes	Sample rate	% of second
InsectWingbeat	50,000	10	16000	16 kHz	100
MosquitoWingbeat	30,000	6	5000	8 kHz	62.5

---

[1] http://www.timeseriesclassification.com.
[2] https://www.kaggle.com/potamitis/wingbeats.

4   Methods

The UCR approach, laid out briefly in Sect. 2, consists of a prepossessing step and a classification step. During the preprocessing step, instances are transformed into the frequency domain via the Fast Fourier Transform algorithm (FFT). Of the resulting 16,000 attribute output vector, indicies 100–2000 are kept and form the data used for classification. In the classification stage, a Nearest neighbour (k-NN) approach is used in conjunction with Euclidean distance. Through leave one out cross validation, k was set as 8.

A second approach, also outlined in Sect. 2 uses the well known neural network AlexNet to derive features that are subsequently classified via a support vector machine (SVM). In this case instances are transformed into spectrogram images prior to classification. The number of FFT bins used is 512 and windows overlapped by 50%.

In Sect. 5, we go on to publish results of approaches that have not been applied to the problem of insect classification. These include: Shapelet Transform (ST) [8] used with the C4.5 decision tree. In this approach, the dataset is transformed under a 48Hr contract, such that it is expressed in terms of intervals which are class discriminant. A C4.5 decision tree is then grown on the training data; Time Series Forest (TSF) [3], in which random intervals are selected and distilled into statistical features that are used to grow C4.5 decision trees; the Bag of SFA Symbols (BOSS) [17], in which instances are first split and compiled into a dictionary of words represented as histograms and classification takes place via a 1-NN used in conjunction with a bespoke distance measure; the contract Random Interval Spectral Ensemble (cRISE), in which random intervals are selected and transformed into spectral and autocorrelation coefficients. These new representations are then combined before being used to grow random decision trees. We also evaluate two ensembles. The first consists of cRISE contracted to 1 h of training, BOSS contracted to 1 h of training and TSF. In this approach (CAWPE) we use a scheme in which constituents are weighted by cross-validated accuracy estimates [5]. The second approach is The hierarchical vote collective of transformation-based ensembles for time series classification (HIVE-COTE). This consists of TSF, cRISE contracted to one hour, BOSS contracted to one hour and ST contracted to 48 h with C4.5.

Furthermore, we evaluate all approaches, other than cRISE which manages transformation internally, in combination with two preprocessing approaches as well as the raw data. The first approach, labelled T-1, sees instances resampled to 6000 Hz prior to transformation and the entire output is used. This reduction in sampling is motivated by evidence that these insects have little to capacity to produce frequencies exceeding 3000 Hz [14]. The second approach, labelled T-2, is the preprocessing step of UCR approach defined at the beginning of this section.

5   Results

In order to produce robust results from which to draw our conclusions, all experiments were subject to a stratified 10-fold cross validation. In the interest of producing reproducible results, all random functions used to produce data folds were seeded.

The rest of this section is organised as follows. In Sect. 5.1 we evaluate the accuracy achieved by benchmark classifiers using just the fundamental frequency attribute. In Sect. 5.2 we investigate the performance of approaches in conjunction with spectral features and in Sect. 5.3 we present and discuss all approaches in respect to timing.

All code used in these experiments is available from the UEA TSC repository[3] and raw results and analysis is available at[4].

5.1   Fundamental Frequency

The fundamental frequencies of the instances in both the MosquitoWingbeat and InsectWingbeat datasets were extracted using a peak finding algorithm in conjunction with the harmonic product spectrum technique. Table 2 displays results from experiments undertaken with these datasets. At just over 50% accuracy, the performance of this feature alone is in-line with results seen in literature evaluating similar datasets.

Table 2. Table showing mean accuracy, the Area under the receiver operator curve (AUROC) and Negative log likelihood (NLL) for 1 Nearest Neighbour with Euclidean distance and Naive Bayes approaches, evaluated over 10 folds on the fundamental frequency attribute of the MosquitoWingbeat (6 classes) and InsectWingbeat (10 classes) datasets.

Dataset	Classifier	Accuracy	AUROC	NLL
InsectWingbeat	ED	**0.56**	0.74	2.93
	NB	0.45	**0.83**	**2.49**
MosquitoWingbeat	ED	**0.56**	0.74	2.95
	NB	0.53	**0.78**	**1.89**

5.2   Spectral Approaches

Table 3 shows the results of cRISE, 8-NN, BOSS, TSF, ST and the ensembles of CAWPE and HIVE-COTE. The results are separated by dataset and ordered with respect to accuracy. All transform/classifier combinations are also published.

---

[3] https://github.com/TonyBagnall/uea-tsc.
[4] https://tinyurl.com/yxqgfl9e.

Table 3. Table showing mean accuracy, AUROC and NLL over 10 folds for ST+C4.5, TSF, cRISE, 8-NN, BOSS and CAWPE ensembles for T-1, T-2 and no spectral transformation.

Dataset	Classifier	Transform	Accuracy	AUROC	NLL
InsectWingbeat	HIVE-COTE	T-1	**0.7951**	**0.9794**	1.1384
	HIVE-COTE	T-2	0.7821	0.9780	1.1258
	TSF	T-1	0.7540	0.9751	0.9667
	TSF	T-2	0.7526	0.9748	**0.9316**
	CAWPE	T-1	0.7482	0.9731	1.2560
	CAWPE	T-2	0.7442	0.9728	1.2316
	cRISE	n/a	0.7172	0.9642	1.3791
	CAWPE	none	0.7138	0.9616	1.9251
	BOSS	T-2	0.6668	0.9496	1.2531
	8-NN	T-2	0.6626	0.9308	1.9543
	BOSS	T-1	0.6620	0.9474	1.2650
	8-NN	T-1	0.6556	0.9275	2.0613
	HIVE-COTE	none	0.6404	0.9616	1.6395
	ST+C4.5	T-2	0.6239	0.8095	2.3152
	ST+C4.5	T-1	0.6229	0.8076	2.3386
	ST+C4.5	none	0.5805	0.7854	2.6107
	BOSS	none	0.5751	0.8962	1.9126
	8-NN	none	0.5639	0.9009	2.7342
	TSF	none	0.4744	0.8647	2.3057
MosquitoWingbeat	HIVE-COTE	T-1	**0.8141**	**0.9680**	0.9874
	TSF	T-1	0.7956	0.9643	**0.8188**
	CAWPE	T-1	0.7881	0.9606	1.0271
	HIVE-COTE	T-2	0.7505	0.9487	1.1580
	TSF	T-2	0.7225	0.9407	1.0180
	CAWPE	T-2	0.7222	0.9395	1.2056
	CAWPE	none	0.7149	0.9327	1.6154
	cRISE	n/a	0.6808	0.9199	1.4349
	BOSS	T-1	0.6768	0.9193	1.2076
	BOSS	T-2	0.6445	0.9035	1.3160
	HIVE-COTE	none	0.6310	0.9287	1.4489
	ST+C4.5	T-1	0.5937	0.7755	2.5144
	ST+C4.5	none	0.5772	0.7808	2.6137
	ST+C4.5	T-2	0.5600	0.7626	2.6760
	TSF	none	0.5189	0.8363	1.8608
	BOSS	none	0.4632	0.7831	2.0347
	8-NN	T-2	0.3829	0.7257	4.3566
	8-NN	T-1	0.2840	0.6402	5.3337
	8-NN	none	0.2539	0.5885	6.1839

The results shown in Table 3 confirm our prior belief, *"that spectral features are most effective"*. This is most obvious when looking at the results of TSF in respect to InsectWingbeat. In this case, we see an increase of 28% in accuracy between spectral and non-spectral features. However, in all cases other than CAWPE and ST accuracy differs by at least 8%.

The effect T-1 and T-2 have on accuracy are confined to the MosquitoWingbeat dataset. Table 3 shows that for the MosquitoWingbeat dataset TSF differs in accuracy by 8%, HIVE-COTE by 6%, ST by 3% BOSS by 4% and CAWPE by 7%. Physical differences used to produce the MosquitoWingbeat dataset result in a larger target area. This results in insects being recorded for a greater duration and ultimately results in signals containing more low energy information, information which the T-2 approach discards.

Overall, HIVE-COTE is the most accurate. On the InsectWingbeat dataset the HIVE-COTE approach is 14% more accurate than the current state of the art approach, 8-NN+T-2. The most effective approach on the MosquitoWingbeat dataset, HIVE-COTE+T-1, is 16% more accurate than the 8-NN+T-2 combination and 9% more accurate than TSF+T-2. Significantly, these results omit powerful time-of-flight information, an attribute that is reported to have significantly increased the accuracy of the 8-NN+T-2 combination to 79.44% [2] on the InsectWingbeat dataset.

5.3 The Relevance of Test Time

The successful application of classification algorithms in real world scenarios also require them to be timely. It is commonly accepted that an algorithm is "real time" if it is able to classify an instance in less time than is represented in the data. Instances from the MosquitoWingbeat represent 620 ms and those from the InsectWingbeat represent 100 ms.

Figure 1 plots mean test time per instance averaged over folds for each approach. The timing data was generated during experiments run on the spectral datasets, the results of which were discussed in Sect. 5.2. Results of non-spectral experiments have been omitted in the interest of brevity.

In all cases TSF performs best and in a timely manner with respect to relative instance length. We note, it also exhibited very little variance across folds. In respect to timing, the UCR transformation approach performs best overall. This is most clear when comparing InsectWingbeat+T-1 and InsectWingbeat+T-2 in respect to TSF. Our observation indicates this is down to the reduced number of instances attributes present for classification.

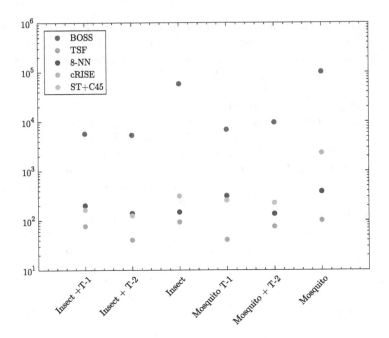

Fig. 1. Figure showing mean test time per instance for all combinations of ST+C4.5, TSF, cRISE, 8-NN, BOSS, CAWPE ensembles, with no spectral transformation, T-1 and T-2 transformations.

6   Conclusion

In conclusion, we have shown that the combination of simple audio features and HIVE-COTE outperforms all approaches evaluated in this paper on both the MosquiotWingbeat and InsectWingbeat datasets. Whilst omitting powerful time of flight information HIVE-COTE in conjunction with spectral features is shown to be 14% more accurate than the benchmark approach on the InsectWingbeat dataset and 16% more accurate on the new MosquitoWingbeat dataset. However, as the slowest constituent of HIVE-COTE, BOSS, does not perform in a timely manner we conclude that even if threaded HIVE-COTE is not timely and therefore would need a considerably faster processor to meet the requirements of an application setting.

We conclude that intra class variance in fundamental frequency prevents its use as a discriminant between species. However, in a real world setting this feature is likely to play a key role in determining candidate intervals in an application setting. Our view is that an appropriate algorithm architecture would consist of layers designed to minimise power consumption, by preventing unnecessary computations. In this context, fundamental frequency could prove an adequate method of solving both *insect|noninsect* and *fly|nonfly* decisions.

Acknowledgements. This work is supported by the Biotechnology and Biological Sciences Research Council [grant number BB/M011216/1], and the UK Engineering and Physical Sciences Research Council (EPSRC) [grant number EP/M015807/1]. The experiments were carried out on the High Performance Computing Cluster supported by the Research and Specialist Computing Support service at the University of East Anglia and using a Titan X Pascal donated by the NVIDIA Corporation.

References

1. Chadwick, L.E.: A simple stroboscopic method for the study of insect flight. Psyche **46**(1), 1–8 (1939)
2. Chen, Y., Why, A., Batista, G., Mafra-Neto, A., Keogh, E.: Flying insect classification with inexpensive sensors. J. Insect Behav. **27**(5), 657–677 (2014)
3. Deng, H., Runger, G., Tuv, E., Vladimir, M.: A time series forest for classification and feature extraction. Inf. Sci. **239**, 142–153 (2013)
4. Greven, H., Kaya, M., Junker, K., Akyuz, L., Amemiya, C.T.: Characterization of tongue worm (pentastomida) chitin supports α-rather than β-chitin. Zoologischer Anzeiger **279**, 111–115 (2019)
5. Large, J., Lines, J., Bagnall, A.: A probabilistic classifier ensemble weighting scheme based on cross-validated accuracy estimates. Data Min. Knowl. Discov., 1–36 (2019)
6. Li, Z., Zhou, Z., Shen, Z., Yao, Q.: Automated identification of mosquito (diptera: culicidae) wingbeat waveform by artificial neural network. In: Li, D., Wang, B. (eds.) AIAI 2005. ITIFIP, vol. 187, pp. 483–489. Springer, Boston, MA (2005). https://doi.org/10.1007/0-387-29295-0_52
7. Lines, J., Taylor, S., Bagnall, A.: Time series classification with HIVE-COTE: the hierarchical vote collective of transformation-based ensembles. ACM Trans. Knowl. Discov. Data **12**(5), 1041–1046 (2018)
8. Lines, J., Davis, L.M., Hills, J., Bagnall, A.: A shapelet transform for time series classification. In: Proceedings of the 18th ACM SIGKDD International Conference on Knowledge Discovery and Data Mining, pp. 289–297. ACM (2012)
9. Moore, A.: Development of a data acquisition system for long-term outdoor recording of insect flight activity using a photosensor. In: 13th Conference on Aerobiology and Biometeorology, American Meteorological Society, Albuquerque, New Mexico (1998)
10. Moore, A., Miller, J.R., Tabashnik, B.E., Gage, S.H.: Automated identification of flying insects by analysis of wingbeat frequencies. J. Econ. Entomol. **79**(6), 1703–1706 (1986)
11. Moore, A., Miller, R.H.: Automated identification of optically sensed aphid (homoptera: Aphidae) wingbeat waveforms. Ann. Entomol. Soc. Am. **95**(1), 1–8 (2002)
12. Potamitis, I.: Classifying insects on the fly. Ecol. Inform. **21**, 40–49 (2014)
13. Potamitis, I., Rigakis, I.: Novel noise-robust optoacoustic sensors to identify insects through wingbeats. IEEE Sens. J. **15**(8), 4621–4631 (2015)
14. Potamitis, I., Rigakis, I.: Large aperture optoelectronic devices to record and time-stamp insects' wingbeats. IEEE Sens. J. **16**(15), 6053–6061 (2016)
15. Reed, S., Williams, C., Chadwick, L.: Frequency of wing-beat as a character for separating species races and geographic varieties of drosophila. Genetics **27**(3), 349 (1942)

16. Sarpola, M., et al.: An aquatic insect imaging system to automate insect classification. Trans. ASABE **51**(6), 2217–2225 (2008)
17. Schäfer, P.: Bag-of-SFA-Symbols in Vector Space (BOSS VS) (2015)
18. Unwin, D., Ellington, C.: An optical tachometer for measurement of the wing-beat frequency of free-flying insects. J. Exp. Biol. **82**(1), 377–378 (1979)
19. Wen, C., Guyer, D.: Image-based orchard insect automated identification and classification method. Comput. Electron. Agric. **89**, 110–115 (2012)
20. Zhang, C., Wang, P., Guo, H., Fan, G., Chen, K., Kämäräinen, J.K.: Turning wingbeat sounds into spectrum images for acoustic insect classification. Electron. Lett. **53**(25), 1674–1676 (2017)

Studying the Evolution of the *Circular Economy* Concept Using Topic Modelling

Sampriti Mahanty[1](✉) ⓘ, Frank Boons[1] ⓘ, Julia Handl[1] ⓘ,
and Riza Batista-Navarro[2] ⓘ

[1] Alliance Manchester Business School, The University of Manchester,
Manchester, UK
sampriti.mahanty@postgrad.manchester.ac.uk,
{frank.boons,julia.handl}@manchester.ac.uk
[2] School of Computer Science, The University of Manchester, Manchester, UK
riza.batista@manchester.ac.uk

Abstract. Circular Economy has gained immense popularity for its perceived capacity to operationalise sustainable development. However, a comprehensive long-term understanding of the concept, characterising its evolution in academic literature, has not yet been provided. As a first step, we apply unsupervised topic models on academic articles to identify patterns in concept evolution. We generate topics using LDA, and investigate topic prevalence over time. We determine the optimal number of topics for the model (k) through coherence scorings and evaluate the topic model results by expert judgement. Specifying k as 20, we find topics in the literature focussing on resources, business models, process modelling, conceptual research and policies. We identify a shift in the research focus of contemporary literature, moving away from the Chinese predominance to a European perspective, along with a shift towards micro level interventions, e.g., circular design, business models, around 2014–2015.

Keywords: Circular economy · Topic modelling · Concept evolution

1 Introduction

In the last 15 years or so the concept of Circular Economy (CE) has gained immense traction amongst academics, practitioners and policy-makers for its perceived capacity to operationalize Sustainable Development [14, 15, 25]. CE is defined as "an economic system that replaces the 'end-of-life' concept with reducing, reusing, recycling and recovering materials in production/distribution and consumption processes" [25]. However, the concept of CE is not new [3] and its theoretical underpinnings stem from a variety of fields such as cleaner production, industrial ecology, and environmental science [27]. Since its inception the definition of CE has been extended to and evolved across a broad spectrum of cross-disciplinary subjects such as industrial ecology, regenerative design, cradle to cradle, performance economy [30]. Given that the

Supported by The University of Manchester.

H. Yin et al. (Eds.): IDEAL 2019, LNCS 11872, pp. 259–270, 2019.
https://doi.org/10.1007/978-3-030-33617-2_27

concept of CE is "not new" [3] and that it "has evolved" [4, 14] it is of interest to understand what is the way by which the concept of CE evolved over the years, emerging from its antecedent concepts to becoming a popular concept in its own right. To enable a careful empirical study, we build on the work of Jiao and colleagues [23] and distinguish the labels of "circular economy" from the textual elements that become attached (or detached) to it over time, which are again distinct from the empirical practices that are referred to with the label of CE.

Although the origins of the concept date back to the 1970s, it is only in the last 15 years that it has been referred to directly and independently in academic literature, distancing itself from its antecedent concepts (e.g., industrial ecology) [34]. CE has been discussed extensively in academic discourse, which is evident in the increasing number of publications. However, its development has not been adequately understood from a temporal perspective characterising the pattern of its evolution using data driven methods. There have been studies in the past addressing CE from a temporal perspective such as the work of Blomsma and Brennan [3] where they adopt a narrative approach in explaining the emergence of CE in prolonging resource productivity. While such qualitative work is valuable, with the vast and rapidly growing literature on CE and limited cognitive capacity of humans [37] it is a difficult task to address all aspects of the CE concept. We thus use quantitative data driven methodologies to complement expert knowledge in the field by providing a holistic picture of the literature, unravel new findings and aid the formulation of new research questions.

As grounding for this work we study the development of CE by conceptualising it as a *process*, drawing inspiration from the work of Boons and colleagues [6, 7]. We consider the development of CE as an *evolutionary process*, adopting the view that there is an analogy between knowledge gain and biological evolution [8]. According to this view, in the course of evolution, species become more adaptive to their natural environment by undergoing natural selection. Likewise, scientific progress is a result of selection mechanisms at the individual level (i.e., scientists) and at the group level (i.e., scientific communities). Scientists perform studies and produce new knowledge, a large part of which is academic literature. They select scientific communities to join or form based on cognitive, social and philosophical grounds [22].

In this paper, we analyse the evolutionary pattern of the CE concept in academic literature over a period of time noting periods of development and major structural changes. Specifically, we make use of CE academic literature as the basis for carrying out such analysis, employing an unsupervised machine learning method, i.e., topic modelling based on Latent Dirichlet Allocation (LDA) [2]. This method generates topics from the CE literature where the topics constitute sets of textual elements that get attached (or detached) to the label of 'Circular Economy' over time enabling our empirical study.

2   Background

We draw inspiration for this paper from the work of Boons and colleagues [6, 7] who addressed process-oriented research questions in studying industrial symbiosis. They put forth the defining feature of process-oriented studies as viewing reality as a stream of events rather than a stable entity. In our study, we employ the backward approach proposed by Boons and colleagues [6] whereby the final outcome of interest is already known and processes leading to it are being uncovered. In our case the popularity of the CE concept in academic discourse addressing questions pertaining to sustainability is the outcome of interest and our aim in this paper is to address: *"What is the way by which the concept of CE evolved in academic literature?"*

In order to define a process we need to define the central subject in the process [38]. A central subject in the process can be any kind of entity such as an individual actor, a group of actors, a lineage, a social movement or a machine [21]. The central subject in our process study is the concept of CE. A concept is defined as a mental representation of cognitive agents, which are crucial to psychological processes such as categorization, inference, memory, learning and decision-making [16, 28]. In order to define these concepts, members of the scientific community use the same language i.e., lexicon kind terms at the very least [26]. Thus, the concept of CE will be associated with certain terms. The evolution of the CE concept will be studied by way of understanding the evolution of its associated terms. This brings us to our approach of using LDA topic modelling where the central entity i.e., the label 'Circular Economy' is fixed and we assess the evolution by the associated terms (generated from the topic model) with the central entity.

LDA has been used as a methodology extensively to assess topics and its evolutionary pattern in the computer science domain such as evolution of topics in the Advanced Computational Linguistics (ACL) articles [19], software development [20], academic articles in the CiteSeer repository [5], source code histories [42, 43]. However, in the area of sustainability research this methodology has not yet been extensively implemented. Shin et al., 2018 [37] in their study to summarise sustainability literature in maritime studies have used LDA topic modelling however they do not assess the evolution of topics. Our application of this methodology on articles related to CE will enable summarising and assessing evolution of the topics over time which has not yet been applied in this field.

3   Methodology

Our approach to understanding the evolution of CE is depicted in Fig. 1. In this section, we describe in detail the steps involved, i.e., (a) Data collection and pre-processing, (b) Topic modelling (c) Topic coherence scoring (d) Expert judgement.

3.1   Data Collection and Pre-processing

The analysis in this paper is based on academic articles published on CE. Data was collected using the Scopus[1] database from 2005 to May 2019. We chose Scopus as it is considered to be one of the largest abstracts and citation databases of peer-reviewed literature, including scientific journals, books and conference proceedings [33].

---

[1] https://www.scopus.com/search/form.uri?display=basic.

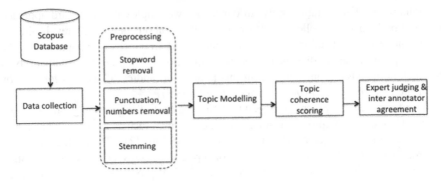

Fig. 1. Diagram depicting our proposed approach

The keyword "circular economy" was used to arrive at selected articles. While there are other terms, which have similar conceptualizations as circular economy and often used interchangeably [13] we limit our search criterion because we are interested in the use of the exact verbiage. We have used abstracts as a proxy for full articles, since abstracts are a compact representation of the whole article normally containing enough key words about research themes [17, 41]. We obtain 3437 results from our search criterion out of which we consider 3300 and omit 137 due to unavailability of abstracts.

The data (3300 abstracts) are read in a CSV format into the 'tm' [12] package in R and standard text mining pre-processing steps are applied over the entire corpus. These steps are as follows:

(1) Lowercasing the corpus, which prevents a word with different capitalisation from being mistaken as two different words.
(2) Removal of standard stopwords such as "a", "an", "and", "the". Such words do not contain information that is required in the analysis and impair the accuracy of the results [37].
(3) Stemming of terms is conducted to remove pluralisation or other suffixes and to normalise different variations of the same word. This technique is often applied in text mining, in order to reduce similar words to a unique term (e.g., "predicting" and "predictive" are transformed into "predict") [24]
(4) Additional pre-processing steps involve removing numbers, punctuation characters, and white spaces to avoid any impairment of the topic model results.

3.2 Topic Modelling

Our analysis is based on Latent Dirichlet Allocation (LDA), [2] a generative probabilistic model of a corpus that represents documents as random mixtures over latent topics where each topic is a distribution of words. Our intent in this paper is to generate topics and then investigate topic prevalence over time. One could argue that LDA does not explicitly model temporal patterns in a text corpus and there are other models, which have been developed to consider the time dimension such as **Dynamic Topic Model** [1] and **Topics over Time (TOT)** [44]. **DTM** represents time as a discrete

Markov process, where topics themselves evolve according to a Gaussian distribution. This model, however, penalizes abrupt changes between successive time periods, discouraging fluctuation in topics over time [11]. Given the nature of our data, which has abrupt fluctuations in terms of sudden rise in the production of articles, we anticipate that there will be fluctuations in topic proportions in successive time periods. Hence, we do not prefer using this model for our analysis. **TOT** represents time as a continuous beta distribution, solving the issue in DTMs. However, the beta distribution is still inflexible since it assumes that evolution of topics will have only a single point of rise and a single point of fall in the entire corpus [11]. This means that the model will not accommodate a situation where a topic has a period of rise followed by a fall and then a subsequent rise.

Hence, we choose LDA, which is a simple and intuitive model, and conduct some simple calculations after generating the topics and use it to assess topic evolution. We use the 'lda [9]' and 'topic models' [18] package in R to generate the topics. We apply LDA over the entire corpus together and then calculate the probability distribution across topics for each document. The topic probabilities for each topic are then summed over each year based on the time stamp associated with each document and visualized graphically in a stacked plot to assess the trend in the topics over the years.

3.3 Automatic Topic Coherence Scorings

For parameterised models such as LDA, the number of topics (k) is a predetermined criterion. There is emphasis on the selection of k since it impacts the interpretability of the topic model, a lower value can divide the corpus into generic semantic contexts while a higher value can generate overlapping or un-interpretable topics [45]. The determination of k can be based on scientific evidence or human judgement [37]. In this section we discuss our approach to determining the optimal value of k.

Topic coherence measures score a single topic by measuring the degree of semantic similarity between high scoring words in the topic thereby distinguishing between semantically interpretable topics and topics that are artefacts of statistical inference [39]. The two coherence measures, which are designed for LDA, matching well with human judgements of topic quality are: (a) the UCI measures [32], which are calculated over an external corpus such as Wikipedia. This metric is an external comparison to known semantic valuations (b) the UMass measure [31], which is an intrinsic score that computes coherence scores over the original corpus that has been used to train the models.

Rosner and colleagues [36] in their study compare the correlation of various coherence measures with human judgment and the UCI metric is shown to be significantly outperforming the UMass metric. Although any-any and one-any metrics slightly outperform the UCI metric but they pose challenges like exponential running time and difficulty in practical application. Hence, we chose the UCI metric to calculate coherence scores and determine the quality of topics.

3.4 Evaluation

It is difficult to judge if the results generated from the topic modelling are entirely meaningful given it's an unsupervised method and there exists no gold standard list of topics for comparison. Whilst we tend to follow scientific evidence in this process, there have been studies showing automatically generated higher coherence scores could also mean lower topic interpretability in some cases [10]. The gold standard for coherence evaluation is human-produced topic rankings [35]. This line of thought leads us to combining automatically generated coherence scores with expert opinion in evaluating the performance of our topic model. We thus conducted a workshop with five participants who are active researchers in different application areas of CE. We consider their expert judgment in validating the topic model results. The workshop aimed to address two questions specifically: (a) Is there semantic coherence in the topics generated? (b) Are all aspects of the CE literature covered in the topic model results?

To further validate the results and the human interpretability of topics we also calculated agreement between our approach and the experts in identifying the topics, based on F-measure (also known as F-score). In calculating the F-score the topic annotations by experts were treated as gold standard and those from the authors as response.

4   Results

In this section, we present the results we obtained upon conducting the evaluation strategy that we discussed in the previous section.

4.1   Optimal Number of Topics from Automatic Coherence Scoring

We compute UCI coherence scores for topics (k) ranging from 1 to 30 and plot it on a graph (Fig. 2). The coherence score is highest until k = 4 before declining. The coherence scores fluctuate until k = 18–21 where it stagnates before dropping again. Coherence scores at k = 18–21 plateau with the coherence scores at k = 7–8. Our rationale for not selecting k as 4, which has the highest coherence value, is that it poses a risk of generating few generic topics [43] and not capturing the wide range of application areas of the CE literature. Along similar lines we do not select k as 7 or 8 and rather select k as 20 which almost has the same coherence score as k = 7–8 and will possibly help in generating more meaningful topics. We determine the optimal number of topics to be 20 based on coherence scoring.

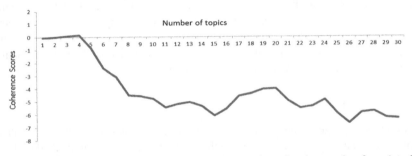

Fig. 2. UCI coherence scores corresponding to the number of topics ranging from 1 to 30

4.2   Expert Judgement

In this section we summarise the results from the workshop to solicit expert judgment on the topic model results. Expert opinion was sought on the following grounds.

(1) *Is there semantic coherence in the topics generated by the model?*
     There was no major disagreement in terms of annotating the topics and the experts reached consensus without much difficulty. There were two topics, which triggered questions.

    (a) *Water, urban, citi, region, land, area, system, environ, resourc, ecology*
          This topic was identified as relating to studies on cities and industrial ecology but the experts believed this could be clubbed with two other topics relating to Industrial ecology/symbiosis rather than being a separate topic.

    (b) *Suppli, chain, compani, studi, barrier, practic, firm, consum, behavior*
          This topic according to the experts could best be represented as two separate topics pertaining to circular supply chain and behavioural studies.

(2) *Are all aspects of the CE literature covered in the topic model results?*
     The experts opined two aspects not being covered in the results. Firstly the aspect of "Digitalisation", in transitioning towards a CE. Secondly articles in the social sciences tend to use "circularity" in verbiage rather than CE although they discuss similar conceptualization. This is a limitation of the study since we are only focusing on the exact usage of the CE terminology and our search criterion of the articles is based on the same.

Overall we can conclude that the topics generated by the model were semantically coherent and could, to a large extent, be matched with human judgment. The experts reached a consensus in assigning topic labels to each of the topics and further grouped the 20 topics into 7 thematic categories to establish broad themes of CE research.

4.3   Performance of Topic Models Against Experts

We obtained an F score of 0.921; which is quite high and is a revalidation of the human interpretability of the topics.

4.4   Topic Model Results

In this section we present a summary of the topic model results. Table 1 shows the ordered list of the terms in each topic, the topic names assigned to each based on the expert judgement, the proportion of the topics in the entire corpus and the broad thematic category i.e., resources, design for circularity, industrial ecology/symbiosis, process modelling, firms, businesses and consumers, meta/conceptual research, policy and governance identified by the experts. Industrial ecology/symbiosis is the topic with the maximum proportion, which is unsurprising since industrial ecology is one of the primary antecedents of the CE concept [30] followed by business models which is rapidly gaining popularity in contemporary CE literature. Topics pertaining to Plastics, simulation based methodologies have the lowest proportion in the entire corpus.

Table 1. Summary of topics

Top 10 terms for each topic	Topic Name	Topic proportion	Thematic category
Wast, manag, recycl, landfill, collect, recoveri, solid, municip, generat, dispos	Waste management	5.03%	Resources
Watewat, treatment, metal, recoveri, sludg, extract, acid, concentr, remov, water	Sludge	4.86%	
Materi, recycl, flow, resourc, metal, raw, mine, product, stock, secondari	Metals	4.78%	
Food, product, digest, agricultur, nutrient, soil, farm, organ, biomass, wast	Food/Agriculture	4.69%	
Energi, plant, fuel, renew, electr, power, gas, technolog, carbon, bioga	Enery	3.91%	
Energ, emiss, consumpt, steel, carbon, china, reduc, reduct, industri, resource	Steel	3.78%	
Ash, properti, cement, materi, concret, test, slag, composit, result, raw	Cement/concrete	3.70%	
Recycl, plastic, packag, materi, wast, chemic, weee, textil, polym, product	Plastics	3.47%	
Product, remanufactur, design, reus, manufactr, compon, consum, diassembl, eol, electron	Design for circularity	4.34%	Design for circularity
Innov, design, project, build, sustain, educ, solut, engin, develop, new	Innovation in design	4.13%	
Develop, economi, circular, china, coal, resourc, enterpris, ecology, industri, mode	Industrial ecology/symbiosis (local)	9.07%	Industrial ecology/symbiosis
Water, urban, citi, region, land, area, system, environ, resourc, ecolog	Urban systems	3.89%	
Industri, park, symbiosi, eco, ecolog, china, chain, develop, chemic, eip	Industrial ecology/symbiosis (regional+national)	3.67%	
Evalu, system, indic, method, effici, circular, economi, index, model, eco	Evaluating CE	5.72%	Process modelling
Environment, impact, life, assess, cycl, product, lca, scenario, result, use	Lifecycle Assessment	5.13%	
System, model, research, design, network, base, optim, data, analysi, simul	Models and simulation based methodologies	3.51%	
Model, busi, circular, economu, valu, system, product, sustain, transit, resourc	Business Models	8.24%	Firms, businesses, consumers
Suppli, chain, compani, studi, barrier, practic, find, firm, consum, behavior	Supply chain + behavioural studies	5.28%	
Research, concept, review, literatur, practic, econom, sustain, framework, discuss	Meta/conceptual research	7.11%	Policy and governance
Polici, govern, european, public, social, regul, countri, econom, develop, environment	European policy and governance	5.69%	

Figure 3 shows the evolution of the topics over time. A significant structural change in the proportion of topics is witnessed in the year 2015. This structural change can be understood from two perspectives. Firstly the shift in focus from macro-level topics such as industrial ecology to micro-level interventions such as circular product design, business models [34]. Secondly, a shift from the Chinese pre-dominance to a more European context. Literature on CE has evolved along with the policy and focus of these economies i.e., 11th, 12th five year plans and CE promotion law in China [40] and advocacy by organisations like the Ellen McArthur Foundation and enactment of the Circular Economy Package in Europe [29].

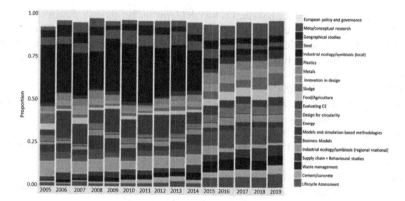

Fig. 3. Topic proportions over time

5 Conclusion and Future Research

Our analyses present a way to summarise the growing literature on the concept of CE from a temporal perspective using unsupervised topic models. We determine the optimal number of topics based on coherence scoring and then revalidate this through human judgement and find the results from the automated coherence scoring and human judgement to be highly correlated. We apply LDA to assess the evolution of topics over time by summing topic proportions across each document for a particular time stamp. This is a simple and intuitive approach and can be easily applied to different kinds of text corpus. Our analysis and visual representations show structural change in the CE literature around the year 2014–2015 shifting towards a European context and micro-level interventions such as circular product design, business models and supply chain from a Chinese predominance and macro level themes.

Whilst this work provides a preliminary understanding of characterizing the evolution of the concept we intend to use these results to further our research in answering how and why the concept of CE evolved in academic discourse.

Acknowledgements. The authors are grateful to Helen Holmes, Wouter Spekkink, Maria Sharmina, Malte Roedl, and Carly Fletcher for serving as the experts to evaluate the topic model results and providing their valuable feedback.

Sampriti Mahanty acknowledges the support from Alliance Manchester Business School.

References

1. Blei, D.M., Lafferty, J.D.: Dynamic topic models. In: Proceedings of the 23rd International Conference on Machine learning, pp. 113–120. ACM, 2006 June
2. Blei, D.M., Ng, A.Y., Jordan, M.I.: Latent Dirichlet allocation. J. Mach. Learn. Res. **3**, 993–1022 (2003)
3. Blomsma, F., Brennan, G.: The emergence of circular economy: a new framing around prolonging resource productivity. J. Ind. Ecol. **21**(3), 603–614 (2017)

4. Bocken, N.M., De Pauw, I., Bakker, C., van der Grinten, B.: Product design and business model strategies for a circular economy. J. Ind. Prod. Eng. **33**(5), 308–320 (2016)

5. Bolelli, L., Ertekin, Ş., Giles, C.L.: Topic and trend detection in text collections using Latent Dirichlet Allocation. In: Boughanem, M., Berrut, C., Mothe, J., Soule-Dupuy, C. (eds.) ECIR 2009. LNCS, vol. 5478, pp. 776–780. Springer, Heidelberg (2009). https://doi.org/10.1007/978-3-642-00958-7_84

6. Boons, F., Spekkink, W., Jiao, W.: A process perspective on industrial symbiosis: theory, methodology, and application. J. Ind. Ecol. **18**(3), 341–355 (2014)

7. Boons, F., Spekkink, W., Mouzakitis, Y.: The dynamics of industrial symbiosis: a proposal for a conceptual framework based upon a comprehensive literature review. J. Clean. Prod. **19** (9–10), 905–911 (2011)

8. Bradie, M.: Assessing evolutionary epistemology. Biol. Philos. **1**(4), 401–459 (1986)

9. Chang, J.: lda: Collapsed Gibbs Sampling Methods for Topic Models. R package version 1.2.3 (2010). http://CRAN.R-project.org/package=lda

10. Chang, J., Gerrish, S., Wang, C., Boyd-Graber, J. L., Blei, D.M.: Reading tea leaves: how humans interpret topic models. In: Advances in Neural Information Processing Systems, pp. 288–296 (2009)

11. Chen, T.H., Thomas, S.W., Hassan, A.E.: A survey on the use of topic models when mining software repositories. Empir. Softw. Eng. **21**(5), 1843–1919 (2016)

12. Feinerer, I.: Introduction to the tm Package. Text Mining in R (2015). ftp://videolan.cs.pu.edu.tw/network/CRAN/web/packages/tm/vignettes/tm.pdf

13. Geisendorf, S., Pietrulla, F.: The circular economy and circular economic concepts—a literature analysis and redefinition. Thunderbird Int. Bus. Rev. **60**(5), 771–782 (2018)

14. Geissdoerfer, M., Savaget, P., Bocken, N.M., Hultink, E.J.: The circular economy–a new sustainability paradigm? J. Clean. Prod. **143**, 757–768 (2017)

15. Ghisellini, P., Cialani, C., Ulgiati, S.: A review on circular economy: the expected transition to a balanced interplay of environmental and economic systems. J. Clean. Prod. **114**, 11–32 (2016)

16. Goertz, G.: Social Science Concepts: A User's Guide. Princeton University Press, Princeton (2006)

17. Griffiths, T.L., Steyvers, M.: Finding scientific topics. Proc. Natl. Acad. Sci. **101**(Suppl. 1), 5228–5235 (2004)

18. Grun, B., Hornik, K.: Topicmodels: an R package for fitting topic models. J. Stat. Softw. **40** (13), 1–30 (2011). https://www.jstatsoft.org/v040/i13, https://doi.org/10.18637/jss.v040.i13. ISSN 1548-7660

19. Hall, D., Jurafsky, D., Manning, C.D.: Studying the history of ideas using topic models. In: Proceedings of the Conference on Empirical Methods in Natural Language Processing, pp. 363–371. Association for Computational Linguistics, 2008 October

20. Hindle, A., Godfrey, M.W., Holt, R.C.: What's hot and what's not: windowed developer topic analysis. In: 2009 IEEE International Conference on Software Maintenance, pp. 339–348. IEEE, 2009 September

21. Hull, D.L.: Central subjects and historical narratives. Hist. Theory **14**(3), 253–274 (1975)

22. Hull, D.L.: Science as a Process: An Evolutionary Account of the Social and Conceptual Development of Science. University of Chicago Press, Chicago (2010)

23. Jiao, W., Boons, F.: Policy durability of Circular Economy in China: a process analysis of policy translation. Resour. Conserv. Recycl. **117**, 12–24 (2017)

24. Kao, A., Poteet, S.R. (eds.): Natural language processing and text mining. Springer, London (2007). https://doi.org/10.1007/978-1-84628-754-1

25. Kirchherr, J., Reike, D., Hekkert, M.: Conceptualizing the circular economy: an analysis of 114 definitions. Resour. Conserv. Recycl. **127**, 221–232 (2017)

26. Kuhn, T.S.: The road since structure. In: PSA: Proceedings of the Biennial Meeting of the Philosophy of Science Association, vol. 1990, no. 2, pp. 3–13. Philosophy of Science Association, 1990 January

27. Lazarevic, D., Valve, H.: Narrating expectations for the circular economy: towards a common and contested European transition. Energy Res. Soc. Sci. **31**, 60–69 (2017)

28. Margolis, E., Laurence, S.: Concepts. In: Zalta, E.N. (ed.) The Stanford Encyclopedia of Philosophy. https://plato.stanford.edu/archives/sum2019/entries/concepts/. (Summer 2019 Edition)

29. Masi, D., Kumar, V., Garza-Reyes, J.A., Godsell, J.: Towards a more circular economy: exploring the awareness, practices, and barriers from a focal firm perspective. Prod. Plan. Control. **29**(6), 539–550 (2018)

30. Merli, R., Preziosi, M., Acampora, A.: How do scholars approach the circular economy? A systematic literature review. J. Clean. Prod. **178**, 703–722 (2018)

31. Mimno, D., Wallach, H.M., Talley, E., Leenders, M., McCallum, A.: Optimizing semantic coherence in topic models. In: Proceedings of the Conference on Empirical Methods in Natural Language Processing, pp. 262–272. Association for Computational Linguistics, 2011 July

32. Newman, D., Noh, Y., Talley, E., Karimi, S., Baldwin, T.: Evaluating topic models for digital libraries. In: Proceedings of the 10th Annual Joint Conference on Digital Libraries, pp. 215–224. ACM, 2010 June

33. Nobre, G.C., Tavares, E.: Scientific literature analysis on big data and internet of things applications on circular economy: a bibliometric study. Scientometrics **111**(1), 463–492 (2017)

34. Prendeville, S., Cherim, E., Bocken, N.: Circular cities: mapping six cities in transition. Environ. Innov. Soc. Transit. **26**, 171–194 (2018)

35. Röder, M., Both, A., Hinneburg, A.: Exploring the space of topic coherence measures. In: Proceedings of the Eighth ACM International Conference on Web Search and Data Mining, pp. 399–408. ACM, 2015 February

36. Rosner, F., Hinneburg, A., Röder, M., Nettling, M., Both, A.: Evaluating topic coherence measures. arXiv preprint arXiv:1403.6397 (2014)

37. Shin, S.H., Kwon, O., Ruan, X., Chhetri, P., Lee, P., Shahparvari, S.: Analyzing sustainability literature in maritime studies with text mining. Sustainability **10**(10), 3522 (2018)

38. Spekkink, W.: Institutional capacity building for industrial symbiosis in the Canal Zone of Zeeland in the Netherlands: a process analysis. J. Clean. Prod. **52**, 342–355 (2013)

39. Stevens, K., Kegelmeyer, P., Andrzejewski, D., Buttler, D.: Exploring topic coherence over many models and many topics. In: Proceedings of the 2012 Joint Conference on Empirical Methods in Natural Language Processing and Computational Natural Language Learning, pp. 952–961. Association for Computational Linguistics, 2012 July

40. Su, B., Heshmati, A., Geng, Y., Yu, X.: A review of the circular economy in China: moving from rhetoric to implementation. J. Clean. Prod. **42**, 215–227 (2013)

41. Sun, L., Yin, Y.: Discovering themes and trends in transportation research using topic modeling. Transp. Res. Part C Emerg. Technol. **77**, 49–66 (2017)

42. Thomas, S.W., Adams, B., Hassan, A.E., Blostein, D.: Modeling the evolution of topics in source code histories. In: Proceedings of the 8th Working Conference on Mining Software Repositories, pp. 173–182. ACM, 2011 May

43. Thomas, S.W., Adams, B., Hassan, A.E., Blostein, D.: Studying software evolution using topic models. Sci. Comput. Program. **80**, 457–479 (2014)

44. Wang, X., McCallum, A.: Topics over time: a non-Markov continuous-time model of topical trends. In: Proceedings of the 12th ACM SIGKDD International Conference on Knowledge Discovery and Data Mining, pp. 424–433. ACM, 2006 August
45. Zhao, W., et al.: A heuristic approach to determine an appropriate number of topics in topic modeling. In: BMC Bioinformatics, vol. 16, no. 13, p. S8. BioMed Central, 2015 December

Mining Frequent Distributions
in Time Series

José Carlos Coutinho[1,2]([✉]), João Mendes Moreira[2], and Cláudio Rebelo de Sá[1]

[1] University of Twente, Enschede, The Netherlands
c.f.pinhorebelodesa@utwente.nl
[2] University of Porto, Porto, Portugal
{up201404293,jmoreira}@fe.up.pt

Abstract. Time series data is composed of observations of one or more variables along a time period. By analyzing the variability of the variables we can reveal patterns that repeat or that are correlated, which helps to understand the behaviour of the variables over time. Our method finds frequent distributions of a target variable in time series data and discovers relationships between frequent distributions in consecutive time intervals. The frequent distributions are found using a new method, and relationships between them are found using association rules mining.

1 Introduction

Time series data is constituted by a set of observations, each of them recorded at a specific time [5]. It can be defined by one or more variables that are measured in each observation. Keeping track of the variable's values along the time allows to study their variability and possibly obtain patterns that repeat or that are correlated.

This paper addresses the problem of finding patterns in time series data, in the form of distributions, and discovering the relationships between them in consecutive time intervals. After dividing the dataset in equal-sized time windows, it measures the distance between the target distributions using the Kolmogorov-Smirnov (KS) distance. When the distance between two distributions is below a threshold, we consider these distributions to be the same pattern. Whenever a distribution is not similar to any previously discovered pattern, we consider it a new pattern. In the end, we only keep the patterns (distributions) that are frequent according to a minimum support defined by the user.

Finally, we use association rules mining to find relationships between patterns on consecutive time windows. After defining a minimum support and confidence for the rules, the method will pair the time windows that are consecutive, labeling the pattern in the previous window as *antecedent* and the pattern in the next window as *consequent*. The method obtains the rules between antecedents and consequents which have values of support and confidence above the minimum.

After running the method, we could obtain distinct patterns of players' speed in football data and of price in electricity and AWS data. In the three cases, the method could find interesting association rules between the patterns found.

H. Yin et al. (Eds.): IDEAL 2019, LNCS 11872, pp. 271–279, 2019.
https://doi.org/10.1007/978-3-030-33617-2_28

2   Background and Related Work

Kolmogorov-Smirnov Statistical Test (KS test) is a statistical test which measures the equality of one-dimensional probability distributions. The KS statistic can be used to quantify the distance between two empirical distribution functions. Jorge *et al.* [7] also used this statistical test in their Distribution Rules method to measure the distance between a distribution of a target variable and a reference distribution.

Association Rules Mining is an area of Data Mining which studies ways of finding relationships between items in a dataset [1]. The relationships come in the form of implications $X \implies Y$, which are referred as *rules*. X and Y are itemsets, where X is the *antecedent* and Y the *consequent*. To measure the relevance of the rules, we check if the itemset $X \cup Y$ is observed frequently enough (*support*) and if the rule is verified frequently enough (*confidence*).

EP-MEANS proposed Henderson *et al.* [6], can be used to find patterns in the variability of one variable. This method clusters probability distributions regarding a target attribute. It is based in the K-means clustering algorithm [4] and the Earth Mover's Distance [8]. EP-MEANS has to pass through the data multiple times in order to get the centroids that will represent each pattern, which can be time consuming.

SPAM (Sequential PAttern Mining), proposed by Ayres *et al.* [3], can be used to find relationships between sequential time intervals. This method finds frequent itemsets sequences by building a lexicographic tree of sequences. Since we are only interested in finding the relationships between patterns in two consecutive timesteps, we chose not to use this method, for simplicity.

3   Proposed Method

We propose Frequent Distributions, a method which discovers frequent distributions of a variable, which we refer to as *profiles*. This method also uses association rule mining to look for relationships between the profiles in consecutive time intervals.

3.1   Finding Frequent Distributions

Let us define a univariate time series dataset ($\mathcal{D}$) as a table with n rows and 3 columns that come in the format $\{value, t, entity\}$, where $t \in T$ and $entity \in E$. We call $E = \{entity_1, \ldots, entity_k\}$ the *entities*, where k is the number of entities represented in $\mathcal{D}$, and we call $T = [t_{first}, t_{last}]$ the *timesteps*, where t_{first} and t_{last} are the first and last timesteps registered in $\mathcal{D}$. Each row $r_{t,entity_i}$ represents an observation for one entity at a specific timestep. It can be formally defined as $r_{t_x,entity_i} = \{value_{(x,i)}, t_x, entity_i\}$.

A profile (pf) is defined as an empirical distribution of the values of a variable during a time interval of size *wsize*. $\mathcal{D}$ will have a fixed set of profiles of the variable ($PF_{variable} = \{pf_1, ..., pf_n\}$) and the variability of variable can switch between the different profiles over time. For example, when looking at records of electricity consumption, we will probably have $PF_{consumption} = \{pf_{day}, pf_{night}\}$. For pf_{day}, the distribution will include lower values, as opposed to pf_{night}, which will include higher values due to the need for artificial lighting during the night.

The Frequent Distribution mining approach iteratively discovers new profiles of a target variable (*target*) in the time series data. It starts by splitting the data into *time windows* of size *wsize*, and then makes one pass through them sequentially. For each time window $tw = [t_0, ..., t_{wsize}]$, we observe $Dist_{target}(tw, entity)$ for each $entity \in E$ and try to assign each $Dist_{target}$ to a profile. This is done by checking the *distance* between the distributions and the discovered profiles. Any distance distribution metric can be used, but, for simplicity we use the Kolmogorov-Smirnov statistical test as in [7]. A distribution is assigned to a known profile pf if $distance(Dist_{target}, pf) < \theta$. The value of θ is in the range $[0, 1]$ and is defined by the user. If $distance(Dist_{target}, pf_i) \geq \theta, \forall pf_i \in PF_{target}$, where PF_{target} is the set of profiles, then the new profile $Dist_{target}$ is added to PF_{target}. Finally, the profiles in PF_{target} which have a support below *minsupp* are discarded. The *minsupp* is decided by the final user. The pseudocode for Frequent Distributions is shown in Algorithm 1.

Input: target, wsize, θ, minsupp
begin

    profiles_list; // Initialize with the initialization method described
    foreach *time window tw of size wsize* **do**
        foreach *entity* **do**
            entity_distribution ⟵ $Dist_{target}(tw, entity)$;
            is_distribution_distinct ⟵ *True*;
            foreach *pf in profiles_list* **do**
                if *distance(entity_distribution, pf)* $< \theta$ **then**
                    is_distribution_distinct ⟵ *False*;
                    pf.count ⟵ *pf.count* + 1;
                else
                    Add *entity_distribution* to *profiles_list*;
                end
            end
        end
    end
    frequent_profiles ⟵ All *profile* in *profiles_list* where
    profile.count > *minsupp*;
    return *frequent_profiles*;
end

Algorithm 1. Frequent Distributions algorithm

3.2   Initialization of the Profiles Set

For this method to work, first we need to initialize PF. The initialization is done by observing $Dist_{target}(tw_{first}, entity)$ for each $entity \in E$. Then, we calculate the *distance* between those distributions and put them in a symmetric matrix, where the column and row indexes correspond to the entity indexes in E. This matrix is then binarized. Values above θ are set to 1, which represent the distributions which were different from each other, and all other values are set to 0. Finally, we group the distributions which have a 0 in the binarized matrix. In each group, the distribution whose entity has the lowest index will be added to PF as a profile.

3.3   Combining with Association Rules Mining

We can combine the Frequent Distributions with Association Rules mining by adding extra steps to the method. The objective is to obtain association rules that, for each entity, measure the transition between profiles in consecutive time windows. We will refer to the consecutive time windows as tw_x and tw_{x+1}, where tw_{x+1} is the window that immediately follows tw_x. For example, it would be interesting to observe that, for a given target attribute, a profile A is always followed by a profile B in the next time window.

In order to do this, first we have to have a record of the profiles observed for each pair $(tw, entity)$. Afterwards, we need to find the frequent itemsets that will be used in the association rules mining. Each itemset will be composed of a pair of consecutive profiles: a previous profile, observed in tw_x, and a next profile, observed in tw_{x+1}. So, we need to scan through the time series and obtain, for each (tw_x, tw_{x+1}), the pair of consecutive profiles. After obtaining all the itemsets in the last step, we use the Apriori algorithm [2] to find the itemsets that are frequent and afterward use association rules mining to find the association rules. An itemset is considered frequent if the support for that itemset is higher than a user-defined minimum support. Also, in our approach, an association rule will only be considered relevant if its confidence value is higher than the user-defined minimum confidence threshold.

We could have used Sequential Pattern Mining [3] to obtain these relationships. However, in order to simplify, we chose to use association rules mining.

4   Results

To test our method, we use data from 3 sources. The first source, which we call Source A, contains football (soccer) spatiotemporal data. The data has multiple datasets, each one representing the xy positions of the players during one match, and from the xy positions we could calculate the speed of the players. The second and third sources are the well-known Electricity and AWS datasets[1]. Given the big number of instance types on the AWS dataset, we decided to use a subset of the data corresponding to the *m4.large* instance type and the Linux/UNIX operative system.

---

[1] https://moa.cms.waikato.ac.nz/datasets/.

In the datasets from Source A, we used the Frequent Distributions to look for speed profiles of players and relationships between the profiles ($target$ = speed). In the Electricity dataset, we focused on the electricity prices ($target$ = price). In the AWS dataset, the focus was the server prices ($target$ = price). For that, we tested different values in the parameters. In all experiments , θ values varied between 0.6-0.9, and $minsupp$ value was 0.01. Also, all rules with lift less or equal to 1 were discarded. In experiments with Source A's data, we used 100, 200 and 600 as the $wsize$ value; with the Electricity dataset, we used 48, 96 and 144; and with the AWS dataset, we used 1440; In Source A, the $wsize$ values correspond to a number of seconds times 10 (for example, $wsize$ = 50 corresponds to a 5-s time window). In the Electricity dataset, the $wsize$ values correspond to the number of half hours in the time window (for example, $wsize$ = 48 corresponds to 48 half hours, which is the same as a day). In the AWS dataset, the $wsize$ values correspond to the number of minutes in the time window.

In all experiments we chose the $wsize$ according to the time length of the profiles we wanted to obtain. We observed that the smaller the $wsize$, the more profiles are found. This can be explained by the fact that having smaller windows implies that each distribution will include fewer examples. Fewer examples lead to more variability between the distributions observed, which in the end translates into finding more profiles.

In the experiments made, the values for θ were chosen empirically. However, since we are looking for distinct profiles, we set the minimum distance to 0.6. On the other hand, when the distances are too big (>0.9) only very few distinct profiles were found. This can be explained by the fact that a higher θ implies that more distributions will be considered similar to each other.

From the results of Source A, we can observe 3 main profiles. (Figure 1). There is one profile for standing still/walking (Fig. 1c), a second profile for slow running (Fig. 1a) and third profile for bursts of speed (Fig. 1b). This third profile has more variability, where the speed observed ranges from 0 to 22 km/h. This can mean that, in this profile, the player is not always running but increases and decreases his speed considerably.

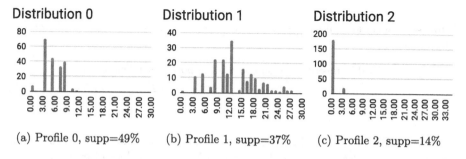

(a) Profile 0, supp=49%   (b) Profile 1, supp=37%   (c) Profile 2, supp=14%

Fig. 1. Speed profiles of players obtained in Source A, with $wsize$ = 200 and θ = 0.7.

Regarding the relationships between profiles, it was found that is common for players to switch from the running profile to the slow running one (Rules 1, 2, 3 and 4 of Table 1). Also, it was found that was common for players to switch from the standing still/walking to the slow running profile (Rules 5 and 6 of Table 1). It was also found that was common for players to keep in the slow running profile (Rules 7, 8 and 9 of Table 1). All rules show that these phenomenons happened more than 50% of the times for multiple players across some of Source A's experiments. This reveals that, in the Source A dataset, players have a tendency to keep in the slow running profile or to return to it after being in a different profile.

Table 1. Most important rules found in Source A's data

Rule_id	Antecedent	Consequent	Support	Confidence	Lift	No. of players	wsize	θ
1	Profile 1	Profile 0	3–16%	50–58%	1.01–1.09	8	100	0.7
2	Profile 1	Profile 0	3%	50%	1.02	1	100	0.8
3	Profile 1	Profile 0	7–21%	50–53%	1.01–1.02	5	200	0.7
4	Profile 1	Profile 0	5%,11%	66%,73%	1.07,1.11	2	600	0.6
5	Profile 2	Profile 0	6–20%	50–53%	1.01	3	100	0.7
6	Profile 2	Profile 0	6%	50%	1.01	1	200	0.7
7	Profile 0	Profile 0	25–31%	51–56%	1.01–1.1	12	100	0.7
8	Profile 0	Profile 0	25%	50%	1.01	1	200	0.7
9	Profile 0	Profile 0	26–38%	51–62%	1.01–1.03	18	600	0.6

In the Electricity dataset, in the experiment with $wsize = 144$ and $\theta = 0.8$, there were found 7 profiles. The profiles are shown in Fig. 2.

The rules found were only relative to the Victoria state, and are shown in Table 2. The rules show that, in more than 60% of the times that Victoria state was in the profile 3, it remained in that profile. This profile is one of the profiles which includes the lowest prices. This means that, when the price of electricity is low for 2 days ($wsize = 144$), it is likely to continue for the next 2 days.

Table 2. Association rules found in the Electricity dataset relative to the Victoria state

Rule_id	Antecedent	Consequent	Support	Confidence	Lift
1	Profile 3	Profile 3	24%	62%	1.58

In the AWS dataset, in the experiment with $wsize = 1440$ and $\theta = 0.9$, there were found 7 profiles. A temporal representation of the values of each profile is shown in Fig. 3.

We found rules relatively to the instances of the two regions that correspond to Canada: *ca-central-1a* and *ca-central-1b*, which are shown in Table 3.

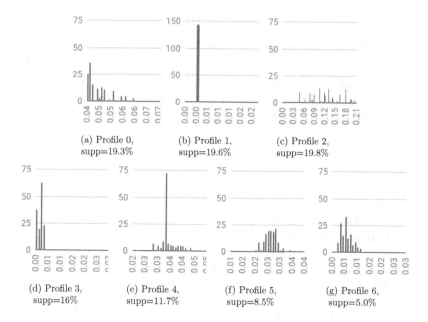

(a) Profile 0,
supp=19.3%

(b) Profile 1,
supp=19.6%

(c) Profile 2,
supp=19.8%

(d) Profile 3,
supp=16%

(e) Profile 4,
supp=11.7%

(f) Profile 5,
supp=8.5%

(g) Profile 6,
supp=5.0%

Fig. 2. Profiles of the electricity prices. Note that the scales may be different between plots.

They show that, in more than 55% of the times that the prices of the canadian servers were in profile 3 or 6, they remained in that profile. These two profiles are the ones that include the lowest prices, which can mean that, when canadian servers have the lowest prices during one day ($wsize = 1440$), it is likely that the prices keep low on the next day.

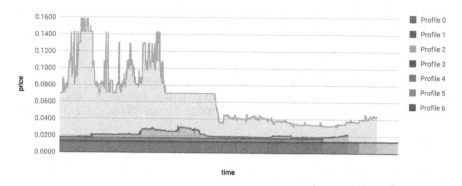

Fig. 3. Variation of the values of each AWS profile of price, relatively to all instances $m4.large$, Linux/UNIX, regardless of the region

Table 3. Association rules found in the AWS dataset relative to the Canadian region

Rule_id	Antecedent	Consequent	Support	Confidence	Lift	Region
1	Profile 3	Profile 3	13%	62%	2.55	ca-central-1a
2	Profile 6	Profile 6	10%	92%	8.52	ca-central-1a
3	Profile 3	Profile 3	13%	55%	2.16	ca-central-1b
4	Profile 6	Profile 6	9%	88%	8.50	ca-central-1b

5   Conclusion

We propose a preliminary study on what we refer as Frequent Distributions, a method to find distributions (profiles) of variables in time series data and relationships between them in consecutive time intervals.

As future we want to improve the initialization and obtain profiles with more than one variable. Also, sequential pattern mining could be used to obtain relationships that are not limited just to two consecutive time windows.

Acknowledgements. This work is financed by National Funds through the Portuguese funding agency, FCT - Fundação para a Ciência e a Tecnologia within project : UID/EEA/50014/2019

References

1. Agrawal, R., Imielinski, T., Swami, A.N.: Mining association rules between sets of items in large databases. In: Buneman, P., Jajodia, S. (eds.) Proceedings of the 1993 ACM SIGMOD International Conference on Management of Data, Washington, DC, USA, May 26–28, 1993. pp. 207–216. ACM Press (1993). https://doi.org/10.1145/170035.170072
2. Agrawal, R., Srikant, R.: Fast algorithms for mining association rules in large databases. In: Bocca, J.B., Jarke, M., Zaniolo, C. (eds.) VLDB 1994, Proceedings of 20th International Conference on Very Large Data Bases, September 12–15, 1994, Santiago de Chile, Chile, pp. 487–499. Morgan Kaufmann (1994)
3. Ayres, J., Flannick, J., Gehrke, J., Yiu, T.: Sequential pattern mining using a bitmap representation. In: Proceedings of the Eighth ACM SIGKDD International Conference on Knowledge Discovery and Data Mining, July 23–26, 2002, Edmonton, Alberta, Canada, pp. 429–435. ACM (2002). https://doi.org/10.1145/775047.775109
4. Bishop, C.M.: Pattern Recognition and Machine Learning, 5th edn. Springer, Information science and statistics (2007)
5. Brockwell, P., Davis, R.: Introduction to Time Series and Forecasting. Springer Texts in Statistics. Springer, New York (2013)
6. Henderson, K., Gallagher, B., Eliassi-Rad, T.: EP-MEANS: an efficient nonparametric clustering of empirical probability distributions. In: Wainwright, R.L., Corchado, J.M., Bechini, A., Hong, J. (eds.) Proceedings of the 30th Annual ACM Symposium on Applied Computing, Salamanca, Spain, April 13–17, 2015, pp. 893–900. ACM (2015). https://doi.org/10.1145/2695664.2695860

7. Jorge, A.M., Azevedo, P.J., Pereira, F.: Distribution rules with numeric attributes of interest. In: Fürnkranz, J., Scheffer, T., Spiliopoulou, M. (eds.) PKDD 2006. LNCS (LNAI), vol. 4213, pp. 247–258. Springer, Heidelberg (2006). https://doi.org/10.1007/11871637_26
8. Rubner, Y., Tomasi, C., Guibas, L.J.: A metric for distributions with applications to image databases. In: Proceedings of the Sixth International Conference on Computer Vision (ICCV-1998), Bombay, India, January 4–7, 1998, pp. 59–66. IEEE Computer Society (1998). https://doi.org/10.1109/ICCV.1998.710701

Time Series Display for Knowledge Discovery on Selective Laser Melting Machines

Ramón Moreno[1]($\boxtimes$)(iD), Juan Carlos Pereira[1](iD), Alex López[1](iD), Asif Mohammed[2], and Prasha Pahlevannejad[3]

[1] IK4-LORTEK, Ordizia, Spain
rmoreno@lortek.es
[2] ADIRA, Canelas, Portugal
[3] SmartFactory-KL, Kaiserslautern, Germany
http://www.lortek.es, http://www.adira.pt, https://smartfactory.de

Abstract. This paper presents a method for displaying industrial time series. It aims to support data and process engineers on the data analytics tasks, specially in the area of Industry 4.0 where data and process joins. The method is entitled SCG, from Splitting, Clustering and Graph making which are its main pillars. It brings two innovations: Samples making and Visualizations. The first one is in charge of build well-suited samples fostered to reach the data exploring objectives, whereas the second one is in charge showing a graph-based view and a time-based view. The final objective of this method is the detection of stable working states on a working machine, which is key for process understanding, while at the same time it enlightens on knowledge discovery and monitoring. The use case in which this work is grounded is the Selective Laser Melting (SLM) industrial process, though the introduced SCG procedure could be applied to any time series collection.

Keywords: Time series · Clustering · Knowledge discovery · Graph · Process monitoring · Selective laser melting · Additive manufacturing

1 Introduction

Knowledge discovery (KD) in industrial processes is a very important topic, specially in the current era of data. It encompasses Data Mining, Machine Learning (ML) and Time Series processing, all of them under the umbrella of Big Data in Industry 4.0 [1]. Nowadays, every device connected to a production pipeline is emitting data, with extraordinary effect in the context of the Industrial Internet of Things (IIOT) [2] where a typical production line has hundreds of sensors and actuators. This overwhelming explosion of data needs to be handled properly in different layers. The first of them is known as *edge* computing, where data must be processed in real time (in terms of milliseconds) and relevant information should be filtered. The second layer is named *fog* computing, and is in charge

© Springer Nature Switzerland AG 2019
H. Yin et al. (Eds.): IDEAL 2019, LNCS 11872, pp. 280–290, 2019.
https://doi.org/10.1007/978-3-030-33617-2_29

of handling the data flow in the production pipeline, connecting cells and interacting with each other (also in terms of milliseconds). Finally, the third layer is well-known as *cloud* computing. This highest layer handles data that is not required for real-time operations and is usually computed on dedicated servers, which may be local or remote, such as Azure, AWS, Google cloud and others. This three level division of data computation for industrial environments has an outstanding research activity of its own.

This work is framed within *fog* computing, focusing on the use case of Selective Laser Melting (SLM) [3] machines. The SLM process is a metal Additive Manufacturing (AM) [4] technology. This process is characterized by layer-by-layer construction. A focused laser beam melts the metal powder according to the 3D design. Metal AM is a real manufacturing alternative for certain type of complex components in short series, particularly in automotive and aerospace sectors. The use of this technology for direct production of functional parts is increasing, beyond its use for obtaining prototypes or casting models. But a lack of knowledge about the process, metallurgy, part performance/behavior and quality control, requires the use of Information and Communications Technologies (ICT) in the SLM production cells to monitor the process. The acquired data may then be used to find correlations with the quality of the product and with the actual process knowledge from engineers and researchers, seeking a robust and stable manufacturing process. Notice that an interesting feature of AM/SLM manufacturing is that it is usually used for short term series, resulting in a wide collection of different pieces, which in terms of modeling means excellent domain sampling.

Time Series clustering [5] has been applied in a variety of domains such as Bioinformatics, Biology, Genetics, Multimedia and Finance. However, at this moment, there is a lack of examples of this technique being applied in the industrial domain. Data sampling and timestamps are the main pillars in this kind of data processing. In controlled or regular processes, the information is saved in regular lapses of time and *ts* thus, depending on the size of *ts*, the information granularity will be finer or coarser, e.g. financial data sets. By contrast, in industrial environments, the manufacturing process is composed of a collection of interacting processes, these processes/data possibly being of different nature (in terms of the physical laws that are involved in the process), therefore the observed variables may be discrete or continuous. It is common for each variable/process to need a different *ts size* to capture data appropriately. In time series processing, all variables are typically merged into unified data sets under common timestamps. Notice that this merging of data implies information loss.

Clustering is known as a statistical family of methods for unsupervised learning which split the multidimensional domain space in many pieces according to a given mathematical function, yielding a labelization of the input data set. Well-known unsupervised learning algorithms include distance based such as c-means, k-means or density based algorithms such as DBCURE, DBSCANN, or DENCLUE [6].

This paper introduces a method for data visualization and process monitoring based on time series clustering which helps to understand an industrial process. It is presented as SCG, as an acronym for the three steps involved (Splitting, Clustering and Graph making).

1. Splitting: a dataset division is carried out transforming the input data on a collection of ordered samples.
2. Clustering: samples are grouped into clusters, where each cluster is going to be identified with a production state.
3. Graph making: a directed graph is made based on the clustering output. First, the cluster sets become in graph nodes. Afterwards, reading sequentially the historical data, changes of nodes are detected and then these ones become edges of the graph.

Onwards, this paper is outlined as follows: Sect. 2 gives a brief description to SLM machinery in the HyProCell project; afterwards, Sect. 3 describes the mathematical time series modeling and clustering. Next, Sect. 4 describes the network building and corresponding views. Section 5 shows the experimental results and finally, Sect. 6 ends with the conclusions and further work.

2 SLM Hybrid Production Cell in HyProCell

The hybrid production cells developed in HyProCell[1] covers the entire Laser Based Additive Manufacturing (LBAM) related process chain. HyProCell project proposes a unique combination of LBAM machines and ICT innovations already available in an integrated multiprocess production cell.

The SLM hybrid cell used in this work is located in Poly-shape (France) using the Adira AddCreator machine. Once the SLM job is finished, the powder extraction is carried out by means of a suction lance and a routine for cleaning the surrounding area after production, therefore minimizing the amount of powder present. Afterwards, the extraction of the piece is done by a robot, and then the substrate and the attached AM parts are placed on the automatic powder removal station. Finally, the cleaned items are then transferred to a CNC milling machine for post-processing. The transfer is performed by a robot and pallet system provided by Erowa. The correct workflow of operations is ensured by the Manufacturing Execution System (MES) developed by TTS. Figure 1 illustrates the different modules and machines of the SLM production cell.

The core SLM machine used as part of this work is the new ADIRA model AC - AddCreator. This machine represents the first in a commercial line of industrial Metal Additive Manufacturing systems for large parts through the modular TLM (Tiled Laser Melting) proprietary technology. This technology divides the existing work area in smaller segments or tiles, which are processed sequentially being able to provide, at the end of the process, very large parts. This is possible given the use of a modular process chamber, which moves around the working area (the powder bed), with integrated gas flow and optical systems, therefore maintaining a stable and inert atmosphere.

---

[1] http://www.hyprocell-project.eu/.

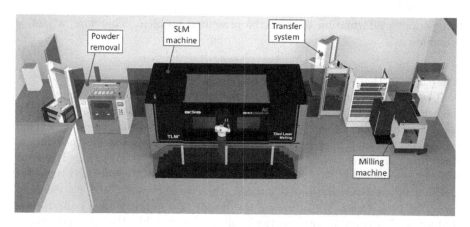

Fig. 1. Overview of the machines and other components in SLM production cell at Poly-Shape

3   Mathematical Background

This section describes the proposed SCG method pipeline. First we give a formal description of time series and after that, we define the concept of sampling and partitioning of time series.

Time Series Modeling

The unit of information of a time series is given by joining a set of measurements in time t, defined with Eq. 1. Given a collection of sensors $\mathfrak{D} = \{s_1, s_2, \ldots, s_n\}$ whose measurements are linked to variables $\mathbf{X} = \{x_1, x_2, \ldots, x_n\}$ that at time t_i take the values $\{x_1^{t_i}, x_2^{t_i}, \ldots, x_n^{t_i}\}$, the collection of measurements $\mathfrak{T}$ is formed as

$$\mathfrak{T} = \{ts_i, x_1^{t_i}, x_2^{t_i}, \ldots, x_n^{t_i}\} \tag{1}$$

for the sake of clarity is re-written as

$$\mathfrak{T}_i = \{ts_i, x_1, x_2, \ldots, x_n\} \tag{2}$$

$\mathfrak{T}_i$ is thence a collection of measurements at time t_i.

A time series is the collection Ω such that

$$\Omega = \{\mathfrak{T}_i, \mathfrak{T}_{i+1}, \ldots, \mathfrak{T}_j\} \tag{3}$$

where $\mathfrak{T}_i$ is the collection of measurements of the sensors $\mathfrak{D}$ at time t_i, $\mathfrak{T}_{i+1}$ the corresponding measurements at time $i + 1$ and so on until time j.

Sample Definition

One single measurement $\mathfrak{T}_i$ by itself does not contain enough information. A collection of continuous measurements from time i to j is necessary as shown in Eq. 3. Hence, the distance between i and j is of interest, in other words, the minimal time series length which guarantees that a collection of $\mathfrak{T}_i$s is going to bring enough information to solve the presented problem. Therefore, a sample is defined as

$$\omega = \{\mathfrak{T}_k, \mathfrak{T}_{k+1}, \ldots, \mathfrak{T}_l\} | k > i, l < j \tag{4}$$

where $\omega \in \Omega$, that is, ω is a subset of Ω.

This point is very important. In most of the works found in the state of the art regarding time series processing, each $\mathfrak{T}$ row of a serial time dataset is taken as a sample. However, this approach is very poor and does not exploit neither Markovian [7] nor Functional Data Analysis [8] approaches (pointing to clustering). In other words, by using a row sample $\mathfrak{T}_i$ as a ω (k = 1 = i), it is being assumed that $\mathfrak{T}_{i-1}$, $\mathfrak{T}_i$ and $\mathfrak{T}_{i+1}$ are independent events and then, the neighboring time-relationship is ignored, thereby it breaks down the essencial aspect of time series. We advocate the use of ω samples with a collection of continuous $\mathfrak{T}$ because it contains the implicit time variable, thus it is possible to analyze the behavior of each x_i along time. Nonetheless, the most important step forward between single $\mathfrak{T}$ and ω, is that with the second one we can use markovian energy functions, functional data analysis or convolutional time based functions, which turns it into a more powerful and useful informative sample description.

Time Series Partitioning

Time series partitioning is key in the SCG. It splits the full time series recorded in the database into samples. Therefore, it is necessary to have defined beforehand the ω size in terms of time, and the corresponding distance $j - i$ which determines the number of $\mathfrak{T}$s that is going to contain each ω. Then, the full TS is split in ωs of the same size. A collection of ωs Ω is defined as

$$\Omega = \{\omega_1, \omega_2, \ldots, \omega_n\} | \forall(\omega_i, \omega_j), \omega_i, \omega_j \in \Omega, \omega_i \cap \omega_j = \emptyset \tag{5}$$

where the above equation means that a measurement $\mathfrak{T}_i$ belongs only to one ω, and ωs are not overlapping each other.

Clustering

Clustering methods group the samples according to a function of similarity such as the Euclidean, Correlation or Cosine distances. There are well-known clustering algorithms, such as the c-means and k-mean and others derived from them, as described in the literature [6,9]. In this work we have chosen the Hierarchical approach [10], due to two main reasons: First, because it is intuitive to use a distance as the splitting parameter. The linkage matrix contains the tree and the distance relationships among samples, so it is not necessary to perform any additional computation. Secondly, the characteristic dendrogram of the hierarchical methods, shows very well suited information about the distribution of the samples in the domain. Hierarchical clustering methods begin from the premise that every sample is its own cluster. After that, following a iterative procedure, the clusters are merged into greater clusters based on distance criteria. The Ward method [11] allows to set an objective function as merging parameter. An overview of the uses and approaches of Ward method can be found in [12].

Working States Discovery

The aim in this work of using clustering over a collection of ω samples, is to discover the most relevant sets. A direct consequence of our clustering using ωs and under the industrial process point of view, is that all ωs included in the same set are working approximately in the same conditions. Obviously, when working with clustering we face the typical question: How many sets are needed to perform the clustering properly? The response is never trivial. In K-means, the number of centroids is an initial parameter. By contrast, when using hierarchical clustering we can obtain different number of set depending on the given cutting distance without the need of any re-computation. This is the key reason why we have opted for hierarchical clustering, as it offers better exploring choices. The optimal number of sets in the clustering process is still an open question whose answer can be put in hands of the expert workers. The process specialist should find out the matching between the clustering output and the process reality. Discover when a data cluster represents an process state. Next point explains the visualizing tools designed to reach this goal.

4 Network Building and Views

This section describes two tools for clustering visualization when working with time series. The main difference between time series and other data collections lies in the *time* variable, which in the case of time series acts as index of all data. Time is a very important variable, it makes the difference between causes and consequences (cause is an event which occurs before the consequence), and it transforms classification in predictions (classification has timeless nature, whereas prediction is a supposition about the future, so it has ephemeral nature).

In order to visualize the cluster and the relationship with time, we assume an equivalence between clusters and directed graphs: A cluster corresponds with a graph node, and a transition between clusters in the historical data corresponds with an edge on the graph. Therefore, from here onwards we will assume the equivalence between graph *node* and cluster *set* and process working *state*. In all cases we are referring to the same information, but seen from different points of view (graphs theory, statistical clustering or industrial processes).

In this part of the work we transform the cluster in a directed graph. As mentioned, a cluster becomes a node, and the directed edges connecting nodes implicitly add the time variable that does not exist on the cluster but exists in the historical data. Hence, network edges re-establish the time dimension. It shows the transitions between sets/nodes/states, and illustrates which relationships are strong and which ones weak.

There are two complementary views of this data. The first one is a directed graph (e.g. Fig. 2b), and the second one a time representation where the changes among states can be observed (e.g. Fig. 2c). Figure 2 shows an illustrating example.

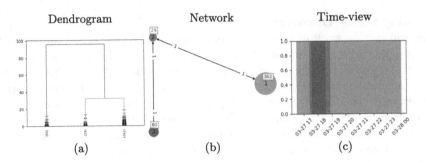

Fig. 2. Different visualizations of a clustering process. On the left (a), shows the den-dogram for 3 sets, it has a view over the data distribution. On the middle is the network visualized as a directed graph with three nodes. On the right (c) a timestamp based representation. It shows the time sequential changes three states 1-2-3-2-1, which correspond with what is represented on the middle view (b).

5   Experimental Results

This section shows the SCG outputs in the above defined order plus an example use case. Firstly, there is a description of the split samples pre-processing (S). Afterwards, the outputs of the clustering processes are shown (C). Later the output network is shown (G).

The experiments have been carried out using our own code developed over python using scipy and networkx libraries. The computer is an Intel(R) Core(TM) i7-8700 CPU @ 3.20 GHz with 32 GB memory. To run a full experiment (clustering and visualizations) takes less than a minute.

Data Setup
Data details: The experimental part is performed using data coming from the SLM Adira machine. The dataset used to accomplish this experiment has a collection of 42000 timestaps and up to 300 variables, that after a data pre-processing step, we have selected 103 variables which are representative of the SLM machine behavior. This data pre-processing step removes variables without information (zero variance, null data or the duplicated ones with fully correlation). The final selected data is a collection of observed variables that include positional parameters, oxygen levels, and temperatures. Therefore the initial Ω contains 42000 rows and 103 columns.

S: Data Pre-processing and Samples Making.
Instead of computing the gross input data, we performed some data transformations. First of all, the input data was normalized to the range [0–1] using the variable ranges provided by the experts in the machine. Secondly, the full input data set is split in ω samples containing the machine registry in slots of 2 min. Third, we add a smoothing operation; instead of using the full set of normalized values, we used the mean of the values in the defined time window of 10 seconds. This transformation is made on each sample, therefore given a sample ω, for all variables x in X, we

compute **mean**(x) as a window-mean on each item x_i in x using a window of size δ, as follow: $\mathbf{mean}(x_i) = \delta^{-1} \sum_{i-(\delta/2)}^{i+(\delta/2)} x_i = mean(\{x_{i-(\delta/2)}, \ldots, x_{i+(\delta/2)}\})$. The output is a collection of ω samples. This data smoothing process implies possible small information loss, nonetheless it avoids spurious outlayers and it reverts in a better robustness. Notice, that in this work we have used only this single transformation, but here we can use whatever other one.

C: Clustering. Going back to the sample definition shown in Eq. 4, $\omega = \{\mathfrak{T}_k, \mathfrak{T}_{k+1}, \ldots, \mathfrak{T}_l\}$, and the set of variables $X = \{x_1, x_2, \ldots, x_n\}$, the matrix representation of a sample takes the shape:

$$\omega = \begin{bmatrix} \mathfrak{T}_0^{x_0} & \mathfrak{T}_0^{x_1} & \mathfrak{T}_0^{x_2} & \cdots & \mathfrak{T}_0^{x_n} \\ \mathfrak{T}_1^{x_0} & \ddots & & & \\ \mathfrak{T}_2^{x_0} & & \ddots & & \\ \vdots & & & \ddots & \\ \mathfrak{T}_m^{x_0} & & & & \ddots \end{bmatrix}$$

A sample ω is the collection of m continuous measurements of the variables X. Therefore, given two samples we can compare differences of a variable x_j at relative time $\mathfrak{T}_i$. From this collection of ω samples the the clustering process is carried out using the Ward Hierarchical clustering[2] method.

The output is a set of clusters C.

G: Network and Time-View. Once the clustering has been accomplished, every sample ω belongs to a unique cluster in C. Indeed, for every cluster in C the number of ωs inside is known. Therefore, with this cluster information in one hand, we can draw the nodes of the network G, and set the node size accordingly with the number of samples that it contains. With the Ω set on the other hand, we run Ω sequentially, ω-by-ω, detecting the changes of cluster in C, which correspond with a change of node in G and draw the edge that is connecting the nodes. Analogously, a historical time-based view is made.

The output is the directed graph G and the time-based view.

Views and Process Exploration

The following views are the visualizing tools for the industrial process engineers, who are free select the number of nodes so as to visualize the corresponding clustering outputs. Thence, in a iterative use of this visualizations (using different number of nodes), the engineers can detect when the most representative process states are reflected in the graph. Here we show three examples of visualizations using 5, 7 and 10 nodes. It is easy to realize that at more nodes used, we get a deeper insight of the industrial process.

At this point, having the clustering C and the visualization tools, we explore the process across the data registry looking for the process' states that best fits

---

[2] https://docs.scipy.org/doc/scipy/reference/cluster.hierarchy.html.

the reality of the process. This exercise must be done with the expert in the process that is the one who understands the industrial process and can add a semantic relationship between the G nodes and corresponding process names.

Figure 3 shows the different outputs depending of the number of states. On the first row the outputs for 5 states clustering is shown. The dendrogram shows the samples distribution. The network shows the machine's state flow between nodes. Indeed, the network shows clearly one bigger node (the green one numbered 2). On the third column is the time-based view where it can be clearly seen that the 2nd state takes most of the working time in the machine until the end of the process. Second and third rows show the corresponding outputs for 7 and 10 states. Increasing the number of states gives a deeper insight into the process. Notice that loops appear on the network view, which means iterative

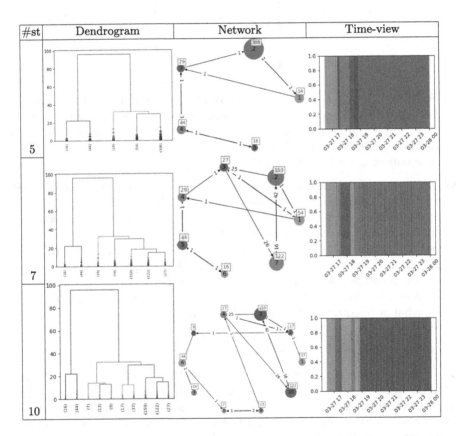

Fig. 3. Different views. The rows correspond to views for 5, 7 and 10 states exploration. The columns show the following information. On the left is the number of states for clustering. The second one shows the dendogram. The third column shows the directed graphs, where the small flag over the network nodes indicates the number of samples inside. Finally, the rightmost one plots the time-based view where on the x-axis is the date-hour tag.

manufacturing states. On the 7-states network there is a loop upon the states 2-3-7 that in the 10-states network corresponds exactly with the 3-4-10 loop.

6   Conclusions

This work takes a step forward in the era of the Industry 4.0 where engineers of different research areas must work together exploring data and understanding the processes. This paper has introduced two tools for time series data visualization: a network-based view and a time-based view. Under the industrial point of view, these tools join process and data, engineering and statistics. In the current era of the data, it makes no sense to explore data only by statistical procedures. Data exploration should be lead by the experts on the process that usually are not experts in data analytics.

As further work, we are going to exploit functional data techniques on ω samples so as to discover the causes of some effects (fabrication defects or errors on SLM processing) in order to learn from faults and support the workers and the MES machinery so as to keep the SLM machine working in the best conditions possible. The natural direction of this works will be addressed towards Support Decision Systems which merges artificial intelligence and human decision.

Authors and Acknowledgments. The work done on this paper is focused in one of the hybrid production Cells that use SLM as AM process, developed in an European Project of Factories of the Future (FoF) which is a public-private partnership (PPP) for advanced manufacturing research and innovation initiative. IK4-LORTEK is a Spanish research center specialized in additive manufacturing, joining processes, and industrial digitization, which is the coordinator of the project and in charge of implementing the self-learning system within the H2020 EU project named HyProCell (Development and validation of integrated multiprocess Hybrid Production Cells for rapid individualized laser-based production), in collaboration with SmartFactory who is in charge of the middleware and OPC-UA adapters, and Adira who is the machine manufacturer.

The part of data generation has been funded by the European Union's Horizon 2020 research and innovation program under Grant Agreement No 723538 (HYPROCELL project). The project is framed in the initiative for advanced manufacturing research and innovation of the Photonics and Factories of the Future Public Private Partnership.

The part of the data engineering has been made under the financial support of the project KK-2018/00104 (Departamento de Desarrollo Económico e Infraestructuras del Gobierno Vasco, Programa ELKARTEK Convovatoria 2018).

References

1. Gokalp, M.O., Kayabay, K., Akyol, M.A., Eren, P.E., Koçyiğit, A.: Big data for industry 4.0: a conceptual framework. In: 2016 International Conference on Computational Science and Computational Intelligence (CSCI), pp. 431–434, December (2016)
2. Wollschlaeger, M., Sauter, T., Jasperneite, J.: The future of industrial communication: automation networks in the era of the internet of things and industry 4.0. IEEE Ind. Electron. Mag. **11**, 17–27 (2017)

3. Yadroitsev, I., Gusarov, A., Yadroitsava, I., Smurov, I.: Single track formation in selective laser melting of metal powders. J. Mat. Process. Technol. **210**(12), 1624–1631 (2010)
4. Frazier, W.E.: Metal additive manufacturing: a review. J. Mat. Eng. Perform. **23**, 1917–1928 (2014)
5. Aghabozorgi, S., Shirkhorshidi, A.S., Wah, T.Y.: Time-series clustering - a decade review. Inf. Syst. **53**, 16–38 (2015)
6. Ben Ayed, A., Ben Halima, M., Alimi, A.M.: Survey on clustering methods: towards fuzzy clustering for big data, pp. 331–336. IEEE, August (2014)
7. DeYoreo, M., Kottas, A.: A bayesian nonparametric markovian model for non-stationary time series. Stat. Comput. **27**, 1525–1538 (2017)
8. Wagner-Muns, I.M., Guardiola, I.G., Samaranayke, V.A., Kayani, W.I.: A functional data analysis approach to traffic volume forecasting. IEEE Trans. Intell. Transport. Syst. **19**, 1–11 (2017)
9. Xu, R., Wunsch, D.: Survey of clustering algorithms. IEEE Trans. Neural Netw. **16**, 645–678 (2005)
10. Legendre, P., Legendre, L.: Chapter 8 - cluster analysis, in numerical ecology. In: Legendre, P., Legendre, L. (eds.) Developments in Environmental Modelling, vol. 24, pp. 337–424. Elsevier, Amsterdam (2012). https://doi.org/10.1016/B978-0-444-53868-0.50008-3
11. Ward Jr., J.H.: Hierarchical grouping to optimize an objective function. J. Am. Stat. Assoc. **58**(301), 236–244 (1963)
12. Murtagh, F., Legendre, P.: Ward's hierarchical agglomerative clustering method: which algorithms implement ward's criterion? J. Classif. **31**, 274–295 (2014)

Special Session on Machine Learning Algorithms for Hard Problems

Using Prior Knowledge to Facilitate Computational Reading of Arabic Calligraphy

Seetah ALSalamah, Riza Batista-Navarro$^{(\boxtimes)}$, and Ross D. King$^{(\boxtimes)}$

Department of Computer Science, University of Manchester, Manchester, UK
Seetah.alsalamah@gmail.com,
riza.batista@manchester.ac.uk, rdking@turing.ac.uk

Abstract. Arabic calligraphy (AC) is central to Arabic cultural heritage and has been used since its introduction, with the first writing of the Holy Quran, up until the present. It is famous for the artistic and complicated ways that letters and words interweave and intertwine to express textual statements – usually quotations from the Quran. These specifications make it probably the hardest of all human writing systems to read. Here, we introduce the challenge of reading Arabic calligraphy using artificial intelligence (AI), a challenge that combines image processing and understanding of texts. We have collected a corpus of 1000 AC images along with annotated quotations from the Quran, pre-processing the images and identifying individual letters using detection methods based on maximally stable extremal regions (MSERs) and sliding windows (SWs). We then collect the identified letters to form bags of extracted letters (BOLs). These BOLs are then used to search for possible quotation from the corpus. Our results show that MSERs outperforms SWs in letter detection. Furthermore, BOL-matching is better than word generation in predicting the correct quotation, with the correct answer found in the list of 10 topmost matches for more than 74% of the 388 test examples.

Keywords: Computational reading · Arabic calligraphy · Pattern recognition · Natural language processing

1 Introduction

Artificial intelligence (AI) has recently made rapid advances in image analysis and understanding of texts, owing to the advent of massive annotated image databases and new developments in machine learning [1]. Arabic is one of the most widely spoken languages in the world. Written Arabic has a number of features that complicate its computational reading: (1) Arabic letters can assume a variety of shapes depending on their position within a word, and (2) some letters come in families that are similar in shape but are distinguished by markers, i.e., dots, that appear above or below the main shape [2].

Arabic calligraphy (AC) is a famous element of Islamic cultural heritage, with an artistic practice based upon the beauty of intersecting script [3]. Arabic quotations, often from the Quran, are beautifully written in AC. In part, because of Islamic an iconism, AC has been a focus of Arabic artistic achievement for over 1400 years,

© Springer Nature Switzerland AG 2019
H. Yin et al. (Eds.): IDEAL 2019, LNCS 11872, pp. 293–304, 2019.
https://doi.org/10.1007/978-3-030-33617-2_30

and many different styles of AC have been developed [4]. To conform to the artistic style of AC, letters are written in a manner where they appear more cursive, are arranged to intersect, interleave, and interweave. Additionally, words are often reordered [5]. These changes in shape and rearrangements of letters and words make AC difficult to read, even for native Arabic readers [4].

Moreover, there are more than 50 different AC styles; the most commonly used are Naskh, Thuluth, Diwani, Nastaliq, Riqa and Kufi [6]. Each AC style has its individual specifications, with different levels of overlap, letter rotation and cursiveness, in addition to different degrees to which marks are drawn around the text. Figure 1 shows examples of six different AC artistic styles expressing the same quotation. These show the variety of representations of the text in stylistic drawing arts, adding to the existing challenge presented by Arabic texts [7].

In this paper, we propose a new method to read the drawn text in the AC images. Our method works at the letter level, to allow for extracting letters from the intersections and interweaving of text in the image. It relies on the availability of calligraphy quotations to simplify the mapping from extracted letters to the target quote.

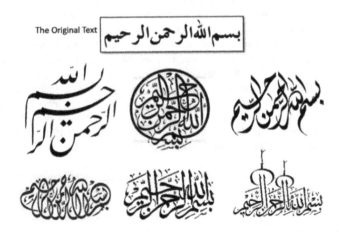

Fig. 1. Original Arabic text (above) and its representation in six different styles of AC, is the famous Arabic phrase translated as "In the name of God, most Gracious, most Compassionate".

2   Related Work

Digitising old documents is a trend in the online era, in which everything is now placed and allocated in order to be available in various media, to be manipulated, translated or learned. This mainly focusses on detecting text from source images; this has been recognised in many types of research that hasbeen conducted in different languages and working on various types of text, whether handwritten or computer-based, or on styles of calligraphic art [8]. However, little of this research has handled calligraphy-based text, since the challenges in this type of text mean dealing with it as shapes rather than letters [9]. On the other hand, there are impressive results from Chinese text recognition, which shares with Arabic the complexity of intersecting letters or characters [10]. However, compared to Arabic letters, which have different shapes according to their location in the word, Chinese characters have constant shapes.

Ye and Doermann [11] summarise research on text detection and recognition in imagery by comparing the methods used and the challenges found in reading text from coloured images in different languages (English, French, Chinese and Korean). They find that, for more than ten different approaches, the results are not very good when stroke detection and word spotting approaches are applied; even with the best end-to-end methods, the final reading accuracy is less than 50%. They cover the remaining problems that reduce the accuracy of reading text from images in three main points as follows: (1) the requirement for a large-vocabulary corpus to obtain the best selection of all probable words; (2) the need to continue improving the correct detection of characters from coloured images; and (3) challenges of building a joint model that can handle multiple languages.

For the Arabic language, previous work done on text recognition achieved more than 90% accuracy [12] on text handwritten in standard Arabic. However, AC cannot be manipulated like standard Arabic text since the intersection of letters needs special segmentation methods [13]. The randomness and challenges in the different styles of AC [4] make its reading difficult, even for those who understand its sophisticated and intersecting styles. Previously reported work focussed on analysing various types of AC [3] and recognising AC text in historical documents [13–15], showing that computational analysis of AC is difficult. As a result, computational reading of AC has been underexplored, and there is a lack of supporting computational resources [4]. To address this gap, we have created our particular dataset to enable the process of reading AC. The initially collected datasets have been published in [16] and will be publicly available after finalising the data collection.

3   Arabic Calligraphy Datasets

To support the development of our computational reading methods, we constructed datasets drawn from AC images. Firstly, more than 1000 AC images were collected from open source websites, mostly from the Free Islamic Calligraphy online platform [17]. These images were then manually annotated to extract: a dataset of letter images and a corpus with the corresponding text. The manual annotation was done by segmenting AC letters and storing them as separate, individual images and, at the same time, by recording the drawn text as a single quotation in the corpus. Figure 2 shows an example of the segmentation of letters into separate images from calligraphy image no.99 and recording of the quotation in the calligraphy corpus.

The resulting A Cletter image dataset contains more than 3200 letter images, with 100 samples for each letter. There are 32 different letter categories, 28 for the main Arabic letters along with four special letters frequently used in the AC context. These letters are the name of God الله, teehmarbuta (ة), alefmaksura (ى) and the short vowel, hamza(ء). All of these images have been saved in their original size upon extraction, with black letters against a white background.

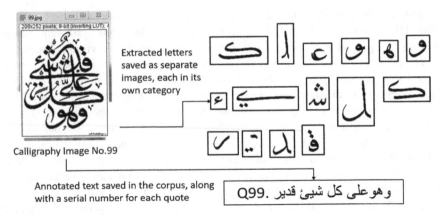

Extracted letters saved as separate images, each in its own category

Calligraphy Image No.99

Annotated text saved in the corpus, along with a serial number for each quote

و هو على كل شيئ قدير .Q99

Fig. 2. Extraction of the letter image dataset and the corresponding corpus from Arabic calligraphy images.

Meanwhile, the calligraphy corpus contains more than 528 different quotations annotated from the calligraphy images. The final annotations were extracted from 1000 images resulting in only 528 unique quotations, with the rest being duplicates written in a different style. The number of words in the quotations ranges from one word to a maximum of 285 words, with an average of four words per quotation. Analysis of the most frequent single character, word and sets of n-words has been conducted to explore common terms in the AC domain. All of the annotation procedures have been approved by a team of three native Arabic readers. Each member of the team completed his/her annotations separately; the final corpus is the result of selecting only matching annotations.

4   Methodology

We decomposed the computational reading of AC images into a series of steps, shown in Fig. 3. The input image is first pre-processed to remove noise and then enhanced by filtering and sharpening. The second step involves the segmentation detection of letters, not words, are extracted to attempt to resolve the disordering of text in AC. Two different methods were used in the detection of letters: maximally stable extremal regions (MSER) and sliding window (SW). With both of these methods, the resulting extracted letters are evaluated separately. In the next stage, the extracted labelled letters are passed to the quotation matching stage to attempt to map them to the probable quotation. The result for each set of extracted letters is a list of likely quotations ranked most likely to least likely. Each of these stages is explained in more detail below.

4.1   Image Pre-processing

In this stage, image enhancement and correction techniques are used to try to improve image quality. The final results are very sensitive to this stage, as AC images generally

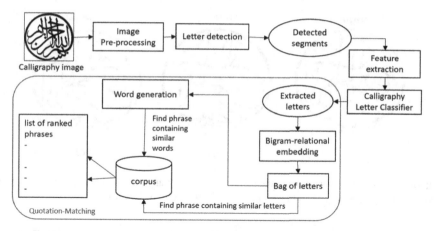

Fig. 3. Diagram depicting our approach to the computational reading of AC, starting from a given input image and ending with the list of probable quotations.

contain significant noise due to their different colours and styles. First, for each image the sharpness of the letter edges is increased, making the letters as clear as possible, capturing as many of the small details as possible. This improves the detail of the image edges and sharpens them. We used the standard deviation of a Gaussian filter to adjust the sharpness of the edges [18]. Secondly, binarisation, also called thresholding, is applied, focussing mainly on removing as much as possible the remaining noise from the image [19]. Finally, small noisy objects are removed; the noise, in this case, was contributed by the artist and the surrounding the text. These are removed from the image by identifying all objects in the image and removing those of a specific size. Based on our manual inspection of 100 images, the smallest object contained in an image typically in the range of 50 to 70 pixels size. We thus eliminated the noise from our images by removing any object that is not bigger than this size. Figure 4 shows an example of an AC image and the results of removing small objects from it in the manner that we have just described. We show the results of using two values of a threshold t, i.e., 50 and 80 pixels, to illustrate the difference between removing objects of size 50 pixels and fewer, and removing objects of size 80 pixels and fewer. With $t = 50$ pixels, some noise remains, but the image is much cleaner than the original image. When $t = 80$ pixels, some of the dots in the letters are removed with the noise (dots in *noon* and *beh*), which is not desirable. This implies that removing noise is a delicate process that could affect the accuracy of letter detection.

4.2   Letter Detection and Recognition

Letter detection involves the extraction of letters from among all of the intersecting and connected shapes within a given AC image. This needs to be carried out without contaminating the letter with noise while selecting the correct location of any related marks (e.g., dots) if they are present. Two detection methods were employed to locate

original   50-Pixel   80-Pixel

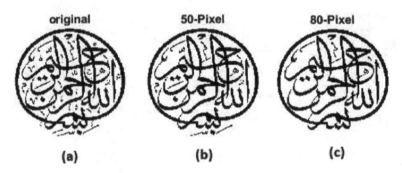

(a)   (b)   (c)

Fig. 4. Example AC image (a) and the results of removing any object whose size is 50 pixels or less (b) and of removing any object whose size is 80 pixels or less (c).

the different letters in an image: maximally stable extremal regions (MSERs) and sliding windows(SWs), which we describe next.

Maximally Stable Extremal Regions (MSERs). This process segments images into regions (corresponding to objects) depending on a grey-level image threshold [20]. The image is scanned according to a threshold, grouping the parts of objects that share the same grey level within their boundaries as one object. Figure 5 shows an example of an AC image and the MSERs detected, shown as coloured segments.

Fig. 5. The MSERs detected in an AC image, shown as coloured segments.

Sliding Windows (SWs). This process uses a rectangular region with a fixed size (64 × 64 pixels) as a window for object detection [21]. This window is used to scan the whole image, moving from left to right and row by row, with the pixels within the window checked continuously for any object that might be contained. The window size is 64 × 64 pixels, similar to the image size that our AC letter classifier was trained on [16]. Every time the window moves to a new position, the features are extracted and sent to the classifier to check whether or not they relate to any letter category.

After an object is identified, it is stored as a separate image and given as input to a classifier for letter recognition. Our AC letter classifier is underpinned by a model trained using support vector machines (SVMs) [16], on two different feature sets: a bag of visual speeded up robust features (SURF) and histogram of oriented gradient (HOG) features. Whereas the former is based on scale-invariant descriptors of image points of interest [22], the latter is counts occurrences of gradient orientation in localised parts of an image [23]. Previous work has shown that the HOG features out perform SURF, with an accuracy of 60% compared with 57% [16]. Thus, the SVM classifier trained on HOG features was selected as our letter recogniser. The classifier was trained with 3200 different image samples for 32 letter categories. The process involved passing an entire calligraphy image through each of the detection methods (described above) and then, for each resulting object segment, having the classifier predict this object's letter category. At the end of this process, the letters are saved as a 'bag of letters': a vector whose elements correspond to the frequency with which each of the 31 Arabic letters of interest was recognised in the given image. The name of God الله category is not placed in the vector as the representative letter (2 *lam*, 1 *hah*) will be separately counted.

4.3   Quotation Matching

It is worth noting that the calligraphic texts that we are trying to read computationally are known, as they are quotations from the Holy Quran and the Sunnah Hadith[1]. This significantly reduces the difficulty of the task and makes the reading of AC in images feasible. We approached quotation matching in two different ways. The first approach made use of a bag of letters (extracted in the previous step) in searching for text in the corpus of quotations. The second approach was based on the generation of all possible words from the extracted letters and then finding quotations containing these words. Meanwhile, for each quotation in the corpus, the frequency of each of the letters it contains is declared to facilitate the search process, described in more detail below.

Bag-of-Letters (BOL) Matching. Firstly, all of the extracted letters are used to form a bag-of-letters vector, which will be used in checking for similarity with the available quotations in the corpus. In filling in the values of this vector, we enforced rules drawn from prior knowledge on letter bigrams that appear most frequently in calligraphic quotations. Such rules were incorporated based on the intuition that accounting for known related letters could compensate for letters that object detection sometimes fails to extract [24]. Table 1 shows the 20 most frequent letter bigrams found in the calligraphy corpus, along with the details of their frequency of occurrence. These are translated into rules which define that if one of these bigram letters is detected (and counted), then the second letter is also counted. These bigrams are restricted in that they occur in only one direction, i.e., from right to left. Therefore, if the right-hand letter is found, then the next letter is added, otherwise there is no addition. This is a unidirectional rule and, for that reason, there is a difference between لا and ال (bigrams with the same letters but in a different order and with different frequencies), for example.

We now describe the process for matching quotations in the corpus. Firstly, the similarity between the bag of letters obtained for a given AC image Bag_{Si} and the bag of letters Bag_{Ti} for every quotation in the corpus, is measured based on the score n which is defined as:

---

[1] The Sunnah Hadith is the second most important textual source of knowledge in Islam after the Holy Quran. It is the report of the Prophet Muhammad's words or actions.

Table 1. The most frequent bigrams found in the calligraphy phrases corpus.

Serial	Bigram	Count	%	Serial	Bigram	Count	%
1	ال	701	6.51%	11	ول	85	0.79%
2	له	264	2.45%	12	حم	83	0.77%
3	لل	212	1.97%	13	ين	79	0.73%
4	لا	125	1.16%	14	مي	77	0.72%
5	لم	122	1.13%	15	لر	76	0.71%
6	وا	110	1.02%	16	لي	70	0.65%
7	من	101	0.94%	17	ما	68	0.63%
8	ان	99	0.92%	18	نا	56	0.52%
9	عل	90	0.84%	19	رب	55	0.51%
10	رح	88	0.82%	20	با	54	0.50%

$$n = \sum_{i \in 0..30}^{let} \begin{cases} 0, & \text{if } Bag_{Si} == Bag_{Ti} \\ 1, & \text{otherwise} \end{cases} \tag{1}$$

Here, i corresponds to the vector element position, whose value ranges from 0 to 30. Hence the two bags are compared element-per-element. If the vectors share the same value for a specific element (i.e., a letter), a score of 0 is returned; otherwise, the score returned is 1. At the end of this element-wise comparison, the value of n will hold the summation of scores for all elements. A smaller value of n thus implies a higher similarity between the input and quotation bags of words. We can then use n to induce a ranking amongst the quotations in the corpus.

Word Generation. In this step, all of the possible context-based words that can be built using the extracted letters, are generated. Given the set of extracted letters $Let_{1..m}$ (together with their related bigrams), possible context-based words containing the maximum number of extracted letters can be predicted according to the naïve Bayes probability, where m is the number of extracted letters:

$$P(W_{ar}|Let_{1..m}) = P(Let_{1..m}|W_{ar}) \times P(W_{ar}) \tag{2}$$

Then, by selecting quotations that contain two or more sequences of the generated words, we obtain a list of quotations ranked according to the number of generated words they contain.

5   Experiments

To evaluate our approach to the computational reading of AC images, we used a test set containing 388 different AC images. This set was labelled with ground truth quotations and contained a wide variety of different styles (Naskh, Thuluth, Diwani, Riqa and Kufi), in various text sizes and representations. Figure 6 shows some samples from the test.

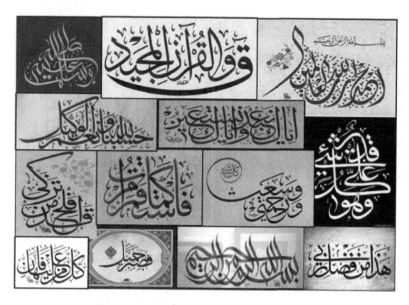

Fig. 6. Samples from the AC image test set.

All 388 images were analysed by the whole sequence of processes described above to predict the quotations they contain. Both MSERs and SWs were employed in the detection phase. Results from both BOL matching and word generation were used in quotation matching. The following parameter values were used in the image pre-processing stage: all objects in the images of size 70 pixels or less were removed; images were binarised with a threshold of 0.4; and, image edges were sharpened with a radius of 2 and strength of 2. All of these parameters have been carefully chosen to optimise performance, based on the result of experiments with small samples of 15 different AC images not from the tested samples.

We evaluate the performance by calculating Top N accuracy [25], which is the percentage having at least one correct quotation in the top-N probable quotations list keeping the number of totals tested images in consideration. Figure 7 shows the Top-10 accuracy results from BOL matching comparing the two methods of detection, MSER and SW. MSER outperforms SW with more correct predictions at a lower level and in the general total. Most of the correct results are found in levels 6 to 10. Altogether, more than 75% for MSER and 68% for SW were correctly located in the top ten prediction levels.

Figure 8 shows the Top-6 accuracy results for word generation from the extracted letters for both methods of detection, MSER and SW. The N level refers to the number of words generated and found in the same predicted quotations with a minimum of 2 words (starting from level 2). In general, the performance of this detection method is inferior to that of BOL matching, as the correct results were only 23%, for SW, and 19%, for MSER, of the total image testing set used. Since SW involves moving around the image and generating more letters than MSER, it is described as having generated slightly more words than MSER.

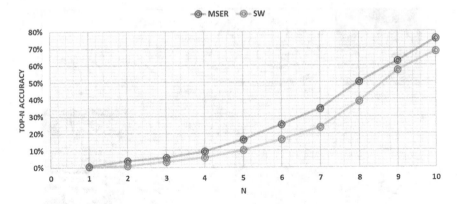

Fig. 7. Top-10 Accuracy results from BOL matching

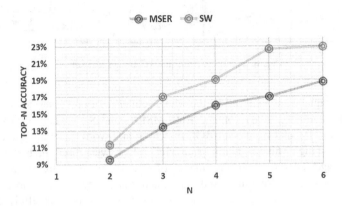

Fig. 8. Top-6 Accuracy results for word generation.

6   Conclusion and Future Work

In this paper, we have compared two methods of detection of AC images, MSER and SW, by using two different search techniques. The results show that BOL matching outperforms word generation from extracted letters, giving an accuracy for all of the top ten suggestions of 75% for MSER and 68% for SW. This is because, in this method, the corpus is directly searched for the bag-of-letter features to present possible suggestions for the text. The conclusion is that MSER gives more precise letter detection, while SW repeats and produces more letters. This helps with word generation, but not with BOL matching, as this is affected by any extra letters. The results so far are promising in terms of further improving the detection and reading of the text in these types of images. In future work, we plan to improve the quotation matching stage by the addition of more methods to map to the correct answer. If our calligraphy dataset could be significantly extended, this would enable the utilisation of state-of-the-art deep learning methods, which have demonstrated their great utility on image analysis problems [26].

Acknowledgement. SA would like to thank King Saud University for funding this research.

References

1. Sun, Y., Zhang, C., Huang, Z., Liu, J., Han, J., Ding, E.: TextNet: Irregular Text Reading from Images with an End-to-End Trainable Network, pp. 1–17 (2018)
2. Azmi, A., Alsaiari, A.: Arabic typography: a survey. Int. J. Electr. Comput. Sci. **9**(10), 16–22 (2010)
3. Bataineh, B., Norul, S., Sheikh, H., Omar, K.: Generating an arabic calligraphy text blocks for global texture analysis. Int. J. Adv. Sci. Eng. Inf. Technol. **1**, 150–155 (2011)
4. Saberi, A., et al.: Evaluating the legibility of decorative Arabic scripts for Sultan Alauddin mosque using an enhanced soft-computing hybrid algorithm. Comput. Human Behav. **55**, 127–144 (2016)
5. Mohamed, N.A., Youssef, K.T.: Arts and design studies utilization of arabic calligraphy to promote the arabic identity in packaging designs. Arts Des. Stud. **19**, 35–49 (2014)
6. Bataineh, B., Abdullah, S.N.H.S., Omar, K.: A novel statistical feature extraction method for textual images: Optical font recognition. Expert Syst. Appl. **39**(5), 5470–5477 (2012)
7. Aburas, A.A., Gumah, M.E.: Arabic handwriting recognition : challenges and solutions electrical and computer engineering dept international islamic university malaysia department of information technology, university technology PETRONAS 2. Pervious related Research Work. In: International Symposium on Information Technology, pp. 1–6 (2008)
8. Bhowmik, S., Sarkar, R., Nasipuri, M., Doermann, D.: Text and non-text separation in offline document images: a survey. Int. J. Doc. Anal. Recognit. **21**(1–2), 1–20 (2018)
9. Nagy, G.: Training a calligraphy style classifier on a non-representative training set. Electron. Imag. **2016**(17), 1–8 (2017)
10. Jiulong, Z., Luming, G., Su, Y., Sun, X., Li, X.: Detecting Chinese calligraphy style consistency by deep learning and one-class SVM. In: 2017 2nd International Conference Image, Vis. Computer ICIVC 2017, pp. 83–86 (2017)
11. Ye, Q., Doermann, D.: Text detection and recognition in imagery: a survey. IEEE Trans. Pattern Anal. Mach. Intell. **37**(7), 1480–1500 (2015)
12. Rabi, M., Amrouch, M., Mahani, Z.: A survey of contextual handwritten recognition systems based HMMs for cursive arabic and latin script. Int. J. Comput. Appl. **160**(2), 31–37 (2017)
13. Azmi, M.S., Omar, K., Nasrudin, M.F., Wan Mohd Ghazali, K., Abdullah, A.: Arabic calligraphy identification for digital jawi paleography using triangle blocks. In: Proceedings 2011 International Conference Electronic Engineering Informatics, ICEEI 2011, July, pp. 1–5 (2011)
14. Bataineh, B., Abdullah, S.N.H.S., Omar, K.: Arabic calligraphy recognition based on binarization methods and degraded images. In: Proceedings 2011 International Conference Pattern Analysis Intelligence Robotics. ICPAIR 2011, vol. 1, pp. 65–70 (2011)
15. Azmi, M.S., Omar, K., Nasrudin, M.F., Muda, A.K., Abdullah, A.: Arabic calligraphy classification using triangle model for Digital Jawi Paleography analysis. In: Proceedings 2011 11th International Conference Hybrid Intelligence System, HIS 2011, pp. 704–708 (2011)
16. AlSalamah, S., King, R.: Towards the machine reading of arabic calligraphy: a letters dataset and corresponding corpus of text. In: 2nd IEEE Workshop on Arabic and Derived Script Analysis and Recognition, ASAR 2018, pp. 19–23 (2018)
17. Bin, P.G., Bin Talal, M.: Free Islamic Calligraphy (2012). https://freeislamiccalligraphy.com/

18. Makandar, A., Halalli, B.: Image enhancement techniques using highpass and lowpass filters. Int. J. Comput. Appl. **109**(14), 21–27 (2015)
19. Chen, Q., Sen Sun, Q., Ann Heng, P., Shen Xia, D.: A double-threshold image binarization method based on edge detector. Pattern Recognit. **41**(4), 1254–1267 (2008)
20. Gui, Y., Bai, X., Li, Z., Yuan, Y.: Color image segmentation using mean shift and improved spectral clustering. EURASIP J. Adv. Signal Process. **2012**(December), 5–7 (2012)
21. Lampert, C.H., Blaschko, M.B., Hofmann, T.: Beyond sliding windows: object localization by efficient subwindow search. In: 26th IEEE Conference on Computer Vision and Pattern Recognition, CVPR (2008)
22. Bay, H., Ess, A., Tuytelaars, T., Vangool, L.: Speeded-Up robust features (SURF) (Cited by: 2272). Comput. Vis. Image Underst. **110**(3), 346–359 (2008)
23. Dalal, N., Triggs, B., Dalal, N., Triggs, B.: Histograms of oriented gradients for human detection to cite this version: histograms of oriented gradients for human detection. IEEE Comput. Soc. Conf. Comput. Vis. Pattern Recognit. **1**, 886–893 (2005)
24. Haroon, M.: Comparative analysis of stemming algorithms for web text mining. Int. J. Mod. Educ. Comput. Sci. **10**(9), 20–25 (2018)
25. Nguyen, N.T.H., Soto, A.J., Kontonatsios, G., Batista-Navarro, R., Ananiadou, S.: Constructing a biodiversity terminological inventory. PLoS ONE **12**(4), 1–23 (2017)
26. Zhang, J., Guo, M., Fan, J.: A novel CNN structure for fine-grained classification of Chinese calligraphy styles. Int. J. Doc. Anal. Recognit. **22**(2), 177–188 (2019)

SMOTE Algorithm Variations in Balancing Data Streams

Bogdan Gulowaty$^{(\boxtimes)}$ and Paweł Ksieniewicz

Department of Systems and Computer Networks,
Wroclaw University of Science and Technology, Wroclaw, Poland
{bogdan.gulowaty,pawel.ksieniewicz}@pwr.edu.pl

Abstract. From one year to another, more and more vast amounts of data is being created in different fields of application. Great deal of those sources require real-time processing and analyzing, which leads to increased interest in streaming data classification field of machine learning. It is not rare, that many of those applications deal with somehow skewed or imbalanced data. In this paper, we analyze usage of SMOTE oversampling algorithm variations in learning patterns from imbalanced data streams using different incremental learning ensemble algorithms.

Keywords: Data streams · Imbalanced learning · Synthetic oversampling · Classifier ensembles

1 Introduction

In 2013 *Twitter* engineers reported, that – on average in one day – their users create over 500 million *tweets* and that they have reached peak of 140 thousand messages in one second [13]. By the end of 2018 *Facebook* noted 1.52 billions daily active users, and over 2 billions monthly [7]. Taking examples from financial domain – the forecast for 2019 is that non-cash transactions count will reach record number of 670 billions [5]. All those applications create vast volumes of data in every minute, which often need to be analyzed in near real-time manner, to provide some kind of business value. Examples may vary from creating accurate advertisement suggestions, collecting and processing data from IOT sensors to preventing money laundering (AML), by analyzing those thousands of banking transactions each hour.

Often, in real-world applications of predictive analysis, there tends to exist phenomena of *data imbalance*. Examples of those might be aforementioned AML, fraud detection or spam detection, where suspicious examples are usually overwhelmed by those insignificant, non-threatening ones. A way to handle this problem is to introduce balance in some certain way, for example by *undersampling* majority or *oversampling* minority patterns, like in SMOTE algorithm.

© Springer Nature Switzerland AG 2019
H. Yin et al. (Eds.): IDEAL 2019, LNCS 11872, pp. 305–312, 2019.
https://doi.org/10.1007/978-3-030-33617-2_31

1.1   Learning from Data Streams

In traditional approach to supervised classification problem, we are dealing with finite dataset $S = \{(X_1, y_1), (X_2, y_2), ..., (X_n, y_n)\}$ of n samples, where X_n is vector consisting of f features $X_n = [x_n^1, x_n^2, ..., x_n^f]$, qualitative or quantitative, and y_n is discrete label assigned to this vector. Our aim is to induce such function $h(x) \rightarrow \hat{y}$ (called hypothesis), using supervised learning algorithm A and knowing that S comes from some unknown distribution D, that will approximate this distribution by minimizing generalization error [1]:

$$\hat{e}(h(x)|S) = \sum_{n=1}^{|S|} 1(h(x_n) \neq y_n),\qquad(1)$$

where

$$1(a \neq b) = \begin{cases} 1 & \text{if } a \neq b \\ 0 & \text{if } a = b. \end{cases}\qquad(2)$$

In case of learning from data streams, we are dealing with possibly infinite amount of samples, which may come from different distributions D. In case the latter case, this phenomena is called *concept drift*. Such conditions require learning algorithms to be able to react to possible changes, for example either by adapting blindly to consecutive processed samples, or by detecting appearance of such drift and reacting appropriately. In other words, algorithm needs to learn incrementally. In this work, we are comparing algorithms, which adapt to changes blindly, by using ensemble of learners and adjusting it accordingly to most recent samples [11].

1.2   Concept Drift

As said in Sect. 1.1, concept drift occurs when, taking consecutive samples or chunks of data, distribution from which patterns were drawn has changed. More precisely, given distribution D_t, from which data was sampled in time t, which was created by some hidden function f_t, drift occurs, when $f_t \neq f_{t+1}$. If the generating function's change was caused by some evolving behaviour – one might imagine data generator with slowly moving in space centroid – then it is incremental drift, as shown in Fig. 1. In case where the next function is completely different with current one and the aspect of change is not connected with distribution's evolving character, there are sudden (Fig. 2), gradual (Fig. 3) and recurring (Fig. 4) drifts.

Fig. 1. Incremental concept drift.

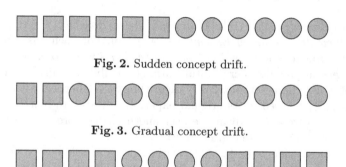

Fig. 2. Sudden concept drift.

Fig. 3. Gradual concept drift.

Fig. 4. Recurring concept drift.

1.3 Imbalanced Data

Problem of imbalanced data was deeply researched in the literature [10]. Generally speaking, dataset is imbalanced, when class with one label greatly overwhelms others. The basic metric for measuring imbalances is ratio. More precisely, data set is imbalanced, when *a priori* probability of labels is unevenly distributed. Hypotheses induced with imbalanced data set tend to *overfit* in favour of majority class, therefore has weak generalization abilities for the minority class. Another problems escalate when it comes to evaluation of such hypothesis. Some metrics tend to distort real performance of classifier. Example of such metric might be accuracy score – given binary dataset with imbalance ratio of 9 to 1, along with majority classifier, accuracy would raise score of 90%, where in fact, this classifier would not recognize single minority class sample.

SMOTE. SMOTE is oversampling algorithm, designed to tackle with data imbalance problem by synthetically generating samples of underrepresented concepts. In its simplest version [6], it starts by randomly choosing sample from processed set. Next step is to select one of k nearest neighbors of this sample, which will be used to generate synthetic pattern. Each of n attributes of the synthetic sample $X_{synth} = [x_{synth}^{(1)}, x_{synth}^{(2)}, ..., x_{synth}^{(n)}]$ is described as

$$x_{synth}^{(n)} = x^{(n)} + gap^{(n)} * (x_{neigh}^{(n)} - x^{(n)}), \tag{3}$$

where:

- $x^{(n)}$ is n-th attribute of chosen sample x,
- $x_{neigh}^{(n)}$ is n-th attribute of randomly chosen neighbor from k neighbors of x sample,
- $gap^{(n)}$ is uniformly distributed random variable from $(0, 1)$ for attribute n.

2 Approach

We have used three, representing different approaches to problem, ensemble-based classification algorithms for non-stationary data streams [12], namely

Streaming Ensemble (SEA) [15], *Accuracy Updated Ensemble 2* (AUE2) [3] and *Modified Weighted Accuracy Ensemble* (WAE) [16]. For each ensemble, *scikit-learn's* [14] implementation of *Gaussian Naive Bayes* was chosen as base classifier. This version provides incremental learning possibility, which was important because of AUE2 characteristic. Every algorithm was either left with its default parameters, as proposed in its original paper, or adjusted to match others. We do not try to prove ones' advantage over another, therefore we did not adjust hyperparameters in more fine grained way.

Four of most widely used SMOTE variations were selected to conduct the comparison – basic SMOTE [6], *Borderline*-SMOTE 1, *Borderline*-SMOTE 2 [9] and *Safe-Level*-SMOTE [4]. Oversampling algorithms were tuned to provide equal balance in terms of created synthetic samples. As for k parameter, 3 nearest neighbors were selected.

All of mentioned algorithms, along with testing environment, except for *Naive Bayes*, were implemented independently. Source code can be found at GIT repository[1].

9 different datasets of size 100 000 samples were created with commonly used, in data streams research, generators, SEA, RBF and HYP. Artificial imbalance of 9 to 1 ratio was introduced, by randomly undersampling minority class. Class labels were also noised by 5%. Every dataset, as described in Table 1, expect SEA modification, was generated by MOA software [2]. To make the paper reproducible, we have made our datasets available to download[2].

Experiments were conducted using *test-then-train* approach, as shown in Algorithm 1. Each classifier was presented with chunk of 250 samples, which – given dataset size of 100 000 samples – created 400 iterations in total. During each iteration, selected metrics were calculated using *Predictive Sequential* approach [8]. As shown in Sect. 1.3, not all metrics are suitable for evaluating classification of imbalanced data. In the experiments, recall, g-mean and balanced accuracy were used.

Algorithm 1. Evaluation procedure with SMOTE

---

function TEST-THEN-TRAIN(*model, stream, metric, batch_size, smote*)
    results ← *Array*[]
    while *stream* is not dry **do**
        chunk ← NEXT(*stream, batch_size*)
        result ← TEST(*metric, model, chunk*)
        results ← ADD(*result*)
        balanced_chunk ← BALANCE(*chunk, smote*)
        TRAIN(*model, balanced_chunk*)
    end while
    return *results*
end function

---

[1] https://github.com/bgulowaty/stream-learn/tree/master-thesis.
[2] https://github.com/bgulowaty/smote-paper.

Table 1. Dataset types used in experiments.

Symbol	Description
SEA^{sudden}	Sudden drift generated with SEA for $\theta = 10$ at both diagonals of decision space. Drift occurs instantly with center being at 50000-th sample
SEA^{gradual}_x	Gradual drift generated with SEA for $\theta = 10$ at both diagonals of decision space. Drift occurs in window of x samples with center being at 50000-th sample
$SEA^{\text{sudden}}_{\text{rec}}$	Recurring sudden drift generated with SEA for $\theta = 10$ at both diagonals of decision space. Drift occurs each 20000 samples
$SEA^{\text{gradual}}_{\text{rec } x}$	Recurring gradual drift generated with SEA for $\theta = 10$ at both diagonals of decision space. Drift occurs each 20000 samples in windows of size x
SEA^{rotating}	Incremental drift modification of SEA. Each 250 samples decision function is rotated by $1°$ around center of decision space, starting from angle of $45°$
HYP	Incremental drift created with shifting hyperplane generator. Samples contain 10 attributes and belong to one of two classes. Plane rotates by 0.01 each one sample and there is 10% probability of rotation direction change
RBF	Incremental drift created with *Radial Basis Function* generator. Samples contain 10 attributes and belong to one of two classes. 50 centroids are used to create samples including 10 drifting by 0.01 each sample

3   Results Discussion

By looking at raw experiments results – for example WAE model evaluated on RBF dataset, as shown in Fig. 5 – one of the first observations that comes out is that usage of SMOTE greatly improves recall of minority class, though it comes at cost of reducing majority class recognizability (Fig. 6).

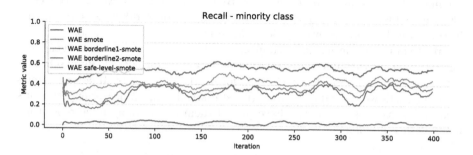

Fig. 5. Minoriy class recall of WAE algorithm evaluated on RBF dataset.

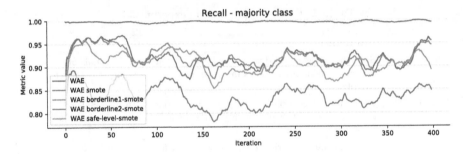

Fig. 6. Majority class recall of WAE algorithm evaluated on RBF dataset.

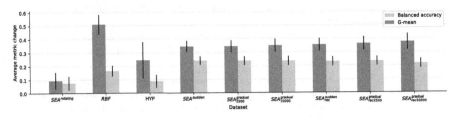

Fig. 7. G-mean and balanced accuracy change for each dataset averaged over SMOTE variations and classification algorithms.

Figure 7 shows change introduced by SMOTE (difference between classifier using SMOTE and unmodified one) for each dataset averaged by all algorithms and oversampling variations. Sudden, gradual and recurring combinations of datasets generated with SEA tend to give similar results. There exists huge standard deviation in case of shifting hyperplane problem, which indicates that either classification or oversampling algorithms react differently to it. Another remark is that improvement in g-mean does not on par with improvement in balanced accuracy, which can be seen by comparing RBF to any of SEA.

Next interesting pattern, seen in Fig. 5, is that – in given example – the original SMOTE variation performs best. By examining Fig. 6, we can see that it also causes biggest drop of majority class recall, although differences are smaller in this case. Figure 8 shows metrics change for each SMOTE variation averaged by algorithms and datasets. It confirms tendency observed before – generally basic variant of oversampling algorithm performs better overall. Research have also shown that said observation is valid for nearly all datasets. *Wilcoxon-Mann-Whitney* tests were conducted to verify statistical significance of experiment's results. For sake of brevity, we have placed them in external online repository, along with some plots[3].

---

[3] https://github.com/bgulowaty/smote-paper.

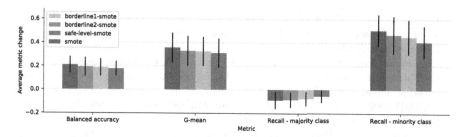

Fig. 8. Metrics change for each SMOTE variation averaged over datasets and classification algorithms.

4 Conclusion and Future Work

In this paper, we have compared different ensemble approaches to non-stationary data streams classification against their combinations with four different variations of SMOTE oversampling algorithm. Experiments were conducted using commonly exploited in evaluation of data stream classifiers generators. Results indicate greatly improvement in minority class detection and overall classification performance introduced by using SMOTE. For both g-mean and balanced accuracy, original SMOTE algorithm variation performed best. Experiments have also shown that, for one data generator (in our case SEA) artificially modified by introducing gradual, sudden and recurring drift, the oversampling algorithm tends to perform similarly.

Further research might span over several directions. Firstly, one could verify if SMOTE would also behave the same way for gradual, sudden and recurring drifts given other data generators. Another interesting topic would be to measure the increase of computational complexity induced by using synthetic oversampling method, as data stream classifiers often need to work under certain bounded performance constraints. Pursuing research in this area would provide answer on SMOTE's usability in more real-life scenarios. Finally, hyperparameters of both oversampling and classification algorithms could be fine-tuned to see, if observed patterns would still be valid.

Acknowledgments. This work is supported by the Polish National Science Center under the Grant no. UMO-2015/19/B/ST6/01597 as well the statutory funds of the Department of Systems and Computer Networks, Faculty of Electronics, Wrocław University of Science and Technology.

References

1. Alpaydin, E.: Introduction to Machine Learning, 2nd edn. The MIT Press, Cambridge (2010)
2. Bifet, A., Holmes, G., Kirkby, R., Pfahringer, B.: MOA: massive online analysis. J. Mach. Learn. Res. **11**(May), 1601–1604 (2010)

3. Brzezinski, D., Stefanowski, J.: Reacting to different types of concept drift: the accuracy updated ensemble algorithm. IEEE Trans. Neural Netw. Learn. Syst. **25**(1), 81–94 (2013)

4. Bunkhumpornpat, C., Sinapiromsaran, K., Lursinsap, C.: Safe-level-SMOTE: safe-level-synthetic minority over-sampling TEchnique for handling the class imbalanced problem. In: Theeramunkong, T., Kijsirikul, B., Cercone, N., Ho, T.-B. (eds.) PAKDD 2009. LNCS (LNAI), vol. 5476, pp. 475–482. Springer, Heidelberg (2009). https://doi.org/10.1007/978-3-642-01307-2_43

5. Capgemini, BNP Paribas: World payments report 2018, October 2018. https://worldpaymentsreport.com/wp-content/uploads/sites/5/2018/10/World-Payments-Report-WPR18-2018.pdf. Accessed 12 Feb 2019

6. Chawla, N.V., Bowyer, K.W., Hall, L.O., Kegelmeyer, W.P.: SMOTE: synthetic minority over-sampling technique. J. Artif. Intell. Res. **16**, 321–357 (2002)

7. Facebook Inc.: Facebook reports fourth quarter and full year 2018 results, January 2019. https://investor.fb.com/investor-news/press-release-details/2019/Facebook-Reports-Fourth-Quarter-and-Full-Year-2018-Results/default.aspx. Accessed 21 Feb 2019

8. Gama, J., Sebastião, R., Rodrigues, P.P.: Issues in evaluation of stream learning algorithms. In: Proceedings of the 15th ACM SIGKDD International Conference on Knowledge Discovery and Data Mining, pp. 329–338. ACM (2009)

9. Han, H., Wang, W.-Y., Mao, B.-H.: Borderline-SMOTE: a new over-sampling method in imbalanced data sets learning. In: Huang, D.-S., Zhang, X.-P., Huang, G.-B. (eds.) ICIC 2005. LNCS, vol. 3644, pp. 878–887. Springer, Heidelberg (2005). https://doi.org/10.1007/11538059_91

10. He, H., Garcia, E.A.: Learning from imbalanced data. IEEE Trans. Knowl. Data Eng. **9**, 1263–1284 (2008)

11. Hoens, T.R., Polikar, R., Chawla, N.V.: Learning from streaming data with concept drift and imbalance: an overview. Prog. Artif. Intell. **1**(1), 89–101 (2012). https://doi.org/10.1007/s13748-011-0008-0

12. Krawczyk, B., Minku, L.L., Gama, J., Stefanowski, J., Woźniak, M.: Ensemble learning for data stream analysis: a survey. Inf. Fusion **37**, 132–156 (2017)

13. Krikorian, R.: New tweets per second record, and how!, August 2013. https://blog.twitter.com/engineering/en_us/a/2013/new-tweets-per-second-record-and-how.html. Accessed 12 Feb 2019

14. Pedregosa, F., et al.: Scikit-learn: machine learning in python. J. Mach. Learn. Res. **12**, 2825–2830 (2011)

15. Street, W.N., Kim, Y.: A streaming ensemble algorithm (sea) for large-scale classification. In: Proceedings of the seventh ACM SIGKDD International Conference on Knowledge Discovery and Data Mining, pp. 377–382. ACM (2001)

16. Woźniak, M., Kasprzak, A.: Data stream classification using classifier ensemble. Schedae Informaticae **23**, 21–32 (2015)

Multi-class Text Complexity Evaluation via Deep Neural Networks

Alfredo Cuzzocrea[1,3]([✉]), Giosué Lo Bosco[2], Giovanni Pilato[3],
and Daniele Schicchi[2]

[1] University of Calabria, Rende, Italy
`alfredo.cuzzocrea@unical.it`
[2] University of Palermo, Palermo, Italy
`{giosue.lobosco,daniele.schicchi}@unipa.it`
[3] ICAR-CNR, Palermo, Italy
`giovanni.pilato@cnr.it`

Abstract. Automatic Text Complexity Evaluation (ATE) is a natural language processing task which aims to assess texts difficulty taking into account many facets related to complexity. A large number of papers tackle the problem of ATE by means of machine learning algorithms in order to classify texts into *complex* or *simple* classes. In this paper, we try to go beyond the methodologies presented so far by introducing a preliminary system based on a deep neural network model whose objective is to classify sentences into more of two classes. Experiments have been carried out on a manually annotated corpus which has been pre-processed in order to make it suitable for the scope of the paper. The results show that a higher detail level of the classification makes the ATE problem much harder to resolve, showing the weaknesses of the model to accomplish the task correctly.

Keywords: Automatic Text Complexity Evaluation · Deep neural network · Text simplification

1 Introduction

Progressive studies in the Natural Language Processing (NLP) field have led the community to develop support systems which can aid people to tackle many different problems. For instance, support systems for business [4,5,7,21], language assistive technologies for students [20] and patients [2] are very common.

In this paper, we present a preliminary study on *text complexity evaluation* topic with the aim of developing a system capable of helping people to assess the complexity of phrases written in the English language at different grades of complexity. The relevance of this study is related to plenty of people communities that encounter problems in daily life actions because they have difficulties in understanding texts. In this regards, the society underestimates the problem, not providing means to help this kind of people which are marginalized.

© Springer Nature Switzerland AG 2019
H. Yin et al. (Eds.): IDEAL 2019, LNCS 11872, pp. 313–322, 2019.
https://doi.org/10.1007/978-3-030-33617-2_32

Tools as that are presented in this paper help writers to elaborate documents suitable for a wider range of people like who have low literacy skills, cognitive impairments, and deafness problems.

The concept of text complexity is strongly related to the reader proficiency; thus, an automatic text complexity evaluation (ATE) system has to be capable of looking into the text from a different point of views relating the peculiarities of complexity with the reading skills. Indeed, many factors affect complexity such as those related to *lexicals, syntactics, morpho-syntactics* and *raw* features which are often difficult to correlate. Since the difficulty of the problem, we have decided to take advantage of the corpora made publicly available by using an approach based on machine learning (ML) algorithms. These resources are specifically created for the topic, and they collect different types of texts which are generally labeled with their grades of complexity. ML methodologies discover what are the competencies of the readers and learn what the characteristics that make a text difficult to comprehend directly analyzing the source data are.

The problem is configured as a classification task; the system is based on Recurrent Neural Network (RNN) algorithms, which are used to classify phrases in three different classes on the basis of their difficulty. The input is structured as a sequence of tokens (i.e., words and punctuation symbols) that are made available for the analysis of the RNN by means of a pre-processing phase. The RNN learns how to identify the complexity of the input analyzing the input sequences detecting most representative features for each grade of difficulty.

The paper is structured as follow: in Sect. 2 we highlight the contributions of our paper, Sect. 3 we report state of the art regarded to the ATE systems, in Sect. 4 it is described the system in a detailed way, experiments and results are explained in Sect. 5 while discussion and conclusion are given in respectively in Sects. 6 and 7.

2   Paper's Contributions

This paper provides the following contributions:

- an innovative approach for supporting the evaluation of text complexity via deep learning techniques;
- the main innovation of our approach consists in supporting *multi-class* text classification, contrary to traditional proposals that consider two classes only;
- a preliminary experimental evaluation that confirms the benefits of our approach.

3   Related Work

The problem of ATE has been analyzed since the 1943rd in which Flesch statistical formula was created to evaluate complexity of a text considering three different language elements: *average sentence length in words, number of affixes* and *number of references to people* [8]. The index has become very common, and it has been applied for evaluating different types of documents like newspaper reports and government publications. Since its relevance, it has been modified to make it more suitable for a specific context. For example, in [14] is proposed a derived index applied to the Navy environment in order to limit understanding difficulties encountered by enlisted personnel and a reworking of Flesch formula has been proposed for measuring complexity of documents written in the Italian language in [9].

Some factors that affect the difficulty of the text are well-known [22] and their use to estimate the text difficulty could be carried out following a rule-based method. Unfortunately, in most of the cases, there are too many factors that have to be linked, making difficult the use of such an approach. This reason has led researchers to use other methodologies to estimate the complexity of texts.

ML algorithms have been tested to exploit ATE corpora for learning from data what distinguishes a text as suitable or not for a target reader. In [24] Simple English Wikipedia and Wikipedia are used to create a corpus which contains sentences labeled respectively as *simple*, and *hard*, then lexical, syntactic and psycholinguistic features are extracted to train SVM model for classifying sentence on the basis of their difficulty. In the medical context, features like word length, sentences length, part-of-speech counts, frequency of common words are used to train ML algorithms for predicting the difficulty of health texts [13]. READ-IT is a sentence classification system based on SVM trained on *hard* and *simple* newspapers written in Italian language. The system exploits lexical, syntactic, raw, and morpho-linguistic features for evaluating the complexity of the input sentence. In [19] a Stochastic Gradient Descent (SGD) classifier is proposed for the binary classification of *complex* and *simple* sentences. The training of the algorithm has been carried out on a specific corpus created aligning sentences of Newsela [26] by means of massAlign [18] system.

The authors have investigated the evaluation of text complexity by using NN models. In [17] is presented a NN based system for evaluating the lexical and syntactic complexity of sentences written in the Italian language. In [15,16], a system based on TreeTagger parser and RNN is presented to examine the syntactic complexity of sentences for Italian and English language. The experiments show the versatility of the system architecture to evaluate the complexity of sentences written in two different languages. Furthermore, in [1] are compared four different ways of representing the elements of the sentence in order to understand how the representation of words and punctuation symbols affect the system performance. The experiments show that the problem of evaluating the text complexity by using RNN model is more related to the architecture than the representation methods.

4   Methodology

We present ATE system for the assessment of sentences at different complexity levels. The evaluation task is tackled as a multi-class classification problem in which the sentences are assigned to three different classes. The architecture of the system is composed of two main modules: *pre-processing* and the *core model*. The pre-processing aims to make sentences suitable for the analysis while the evaluation of sentences complexity is carried out by an RNN which represent the core model.

4.1 Pre-processing

In the NLP field texts are often evaluated as a sequence of tokens like words and punctuation symbols. This kind of representation allows exploiting methodologies based on sequence analysis, which can extract common pattern and regularities that help the classification task. In our case, the pre-processing phase deals with the tokenization of sentences in words and punctuation symbols, and it maps tokens into vectors of real numbers, which make the sentence analyzable by the RNN. The mapping is accomplished by means of a dictionary for the English language that associate words and punctuation symbols to a 300-dimensional vector of real numbers. The dictionary is created by means of Fast-Text, a library for efficient learning of word representations and sentence classification, and it is made publicly available [11]. After the pre-processing phase, each sentence is structured as a matrix of real numbers in which rows are the vector representation of tokens.

4.2 System Architecture

The architecture of the system is an RNN based on Long Short Term Memory (LSTM) neural units. The LSTM unit architecture has been created by researchers with the aim of overcoming the *vanish gradient* problem [10] which affects the training of the classic RNN by making it difficult for long sequences. The formulation of this type of neural units is more complex than the classic one and it introduce the concept of *gate* which act during the elaboration of the sequence. Indicating x_t as the the input sequence at time t, W_h and U_h the weight matrices for the input and the hidden state and b_h the bias vector, the functioning of a LSTM unit is described by the follow equations:

$$f_t = \sigma(W_f x_t + U_f h_{t-1} + b_f)$$
$$i_t = \sigma(W_i x_t + U_i h_{t-1} + b_i)$$
$$o_t = \sigma(W_o x_t + U_o h_{t-1} + b_o)$$
$$c_t = tanh(W_c x_t + U_c h_{t-1} + b_c)$$
$$s_t = f_t \odot s_{t-1} + i_t \odot c_t$$
$$h_t = tanh(s_t) \odot o_t$$

where f_t, i_t, o_t are respectively the input, forget and output gates, the $\odot$ is the element-wise multiplication while tanh is the hyperbolic tangent and sigma is the sigmoid function.

The network is composed of two layers. The first one is a recurrent layer made up of 512 LSTM units that deal with the analysis of the sequence of vectors representing the sentence tokens. The second one is a fully connected layer with three outputs that are stimulated by the previous layer and activated by using `softmax` function [10]. The output of the LSTM layer represents a cumulative result of the analysis of the sentence complexity which stimulating the next layer to allow the network to classify sentence to three different classes on the basis of their difficulties. The graphical representation of the network architecture is illustrated in Fig. 1.

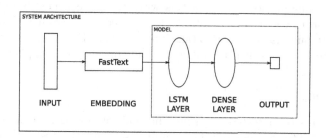

Fig. 1. System description. The input sequence is represented as sequence of vectors by means of the embedding. The LSTM and DENSE layers analyzes the sequence in order to classify sentences as *hard* or *easy* to understand.

The training of the network has been carried out, minimizing the *categorical cross-entropy* loss function by means of *rmsprop* [12] optimization algorithms applied to batches of size 50. In order to prevent over-fitting, it has been applied to the LSTM layer, a dropout regularization technique of 30%. The network has been trained for a variable number of epochs in the range between 1 and 30.

5   Experiments

In this section, we provide our experimental assessment. Due to the preliminary nature of our work, we only consider our technique, with no comparison with existent methods. We left as future work comparison with other proposals.

5.1   Corpus

There is a lack of resources for training ATE systems. To the best of our knowledge, there is not exist a multi-class corpus useful for reaching the objective of this paper; thus the corpus used for training and evaluating the system is extracted from the Newsela [26] resource. Newsela is a collection of 1130 articles which have been manually simplified by a pool of experts. Each article is simplified four times, and it is labeled with an increasing integer number that represents its difficulty. Label *0* indicates the not simplified resource while increasing numbers from *1* to *4* represent an increasing level of simplification. Unfortunately, Newsela is not appropriate for the aims of paper since it is organized in articles and there is no information about the complexity of the sentences inside them. To overcome this trouble, by taking inspiration of [19], we have used a sentences alignment algorithm presented in [18] called MASSAlign. MASSAlign is capable of coupling paragraphs and sentences in comparable documents by exploiting vicinity-driven approach. The tool is used to create a corpus of sentences at different grades of complexity by aligning sentences and taking into account only ones which have been simplified between a document and its successive simplification. The complexity label associated to each sentence is coherent with that one of the belonging document. In this way, it has been possible to create a multi-class corpus containing about 500.000 labeled sentences useful for training the system. In our case, we have to take into account only sentences labeled with *0,2* and *4* which represent three different grade of sentence complexity.

5.2 Experiments

The system has been tested by using the 66% of corpus as training-set and the remaining 33% as test-set. The training-set and the test-set are created in a balanced way taking randomly the same percentage of elements for each class. The system performance evaluation is measured in two different manners. First, we measure cumulative scores by means of averaged recall and averaged precision, then recall and precision is calculated for single classes. This method allows us to map multiple classes into two classes only, thus devising a suitable experimental methodology for multi-class problems, via an ad-hoc 3 x 3 *confusion matrix*, like in classical approaches [23].

Average recall and precision give us information about how well the system is capable of recognizing elements of the five classes and how many mistakes makes during the classification process. The recall and precision measures calculated for each class showing the performance of the system limited to a single class. The Table 1 contains the results achieved by the system for each epoch.

6 Discussion

The evaluation of text complexity has been tackled by using different methodologies. In the context of ML algorithms, state of the art proposes this task as a binary classification problem in which the system is trained to identify *complex* and *simple* sentences. This approach has been showed to be effective but quite rough since the complexity of a text is characterized by many traits which make it comprehensible for many readers classes. In this paper, we configure the problem as a multi-class classification task in order to assign sentences to three different classes on the basis of their difficulty.

Dealing with such a problem are harder than a binary classification task since a higher number of classes requires that the network extracts a greater amount of details in order to carry out the classification in the best possible way. Furthermore, in the context of text complexity evaluation, the subdivision of sentences into a specific number of classes could be difficult also for a human

Table 1. Measurement of system performance. From left to the right: number of epochs utilized for the training process, average recall, average precision, f1 score for each class, average values from multi-class settings.

Epochs	Avg Rec.	Avg Prec.	Class 0	Class 1	Class 2
1	0.497	0.456	0.573	0.070	0.615
2	0.524	0.521	0.598	0.325	0.608
3	0.553	0.544	0.600	0.383	0.657
4	0.521	0.517	0.555	0.209	0.640
5	0.534	0.530	0.497	0.375	0.666
6	0.552	0.565	0.531	0.474	0.657
7	0.580	0.568	0.618	0.400	0.688
8	0.565	0.561	0.633	0.375	0.653
9	0.551	0.551	0.631	0.304	0.643
10	0.555	0.553	0.519	0.396	0.691
11	0.580	0.568	0.627	0.330	0.698
12	0.525	0.544	0.612	0.222	0.614
13	0.576	0.580	0.568	0.470	0.689
14	0.572	0.571	0.548	0.456	0.696
15	0.574	0.581	0.573	0.477	0.679
Epochs	Avg Rec.	Avg Prec.	Class 0	Class 1	Class 2
16	0.581	0.572	0.572	0.439	0.705
17	0.571	0.577	0.542	0.477	0.692
18	0.575	0.565	0.582	0.422	0.696
19	0.554	0.571	0.436	0.492	0.699
20	0.576	0.563	0.596	0.400	0.697
21	0.567	0.568	0.547	0.460	0.689
22	0.573	0.563	0.564	0.426	0.704
23	0.564	0.556	0.607	0.375	0.679
24	0.572	0.562	0.597	0.394	0.699
25	0.565	0.564	0.569	0.438	0.686
26	0.564	0.557	0.580	0.407	0.691
27	0.550	0.563	0.536	0.461	0.666
28	0.536	0.559	0.529	0.456	0.645
29	0.559	0.554	0.585	0.391	0.684
30	0.561	0.548	0.600	0.344	0.692

operator. Indeed, their might exists sentences that do not completely belong to a specific class suggesting to increase the detail levels adding more classes.

In our case, does not exist a handmade corpus that associate sentences to classes, thus, we have relied on an extraction procedure based on MASSAlign tool and Newsela to create a corpus of sentences labeled according to the document complexity to which they belong. Unfortunately, the produced corpus needs to be refined, and it can be used only for preliminary work. Indeed, the difficulty of Newsela documents is measured in a cumulative way, and the complexity level of a sentence inside a document could be different from that one of the belonging document. Thus, two sentences inside the corpus marked with the same label might not be at the same difficulty level since the labels represent only how many time it has been simplified. The uncertainties in the corpus creation can affect the training of the NN, which in an ambiguous environment is prone to make classification errors. Furthermore, the reason why the functioning of the network is not effective might be attributed to either the architecture or the corpus.

Considering the above premise, the results show the limited performance of the system that achieves higher results in the classification of sentences identified as most *complex* and *simplest*. The classification of an intermediate level of complexity the network shows low performances, which are improved by increasing the number of epochs.

7   Conclusions and Future Work

We have introduced a system whose aim is to improve current ATE binary systems taking into account different levels of text complexity. The system learns how to classify sentences taking into account the proficiency of the reader directly from the data set by using the raw text only. Nonetheless, the difficulty of the problem the system shows good abilities to understand different complexity levels by suggesting that both a more reliable corpus and a powerful architecture might lead to improving its performance. The future works will be focused on further experiments which will exploit different types of models and corpora that are under development.

Future work is mainly oriented to extend our approach as to consider big data characteristics (e.g., [3,6,25]).

References

1. Bosco, G.L., Pilato, G., Schicchi, D.: A neural network model for the evaluation of text complexity in Italian language: a representation point of view. Procedia Comput. Sci. **145**, 464–470 (2018)
2. Alfano, M., Lenzitti, B., Lo Bosco, G., Perticone, V.: An automatic system for helping health consumers to understand medical texts, pp. 622–627 (2015)
3. Braun, P., Cameron, J.J., Cuzzocrea, A., Jiang, F., Leung, C.K.: Effectively and efficiently mining frequent patterns from dense graph streams on disk. In: 18th International Conference in Knowledge Based and Intelligent Information and Engineering Systems, KES 2014, Gdynia, Poland, 15–17 September 2014, pp. 338–347 (2014)

4. Chiavetta, F., Lo Bosco, G., Pilato, G.: A lexicon-based approach for sentiment classification of Amazon books reviews in Italian language, vol. 2, pp. 159–170 (2016)
5. Chiavetta, F., Lo Bosco, G., Pilato, G.: A layered architecture for sentiment classification of products reviews in Italian language. Lect. Notes Bus. Inf. Process. **292**, 120–141 (2017)
6. Cuzzocrea, A., Bertino, E.: Privacy preserving OLAP over distributed XML data: a theoretically-sound secure-multiparty-computation approach. J. Comput. Syst. Sci. **77**(6), 965–987 (2011)
7. Di Gangi, M., Lo Bosco, G., Pilato, G.: Effectiveness of data-driven induction of semantic spaces and traditional classifiers for sarcasm detection. Nat. Lang. Eng. **25**(2), 257–285 (2019)
8. Flesch, R.: Marks of Readable Style; A Study in Adult Education. Teachers College Contributions to Education (1943)
9. Franchina, V., Vacca, R.: Adaptation of flesh readability index on a bilingual text written by the same author both in Italian and English languages. Linguaggi **3**, 47–49 (1986)
10. Goodfellow, I., Bengio, Y., Courville, A.: Deep Learning. MIT Press, Cambridge (2016)
11. Grave, E., Bojanowski, P., Gupta, P., Joulin, A., Mikolov, T.: Learning word vectors for 157 languages. CoRR abs/1802.06893 (2018). http://arxiv.org/abs/1802.06893
12. Hinton, G., Srivastava, N., Swersky, K.: Neural networks for machine learning lecture 6a overview of mini-batch gradient descent (2012)
13. Kauchak, D., Mouradi, O., Pentoney, C., Leroy, G.: Text simplification tools: using machine learning to discover features that identify difficult text. In: 2014 47th Hawaii International Conference on System Sciences, pp. 2616–2625, January 2014. https://doi.org/10.1109/HICSS.2014.330
14. Kincaid, J.: Derivation of New Readability Formulas: (automated Readability Index, Fog Count and Flesch Reading Ease Formula) for Navy Enlisted Personnel. Research Branch report, Chief of Naval Technical Training, Naval Air Station Memphis (1975). https://books.google.it/books?id=4tjroQEACAAJ
15. Lo Bosco, G., Pilato, G., Schicchi, D.: A sentence based system for measuring syntax complexity using a recurrent deep neural network. In: 2nd Workshop on Natural Language for Artificial Intelligence, NL4AI 2018, vol. 2244, pp. 95–101. CEUR-WS (2018)
16. Schicchi, D., Lo Bosco, G., Pilato, G.: Machine learning models for measuring syntax complexity of English text. In: Samsonovich, A.V. (ed.) BICA 2019. AISC, vol. 948, pp. 449–454. Springer, Cham (2020). https://doi.org/10.1007/978-3-030-25719-4_59
17. Lo Bosco, G., Pilato, G., Schicchi, D.: A recurrent deep neural network model to measure sentence complexity for the Italian language. In: International Workshop on Artificial Intelligence and Cognition, 6th Edition, Palermo, Italy (2018, in press)
18. Paetzold, G., Alva-Manchego, F., Specia, L.: Massalign: alignment and annotation of comparable documents. In: Proceedings of the IJCNLP 2017, System Demonstrations, pp. 1–4 (2017)
19. Scarton, C., Paetzold, G., Specia, L.: Text simplification from professionally produced corpora. In: Proceedings of the Eleventh International Conference on Language Resources and Evaluation (LREC-2018). European Languages Resources Association (ELRA), Miyazaki, Japan, May 2018. https://www.aclweb.org/anthology/L18-1553

20. Schicchi, D., Pilato, G.: A social humanoid robot as a playfellow for vocabulary enhancement. In: 2018 Second IEEE International Conference on Robotic Computing (IRC), pp. 205–208. IEEE Computer Society, Los Alamitos, February 2018

21. Schicchi, D., Pilato, G.: WORDY: a semi-automatic methodology aimed at the creation of neologisms based on a semantic network and blending devices. In: Barolli, L., Terzo, O. (eds.) Complex, Intelligent, and Software Intensive Systems. AISC, vol. 611, pp. 236–248. Springer International Publishing, Cham (2018). https://doi.org/10.1007/978-3-319-61566-0_23

22. Siddharthan, A.: A survey of research on text simplification. ITL Int. J. Appl. Linguist. **165**(2), 259–298 (2014)

23. Subramani, S., Michalska, S., Wang, H., Du, J., Zhang, Y., Shakeel, H.: Deep learning for multi-class identification from domestic violence online posts. IEEE Access **7**, 46210–46224 (2019)

24. Vajjala, S., Meurers, D.: Assessing the relative reading level of sentence pairs for text simplification. In: Proceedings of the 14th Conference of the European Chapter of the Association for Computational Linguistics, pp. 288–297 (2014)

25. Wu, Z., Yin, W., Cao, J., Xu, G., Cuzzocrea, A.: Community detection in multi-relational social networks. In: Web Information Systems Engineering - WISE 2013–14th International Conference, Nanjing, China, October 13–15, 2013, Proceedings, Part II, pp. 43–56 (2013)

26. Xu, W., Callison-Burch, C., Napoles, C.: Problems in current text simplification research: new data can help. Trans. Assoc. Comput. Linguist. **3**, 283–297 (2015)

Imbalance Reduction Techniques Applied to ECG Classification Problem

Jędrzej Kozal and Paweł Ksieniewicz$^{(\boxtimes)}$

Department of Systems and Computer Networks,
Wroclaw University of Science and Technology, Wroclaw, Poland
`pawel.ksieniewicz@pwr.edu.pl`

Abstract. In this work we explored capabilities of improving deep learning models performance by reducing the dataset imbalance. For our experiments a highly imbalanced ECG dataset MIT-BIH was used. Multiple approaches were considered. First we introduced mutliclass UMCE, the ensemble designed to deal with imbalanced datasets. Secondly, we studied the impact of applying oversampling techniques to a training set. SMOTE without prior majority class undersampling was used as one of the methods. Another method we used was SMOTE with noise introduced to synthetic learning examples. The baseline for our study was a single ResNet network with undersampling of the training set. Mutliclass UMCE proved to be superior compared to the baseline model, but failed to beat the results obtained by a single model with SMOTE applied to training set. Introducing perturbations to signals generated by SMOTE did not bring significant improvement. Future work may consider combining multiclass UMCE with SMOTE.

Keywords: Machine learning · ECG classification · Imbalanced data

1 Introduction

When dealing with imbalanced datasets we are faced with a significant disparity in number of learning examples from different classes. We call the class with higher number of samples the majority class, and the class with lower number of samples the minority class. Majority class can dominate the predictions made by models with ease. This influences the performance of our machine learning algorithms. It is easy for neural network to overfit the imbalanced dataset, just by adjusting biases of last layer to always point the majority class. In this case accuracy of the model can be as high as the proportion of the majority class to the all learning examples.

There are many recent papers regarding the ECG signal classification with deep neural networks [5,8,12,13]. Some of them introduce innovative ideas or report very high value of metrics. However only few of them mention the problem of imbalance of the dataset. In this work, we argue, that as in many medical problems, the imbalance of the ECG dataset is the key factor and addressing it can bring significant improvement to the model performance.

© Springer Nature Switzerland AG 2019
H. Yin et al. (Eds.): IDEAL 2019, LNCS 11872, pp. 323–331, 2019.
https://doi.org/10.1007/978-3-030-33617-2_33

One of basic methods of dealing with imbalanced data is undersampling i.e. drawing learning examples randomly from a majority class in order to equalise its size with the minority class. This is not sufficient when dealing with deep learning models, which require many learning examples. Purpose of this study is to examine the effect of imbalance reduction methods on the quality of deep networks classification metrics.

The single elements of a datasets are sometimes referred to as a samples. Due to a temporal character of our data (each element of a dataset is signal consisting of many samples) this may be confusing. To avoid misconceptions from now on we will refer to a single element of dataset as a learning example (or just example) and to an elements of learning example as samples.

2   Related Work

Our work is heavily influenced by [8]. In that paper little emphasis was put on the imbalance of the data. In order to equalise the number of samples, a simple data augmentation technique was applied. Taking imbalance of a dataset into consideration can greatly improve a performance of the model.

2.1   Residual Networks

Residual networks were proposed in [3]. They utilise skip connections to ease a learning task. Due to the usage of skip connections a neural network layers must learn only the residual mapping $\mathcal{F}$:

$$y = x + \mathcal{F}(x) \tag{1}$$

instead of learning the whole mapping from x to y. Most common application of residual learning are convolutional neural networks (CNNs). CNNs can be applied to any data with grid-like structure [7], including a time series. Some papers studied the impact of applying a residual learning to the optimisation process [2]. This study showed that ResNets performance does not degrade when using deeper architectures.

2.2   UMCE

Undersampled Majority Class Ensemble (UMCE) for a binary classification task was introduced in [9]. UMCE is an ensemble classifier that changes an unbalanced learning problem into a set of smaller, balanced problems. This can be achieved by dividing the set of majority class examples into k folds, containing a number of examples equal to the number of examples of the minority class. Next, a pool of base classifiers Ψ is trained on a training set consisting of one of k majority class folds and all minority examples. Using the supports of Ψ a fussing method can be applied to obtain the final decision of an ensemble.

2.3 SMOTE

Synthetic Minority Over-sampling Technique (SMOTE) was introduced in [4]. It is proven to be an effective technique. SMOTE generates additional examples by using distances in the feature space. First it chooses k nearest neighbours of a given example. One of neighbours is chosen randomly. An additional sample is generated by selecting a random point on the line in the feature space connecting the neighbour and the base example. Sometimes SMOTE can choose more than one neighbour, depending on the desired number of a synthetic examples. Also it can be combined with undersampling of the majority class.

3 Methods

3.1 Undersampling and the Baseline Model

As the most simple method of solving this task we employed undersampling to the lowest cardinality class. This solution served as a baseline for the comparison of all methods performance. Also, for data with applied undersampling, the structure of ResNet network was fine tuned. Networks with the same structure were used later in experiments. Of course structure of the network can be obtained for each of the methods separately, but the sole purpose of this study is to compare an impact of the imbalance reduction methods, not to obtain the highest possible value of metrics.

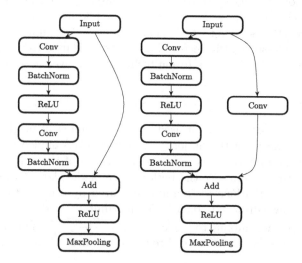

Fig. 1. (Left) the schema of a block with no conv on skip connections - block A of ResNet. (Right) the schema of a block with conv on skip connection - block B of ResNet.

Base Model. As a base model, a ResNet with 1D convolution was utilised. Used ResNet consists of two kinds of blocks: A and B (see Fig. 1). Block B is used when the number of filters is increased. In that case convolution is applied to the skip connection. This changes the tensor shape of a skip connection (its depth) and enables adding it to an output of the two convolutional layers with an increased number of filters.

The number of kernels in each convolutional layer was, respectively 32-32-64-64-128. All kernels were of size 5 and were applied with stride 1. MaxPooling was applied with stride 2. Skip connections were applied to two convolutional layers with batch normalisation. The input of the whole model is passed to a single convolution layer with 32 filters. Output of this layer is feed into five blocks, respectively: A, A, B, A, B. After these blocks three fully connected layers were applied as the classifier part of the network. The first two with the ReLU activation function and 512 neurons, and the last one with softmax.

3.2 Multiclass UMCE

For the purpose of this paper, a multiclass UMCE variant was developed based on the solution for binary classification. The algorithm is given in 1. The majority and minority classes here are classes with highest and lowest number of samples respectively. Please note that we limit the maximum number of base classifiers to 10 in order to restrict the training time of the whole ensemble. This limitation can be omitted or tuned when necessary. The number of folds for each class is different, therefore during the training of base classifiers we draw one of all folds available for each class. For each training set, the fold for a class is drawn once, so there is no risk that training set will contain repeated examples. After training each of the base classifiers, the vector of all supports is used to provide the final decision. In the case of this work fusion technique was average support.

3.3 SMOTE

In our experiments SMOTE was utilised with the number of neighbours $k = 5$. All classes with except of the majority class were oversampled to the number of examples of the majority class. No undersampling of the majority class was performed. In order to improve the performance of the classification algorithm we introduced perturbations to samples generated by SMOTE. First we added random noise drawn from $U(-0.05, 0.05)$ distribution, and later we stretched signals by applying resampling to the length of $1 + \frac{U(-0.5, 0.5)}{3}$. This augmentation method was used in [8]. Later we will refer to this method as SMOTE with data augmentation.

4 Experiment Setup

In order to reduce the impact of variation, each model was trained with 10 fold cross validation. Tuning of hyperparameters was performed using 10-fold cv,

#min = number of learning examples of minority class;
#max = number of learning examples of majority class;
IR = $\frac{\#max}{\#min}$;
number of classifiers = $\min(10, \lfloor IR \rfloor)$;
for *i from 0 to number of classes* **do**
    #i = number of samples in class;
    $k_i = \left\lfloor \frac{\#i}{\#min} \right\rfloor$;
    k_i-fold division of all classes;
end
for *j from 0 to number of classifiers* **do**
    draw random fold for each class;
    combine all folds in training set;
    sample base classifier Ψ_j;
    train Ψ_j on training set
end

Algorithm 1. Multiclass UMCE training algorithm

but without test set examples from the standard train-test split of a dataset. Precision, recall and f1-score were reported to compare the performance of each model.

Statistical analysis was performed using Kruskal test with Conover posthoc tests. We were considering precision, recall and f1-score separately. Tests were performed using results from folds averaged over all labels.

4.1 Dataset

In our study MIT-BIH Arrhythmia Dataset was used. It is available online[1]. ECG signals in this dataset are normalised to fit $[0, 1]$ interval, and each training example is 187 samples long. There are 109446 training examples with 5 classes. Details of class cardinality are given in Table 1. IR in this table is calculated with regard to the majority class (label 0).

Table 1. Number of samples and IR for each class.

Class	0	1	2	3	4
Number of samples	72470	2223	5788	641	6431
IR	-	32.6:1	12.52:1	113.06:1	11.27:1

---

[1] https://www.kaggle.com/shayanfazeli/heartbeat.

4.2   Tools

All computations were performed using computer with IntelCore i7 8700 CPU and GTX 1070Ti GPU. The dataset used for this study was downloaded from kaggle service. The models were implemented using keras [6] library with tensorflow backend [1]. To develop experiment framework packages scikit-learn [10], imb-learn, numpy [11] and scipy were utilised.

4.3   Neural Network Training Process

All Neural Networks were trained for 20 epochs with initial learning rate 0.001 decreased at epochs 8, 13 and 18 by factor of 10. Adam optimiser algorithm was employed with categorical cross entropy loss function. Batch size was 32.

5   Results

Table 2. Test set results averaged over all folds.

Model	Label	Precision	Recall	f1-score
ResNet with undersampling	0	0.994	0.892	0.940
	1	0.306	0.890	0.455
	2	0.822	0.934	0.874
	3	0.225	0.927	0.361
	4	0.933	0.982	0.957
	avrg	0.656	0.925	0.717
Multiclass UMCE	0	1.000	0.918	0.955
	1	0.369	0.919	0.524
	2	0.878	0.951	0.913
	3	0.25	0.94	0.394
	4	0.967	0.989	0.977
	avrg	0.693	0.943	0.753
ResNet with SMOTE	0	0.994	0.996	0.995
	1	0.917	0.893	0.905
	2	0.976	0.972	0.974
	3	0.871	0.818	0.843
	4	0.993	0.993	0.993
	avrg	0.950	0.934	0.942
ResNet with SMOTE (with data augmentation)	0	0.993	0.996	0.995
	1	0.927	0.884	0.905
	2	0.978	0.969	0.974
	3	0.867	0.817	0.841
	4	0.995	0.994	0.994
	avrg	0.952	0.932	0.942

Table 3. H-values and p-values for Kurskal test

Metric	H-value	p-value
Precision	33.354	2.710e−07
Recall	17.148	0.001
f1-score	32.975	3.258e−07

Table 4. Conover post-hoc test results. SwA is short for SMOTE with augmentation.

precision

Undersampling UMCE	Undersampling SMOTE	Undersampling SwA	UMCE SMOTE	UMCE SwA	SMOTE SwA
2.492e−05	2.089e−13	5.056e−15	1.606e−07	1.562e−09	0.130

recall

Undersampling UMCE	Undersampling SMOTE	Undersampling SwA	UMCE SMOTE	UMCE SwA	SMOTE SwA
6.985e−06	0.022	0.064	0.006	0.001	0.634

f1-score

Undersampling UMCE	Undersampling SMOTE	Undersampling SwA	UMCE SMOTE	UMCE SwA	SMOTE SwA
3.990e−05	1.011e−13	6.389e−14	4.120e−08	2.341e−08	0.852

Results averaged over all folds for separate labels and averaged over all labels are given in Table 2. As expected, for all models, the most problematic classes were the ones with the lowest number of samples (1 and 3).

Based on the results of Kruskal tests presented in Table 3 with $\alpha = 0.005$ significance level, we can conclude that all algorithms differ in a significant way. We analyse the post-hoc tests results provided in Table 4 with the same significance level. In case of precision and f1-score all the algorithms differ in a significant way, with the exception of SMOTE and SMOTE with augmentation. For recall all algorithms differ with the exception of undersampling and SMOTE with augmentation and SMOTE and SMOTE with augmentation.

All of the analysed algorithms achieved the high value of recall. This property is desired in all medical applications. Of all methods multiclass UMCE obtained highest average recall and recall for classes 1 and 3. Considering precision and f1-score, UMCE proposed in this paper brings improvement compared to a single model with undersampling. However SMOTE with and without data augmentation obtain the best results, without significant difference between these two approaches. Introducing augmentation in this case had only a small impact on the precision-recall tradeoff. F1-score remains the same for SMOTE with and without augmentation.

6   Conclusions

This research is quite compelling because it employs classic machine learning techniques to push forward the performance of deep learning models. By using this approach we were able to improve on the values of precision, recall and f1-score metrics for ResNets.

More often than not SMOTE is used after applying undersampling to the majority class. That was not the case in our experiments. Deep learning models in general benefit greatly from the increased number of the learning examples in a dataset. This dependence on the size of the dataset also may contribute to worse performance of multiclass UMCE compared to SMOTE. Future work may study the influence of applying SMOTE to reduce IR to some degree, and then constructing multiclass UMCE. Another possible direction of research is to increase the capacity of the ResNet model while applying SMOTE or to use a loss function designed to deal with the imbalanced problems.

Also, one can find surprising that SMOTE applied to temporal signals can yield such good results. SMOTE does not take into consideration temporal dependence of signal samples. In this case we assume that the success of SMOTE can be attributed to a great number of learning examples in the dataset and to normalised value of each signal to [0,1] interval. The code used for experiments is available online[2].

Acknowledgments. This work is supported by the Polish National Science Center under the Grant no. UMO-2015/19/B/ST6/01597 as well the statutory funds of the Department of Systems and Computer Networks, Faculty of Electronics, Wrocław University of Science and Technology.

We also wanna thank Michał Leś for lending his computing power resources. Thanks to him this results could be collected and presented.

References

1. Abadi, M.: TensorFlow: Large-scale machine learning on heterogeneous systems (2015). http://tensorflow.org/
2. Li, H., et al.: Visualizing the loss landscape of neural nets. CoRR abs/1712.09913 (2017). http://arxiv.org/abs/1712.09913
3. He, K., et al.: Deep residual learning for image recognition. CoRR abs/1512.03385 (2015). http://arxiv.org/abs/1512.03385
4. Bowyer, K.W., et al.: SMOTE: synthetic minority over-sampling technique. CoRR abs/1106.1813 (2011). http://arxiv.org/abs/1106.1813
5. Jun, T.J., et al.: ECG arrhythmia classification using a 2-D convolutional neural network. CoRR abs/1804.06812 (2018). http://arxiv.org/abs/1804.06812
6. Chollet, F.E.A.: Keras (2015). https://keras.io
7. Goodfellow, I., Bengio, Y., Courville, A.: Deep Learning. MIT Press, Cambridge (2016). http://www.deeplearningbook.org

---

[2] https://github.com/jedrzejkozal/ecg_oversampling.

8. Kachuee, M., Fazeli, S., Sarrafzadeh, M.: ECG heartbeat classification: a deep transferable representation. CoRR abs/1805.00794 (2018). http://arxiv.org/abs/1805.00794

9. Ksieniewicz, P.: Undersampled majority class ensemble for highly imbalanced binary classification. Proc. Mach. Learn. Res. 1, 1–13 (2010)

10. Pedregosa, F.E.A.: Scikit-learn: machine learning in Python. J. Mach. Learn. Res. 12, 2825–2830 (2011)

11. van der Walt, S., Colbert, S.C., Varoquaux, G.: The numpy array a structure for efficient numerical computation. Comput. Sci. Eng. 13(2), 22–30 (2011)

12. Xiong, Z.E.A.: ECG signal classification for the detection of cardiac arrhythmias using a convolutional recurrent neural network. Physiol. Meas. (2018). https://doi.org/10.1088/1361-6579/aad9ed

13. Xu, S.S., Mak, M.W., Cheung, C.C.: Towards end-to-end ECG classification with rawsignal extraction and deep neural networks. IEEE J. Biomed. Health Inform. (2019)

Machine Learning Methods for Fake News Classification

Paweł Ksieniewicz[1]([✉]) [iD], Michał Choraś[2], Rafał Kozik[2] [iD],
and Michał Woźniak[1] [iD]

[1] Department of Systems and Computer Networks, Faculty of Electronics,
Wrocław University of Science and Technology, Wybrzeże Wyspiańskiego 27,
50-370 Wrocław, Poland
{pawel.ksieniewicz,michal.wozniak}@pwr.edu.pl
[2] Department of Teleinformatics Systems,
UTP University of Science and Technology, Bydgoszcz, Poland
{chorasm,kozikr}@utp.edu.pl

Abstract. The problem of the fake news publication is not new and
it already has been reported in ancient ages, but it has started having
a huge impact especially on social media users. Such false information
should be detected as soon as possible to avoid its negative influence
on the readers and in some cases on their decisions, e.g., during the
election. Therefore, the methods which can effectively detect fake news
are the focus of intense research. This work focuses on fake news detection
in articles published online and on the basis of extensive research we
confirmed that chosen machine learning algorithms can distinguish them
from reliable information.

Keywords: Fake news · Online disinformation · Classification ·
Classifier ensebles · Random Subspace

1 Introduction

The problems related to the fake news are not new, but such type of misinformation has become very popular when it has been used during 2016 U.S.
Presidential Campaign. Nowadays, fake news can easily spread mainly using
social media but occasionally the mainstream media as well. It is worth noting
that its popularity can rapidly growth according the rule that *false news spreads
faster and wider*. Its extensive spread has a serious negative impact on media
users and society.

Sabrina Tavernise defines fake news as *"a made-up story with an intention to
deceive"*[1]. Nevertheless, it is very hard to define however fake news, especially

---

[1] Sabrina Tavernise, *As Fake News Spreads Lies, More Readers Shrug at the Truth*,
The New York Times, Dec. 6, 2016, https://www.nytimes.com/2016/12/06/us/fake-
news-partisan-republican-democrat.html.

© Springer Nature Switzerland AG 2019
H. Yin et al. (Eds.): IDEAL 2019, LNCS 11872, pp. 332–339, 2019.
https://doi.org/10.1007/978-3-030-33617-2_34

because there is no agreed upon its definition and such formalization still remains a challenging problem.

The main goal of publishing such information with false content is to attract readers, what could increase publisher rank and popularity, which consequently increases revenues form adds. It could be also used as the political weapon, which may impact on individual decision, e.g., during election, but also to promote a desirable ideas or behaviours on the ground. Because the definition of fake news is not clear, but experienced human experts can labelled incoming bits of news (usually with the slight delay). It is worth mentioning Lithuania which is on the frontline of a Russian offensive to sow misinformation in the western world. There it is allowed to order shutting down servers for 48 h without a court order if they are used to spread fake news[2]. In Lithuania the net of elves has been established, which gathers volunteers which manually detect the fake news appearances and report them to the authorities. Unfortunately, it is impossible to use human power to label all the messages then the machine learning methods have been starting to be used as the fake news detectors.

The fake news problem is the emerging one for the researchers working on the computer science. Even though, there are plenty of online tools designed for fake news detecting, they are not sufficiently successful [10]. Thus, some project founded by European Commission were introduced eg. SocialThruth [7].

The fake news definition is unclear, it is also worth enumerating what is not fake news [17]:

- satire news with proper context and hoaxes,
- rumors,
- conspiracy theories,
- misinformation that is created unintentionally.

In this work we will focus on finding solution to recognize fake news collected in a popular *Getting Real about Fake News*[3] dataset, being container of text and metadata scraped from 244 websites by the *BS Detector* Chrome Extension. Let us shortly present representative works from this domain.

Shu et al. [17] presented interesting study, where types of features which could be use to fake news detection has been discussed. Zhang et al. [18] enumerated the following groups of methods which may be employed to this analysis: creator/reader analysis, news content analysis, style-based analysis, add social context analysis. In [6] image analysis is presented as the possible approach to fake news detection. An interesting survey on the state-of-the-art technologies for fake news detection has been presented by Conroy et al. [8]. The authors have identified two groups of methods to classify fake news, namely linguistic (such as deep syntax analysis, semantic analysis, rhetorical structure and discourse analysis, automated numerical analysis, based on SVM or Bayesian models) and

---

[2] Michael Peel, Fake news: How Lithuani's 'elves' take on Russian trolls, Financial Times, Feb. 4, 2019, https://www.ft.com/content/b3701b12-2544-11e9-b329-c7e6ceb5ffdf.

[3] https://www.kaggle.com/mrisdal/fake-news.

social network analysis approaches (e.g. analysis of poster's network behaviour, and analysis of data linked to the original news/content). Ferrra et al. [9] proposed to employ n-grams or bag-of-words representation, while Afroz et al. [1] used stylometry methods to analyse articles. Castilio et al. [4] used features from users' posting and re-posting behavior, from the text of the posts, and from citations to external sources to assess credibility of tweets. Sharma et al. [16] discussed several topics related with the problem under consideration and pointed out the possibility of using SCAN (*Scientific Content Analysis*) to this problem. An interesting approach was proposed by Jin et al. [13], who turned away from text analysis and used an image content in the task of automatic news verification. Zhang et al. [18] noticed the dynamic nature of the social media messages and proposed to analysed them as streaming data. Horne and Adali [12] used SVM to distinguished among fake, real, and satire messages. SVM has been also used in [5] to detect potentially fake online posts by classifying users using semantic analysis and behavioural features descriptors. An interesting comparative study was presented in [11], where authors tested several classification methods based on linguistic features. They concluded that many well-known classifiers, especially classifier ensembles, can be used as valuable and effective fake news detectors. Bondielli et al. [3] noted that most of the approaches treated fake news detection task as classification problem, but also other approaches as anomaly detection or cluster analysis could be use. Atodiresi et al. [2] discussed tweet analysis system based on NLP tools. Authors treated this task as the regression problem and thanks that they were able to assign credibility value to each given message.

In a nutshell, the main contributions of this paper are as follows:

- The proposition of employing an classifier ensemble methods to the fake news classification task.
- Unlike the most popular NLP-based features, a simple word frequency analysis is used to message representation.
- An extensive experimental analysis backed-up by the statistical tests.

2   Proposed Methods

Most of the work on the topic of pattern recognition in the context of fake news attempts to extract linguistic factors from text data, whereas classic NLP solutions have resulted from simple methods such as *Count Vectorizer*. This approach analyzes the texts, searching for the most frequently appearing words in them (apart from the so-called *stop-word dictionary* containing conjunctions, pronouns and prepositions of the language being processed) and converts them into feature vectors, containing the number of their occurrences in each text, or the frequency of their occurrence.

The approach proposed in this paper is based on the basic *Count Vectorizer*, extracting data from the *Getting Real about Fake News* collection. It contains 13,000 texts marked as fake news by *BS Detector Chrome Extension* users. To create the possibility of performing binary classification experiments, the dataset has been supplemented with the same number of texts from sources considered reliable, such as the *New York Times, Reuters, Washington Post*, or *CNN*.

Each object in the analyzed data consists of the title of the article and its content. For both of these attributes, independently, a thousand of the most common words have been designated, which together gives a base of 26,000 patterns of 2,000 features. A set with such a large number of final attributes should be a subject to any problems resulting from the *curse of dimensionality*, which means that typical approaches to the solution may prove to be inefficient [15]. Nowadays, *deep convolutional networks* are commonly most beloved solution in such cases, but the following work tries to implement an effective way of classifying this type of data using only basic methods, so the choice of the classifier fell on a *decision tree* of the CART type.

The research is intended to compare two approaches. The first is to train the base classifier on the full available space of features, so as to gain a reference point for the second, ensemble approach. The essence of the latter is the construction of a homogeneous pool of classifiers diversified with the *Random Subspace* method, which for each member classifier selects a random subset of the problem features of a given size.

In this case, it is also important to find the *combination rule*, which works best in the task of generalizing many decisions available in the pool of classifiers to the final decision of the entire committee. Two solutions will be tested here. The first is the classic *majority voting*, where the committee's decision falls on the class most commonly present in the decision of the base classifiers [14]. The second is the use of probabilistic interpretation of CART trees and the accumulation of support obtained for each class in order to determine the largest.

Each of the approaches has been implemented in two variations: regular and weighted. In a weighted variant, we assign a weight to each classifier, consistent with the quality achieved by it. In order to obtain it, a validation subset of 20% of available patterns is extracted from the training set, on which we calculate the accuracy measure (Fig. 1).

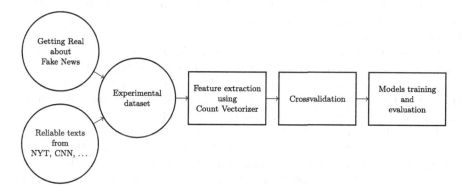

Fig. 1. Overview of processing.

3   Experimental Study

The implementation of the proposed method, together with the method of developing a data set and extraction of features, has been realized in Python, according to the scikit-learn framework API and made available in the public GIT repository[4]. The standard implementation of the CART base classifier available there was also used.

All experiments were carried out based on stratified k-fold cross-validation with five folds. The statistical dependence of the obtained results was examined in accordance with the student's T-test, and the role of the metric – due to the balanced distribution of dataset classes – was accuracy.

According to the default values for this type of method, the population of each classifier consisted of one hundred members. In the first phase of the experiment, the ability to discriminate classifiers, built successively on 20, 100, 200, 1000 and 2000 features (half divided into keywords from the title and content of the article) was tested, and for the size of the subspace – the default – square root from the number of features was selected. The results of this phase of the experiment are presented in Table 1.

As can be seen, the selection of only twenty features from the available data set with the ensemble structured on four-dimensional subspaces leads, in each considered approach, to solution oscillating around the random classifier, which does not allow to draw significant conclusions in the context of differences between them. However, even at that time statistically significantly worse methods are based on majority voting.

With one hundred attributes, the differences between classifiers begin to become noticeable and *support accumulation* methods start to have a slight advantage over the base approach, but this is not a statistically significant difference. A similar situation occurs with two hundred features. With a thousand features, it is even more visible, but the significant difference between the basic method and majority voting disappears.

The case of two thousand features is interesting. The difference between the SUPPORT ACCUMULATION methods and the basic CART becomes big enough to gain statistical significance, and approaches based on majority voting achieve higher results than it.

The observed trend in this case led to the implementation of the second phase of the experiment, where we abandon subspace strategy with the size of the number of features, in favour of using half the available space in each member classifier. The results are presented in Table 2, again for two thousand features, but with a subspace containing thousands of attributes.

The results obtained in this way are quite surprising. The difference between the various combination rules is almost completely eliminated and the simplest, regular MAJORITY VOTING allows achieving the optimal result in the vicinity of 89% accuracy of the classification. It exaggerates the discriminating abilities

---

[4] https://github.com/xehivs/fakenews.

Table 1. Average accuracy obtained during first phase of experimental evaluation. The information on those methods that performed poorly in comparison with the method named in the column, placed directly below the score.

CART	Random Subspace CART ensemble			
	SUPPORT ACC.		MAJORITY VOTING	
	REGULAR	WEIGTED	REGULAR	WEIGTED
1	2	3	4	5
20 features, subspace 4				
0.560	0.561	0.560	0.485	0.485
4, 5	4, 5	4, 5	—	—
100 features, subspace 10				
0.643	0.673	0.676	0.555	0.558
4, 5	4, 5	4, 5	—	—
200 features, subspace 14				
0.683	0.715	0.717	0.578	0.578
4, 5	4, 5	4, 5	—	—
1000 features, subspace 31				
0.683	0.715	0.717	0.578	0.578
—	—	4	—	—
2000 features, subspace 44				
0.743	0.780	0.781	0.757	0.757
—	1	1	—	—

Table 2. Average accuracy obtained during experimental evaluation. The information on those methods that performed poorly in comparison with the method named in the column, placed directly below the score.

CART	Random Subspace CART ensemble			
	SUPPORT ACC.		MAJORITY VOTING	
	REGULAR	WEIGTED	REGULAR	WEIGTED
1	2	3	4	5
2000 features, subspace 1000				
0.743	0.891	0.892	0.891	0.892
—	1	1	1	1

of the basic CART (74%) and, comparing the results with those available in the literature, allows for the construction of a competitive classification system for much more computationally demanding *deep learning models*.

4 Conclusions

The main aim of this work was to employ the classifier ensemble algorithms to the problem of fake news classification. We proposed homogeneous decision tree ensembles diversified using the *Random Subspace* method. The computer experiments confirmed the usefulness of the proposed approach and on the basis of statistical analysis we may conclude that the proposed approach is statistically significantly better than state-of-art individual models, e.g., used by the mentioned comparative study [11].

The results presented in this paper are quite promising therefore they encourage us to continue our work on employing classification methods to the problem under consideration. In the future we are going to extend our research by employing methods dedicated to data stream classification, especially detecting and reacting to the concept drift appearance, as well employing algorithm dedicated to analyse serious difficulties in data as data balancing, data cleaning etc.

Nevertheless, we have also to remember that the related topics, as rumour classification, clickbait detection, truth discovery, or spammer and bot detection remain important topics for assessing credibility of social media publications [17].

Acknowledgement. This work is funded under SocialTruth project, which has received funding from the European Union's Horizon 2020 research and innovation programme under grant agreement No. 825477.

References

1. Afroz, S., Brennan, M., Greenstadt, R.: Detecting hoaxes, frauds, and deception in writing style online. In: Proceedings of the 2012 IEEE Symposium on Security and Privacy, SP 2012, Washington, DC, USA, pp. 461–475. IEEE Computer Society (2012). https://doi.org/10.1109/SP.2012.34
2. Atodiresei, C.S., Tănăselea, A., Iftene, A.: Identifying fake news and fake users on Twitter. Procedia Comput. Sci. **126**, 451–461 (2018). https://doi.org/10.1016/j.procs.2018.07.279. http://www.sciencedirect.com/science/article/pii/S1877050918312559. Knowledge-Based and Intelligent Information & Engineering Systems: Proceedings of the 22nd International Conference, KES-2018, Belgrade, Serbia
3. Bondielli, A., Marcelloni, F.: A survey on fake news and rumour detection techniques. Inf. Sci. **497**, 38–55 (2019). https://doi.org/10.1016/j.ins.2019.05.035. http://www.sciencedirect.com/science/article/pii/S0020025519304372
4. Castillo, C., Mendoza, M., Poblete, B.: Information credibility on Twitter. In: Proceedings of the 20th International Conference on World Wide Web, WWW 2011, pp. 675–684. ACM, New York (2011). https://doi.org/10.1145/1963405.1963500, http://doi.acm.org/10.1145/1963405.1963500
5. Chen, C., Wu, K., Venkatesh, S., Zhang, X.: Battling the internet water army: detection of hidden paid posters. In: 2013 IEEE/ACM International Conference on Advances in Social Networks Analysis and Mining (ASONAM 2013), pp. 116–120 (2011)

6. Choraś, M., Giełczyk, A., Demestichas, K., Puchalski, D., Kozik, R.: Pattern recognition solutions for fake news detection. In: Saeed, K., Homenda, W. (eds.) CISIM 2018. LNCS, vol. 11127, pp. 130–139. Springer, Cham (2018). https://doi.org/10.1007/978-3-319-99954-8_12

7. Choraś, M., Pawlicki, M., Kozik, R., Demestichas, K., Kosmides, P., Gupta, M.: Socialtruth project approach to online disinformation (fake news) detection and mitigation. In: Proceedings of the 14th International Conference on Availability, Reliability and Security, p. 68. ACM (2019)

8. Conroy, N., Rubin, V.L., Chen, Y.: Automatic deception detection: methods for finding fake news. Proc. Assoc. Inf. Sci. Technol. **52**, 1–4 (2015). https://doi.org/10.1002/pra2.2015.145052010082

9. Ferrara, E., Varol, O., Davis, C., Menczer, F., Flammini, A.: The rise of social bots. Commun. ACM **59**(7), 96–104 (2016). https://doi.org/10.1145/2818717. http://doi.acm.org/10.1145/2818717

10. Giełczyk, A., Wawrzyniak, R., Choraś, M.: Evaluation of the existing tools for fake news detection. In: Saeed, K., Chaki, R., Janev, V. (eds.) CISIM 2019. LNCS, vol. 11703, pp. 144–151. Springer, Cham (2019). https://doi.org/10.1007/978-3-030-28957-7_13

11. Gravanis, G., Vakali, A., Diamantaras, K., Karadais, P.: Behind the cues: a benchmarking study for fake news detection. Expert Syst. Appl. **128**, 201–213 (2019). https://doi.org/10.1016/j.eswa.2019.03.036. http://www.sciencedirect.com/science/article/pii/S0957417419301988

12. Horne, B.D., Adali, S.: This just. In: Fake News Packs a Lot in Title, Uses Simpler, Repetitive Content in Text Body, More Similar to Satire than Real News. CoRR abs/1703.09398 (2017). http://arxiv.org/abs/1703.09398

13. Jin, Z., Cao, J., Zhang, Y., Zhou, J., Tian, Q.: Novel visual and statistical image features for microblogs news verification. IEEE Trans. Multimedia **19**(3), 598–608 (2017). https://doi.org/10.1109/TMM.2016.2617078

14. Ksieniewicz, P.: Combining *Random Subspace* approach with smote oversampling for imbalanced data classification. In: Pérez García, H., Sánchez González, L., Castejón Limas, M., Quintián Pardo, H., Corchado Rodríguez, E. (eds.) HAIS 2019. LNCS, pp. 660–673. Springer, Cham (2019). https://doi.org/10.1007/978-3-030-29859-3_56

15. Ksieniewicz, P., Woźniak, M.: Dealing with the task of imbalanced, multidimensional data classification using ensembles of exposers. In: First International Workshop on Learning with Imbalanced Domains: Theory and Applications, pp. 164–175 (2017)

16. Sharma, K., Qian, F., Jiang, H., Ruchansky, N., Zhang, M., Liu, Y.: Combating fake news: a survey on identification and mitigation techniques. ACM Trans. Intell. Syst. Technol. **10**(3), 21:1–21:42 (2019). https://doi.org/10.1145/3305260. http://doi.acm.org/10.1145/3305260

17. Shu, K., Sliva, A., Wang, S., Tang, J., Liu, H.: Fake news detection on social media: a data mining perspective. SIGKDD Explor. Newsl. **19**(1), 22–36 (2017). https://doi.org/10.1145/3137597.3137600. http://doi.acm.org/10.1145/3137597.3137600

18. Zhang, X., Ghorbani, A.A.: An overview of online fake news: characterization, detection, and discussion. Inf. Process. Manag. (2019). https://doi.org/10.1016/j.ipm.2019.03.004. http://www.sciencedirect.com/science/article/pii/S0306457318306794

A Genetic-Based Ensemble Learning Applied to Imbalanced Data Classification

Jakub Klikowski, Paweł Ksieniewicz$^{(\boxtimes)}$, and Michał Woźniak

Department of Systems and Computer Networks,
Wrocław University of Science and Technology, Wrocław, Poland
{jakub.klikowski,pawel.ksieniewicz,michal.wozniak}@pwr.edu.pl

Abstract. Imbalanced data classification is still a focus of intense research, due to its ever-growing presence in the real-life decision tasks. In this article, we focus on a classifier ensemble for imbalanced data classification. The ensemble is formed on the basis of the individual classifiers trained on supervise-selected feature subsets. There are several methods employing this concept to ensure a high diverse ensemble, nevertheless most of them, as *Random Subspace* or *Random Forest*, select attributes for a particular classifier randomly. The main drawback of mentioned methods is not giving the ability to supervise and control this task. In following work, we apply a genetic algorithm to the considered problem. Proposition formulates an original learning criterion, taking into consideration not only the overall classification performance but also ensures that trained ensemble is characterised by high diversity. The experimental study confirmed the high efficiency of the proposed algorithm and its superiority to other ensemble forming method based on random feature selection.

Keywords: Machine learning · Classification · Imbalanced data · Feature selection · Genetic algorithm

1 Introduction

The most of classification data sets do not have exactly the same number of instances for each class. The serious problem appears when the data is significantly imbalanced, when one of the classes quantitatively dominates over other ones. Data imbalance may have a negative effect on the efficiency of standard classification algorithms, thus this problem requires a proper approach and the use of dedicated techniques [10].

Data pre-processing is one of the solutions to this problem. The operation of such methods consists in the prior preparation of data, where algorithm equalizes the number of instances in particular classes. There are tree main approaches: creating additional objects in the minority class (*oversampling*), removing objects in the majority class (*undersampling*), or combine both techniques.

© Springer Nature Switzerland AG 2019
H. Yin et al. (Eds.): IDEAL 2019, LNCS 11872, pp. 340–352, 2019.
https://doi.org/10.1007/978-3-030-33617-2_35

Another approach focuses on designing algorithms dedicated to imbalanced data analysis. Appropriate procedures and modifications change the operation of standard classifiers. The introduced changes are aimed at improving the obtained classification quality by increasing the weight of minority data in the imbalanced set. One of such modifications is the feature selection. Lee et al. [12] presented an classifier with feature selection based on fuzzy entropy for pattern classification. The pattern space is partitioned into non-overlapping fuzzy decision regions. It allows to obtain smooth boundaries and achieve better classification performance. Koziarski et al. [9] came up with idea of feature selection based ensemble learning method, which creates subspaces in guided incremental manner according to parameters setting. One of the parameters decides about preference toward feature quality or diversity. Canuto and Nascimento [4] proposed the genetic-based feature selection method to create ensemble classifiers. They use an hybrid and adaptive fitness function consists of the features correlation measure and the ensemble accuracy. Du et al. [6] invented an feature selection based method for imbalanced data with genetic algorithm. It improves the quality of a single model by using geometric mean score as fitness function. Haque et al. [8] proposed an genetic algorithm method for learning heterogeneous ensembles on imbalanced data. This idea is based on the search for possible combinations of different types of classifiers in the ensemble with selecting the best compositions using genetic algorithm with the Matthews correlation coefficient score as the fitness function.

In this work, we present the *Genetic Ensemble Selection* (GES) classifier, being a novel ensemble learning method which trains diverse individuals on the basis of supervised selected features using a genetic algorithm. Base genetic approach is extended by an original regularisation component to the learning criterion, aiming to ensure a high diversity of the ensemble and to protect it against the hazard of overfitting. The experiments conducted on the collection of benchmark datasets prove the validity and a good performance of the proposed method. The solution refers to the method proposed by Ksieniewicz and Woźniak [11] who applied random search to find the best ensemble in the pool using (*balanced accuracy score*) [3]. Following work expands research on this idea by implementing a genetic algorithm to search optimal subspaces of features and modifies the proposed regularisation criterion for alternatives solutions.

In a nutshell, the main contributions of this works state as follows:

- Proposition of a genetic-based classifier ensemble forming algorithm employing feature selection.
- Proposition of an original fitness function consisting of a classification performance metric with the regularisation component.
- Evaluation of the proposed method on the basis of the wide range benchmark datasets.

2   Proposed Method

The aim of the classification algorithm is to assign an observed object to one of the predefined set of labels $\mathcal{M} = \{1, 2, ..., M\}$. The decision is made on the basis of an analysis of the selected attributes, which are gathered in the feature vector x

$$x = [x^1, x^2, \ldots, x^d]^T \in \mathcal{X}. \tag{1}$$

Let us define a pool of individual classifiers as

$$\Pi = \{\Psi_1, \Psi_2, \ldots, \Psi_K\}, \tag{2}$$

where Ψ_k denotes the k-th elementary classifier.

In this work the individuals use subsets of attributes to ensure the high diversity of the ensemble [14], therefore let's propose the following representation of Π as a bit word

$$\Pi = \left[\left[b_1^1, b_1^2, \ldots, b_1^d \right] \left[b_2^1, b_2^2, \ldots, b_2^d \right] \ldots \left[b_K^1, b_K^2, \ldots, b_K^d \right] \right], \tag{3}$$

where b_i^j denotes if the jth feature is used by the ith classifier.

As the optimisation criterion we propose

$$Q(\Pi) = metric(\Pi) - \alpha * \frac{no - features(\Pi)}{d} + \beta * \frac{av - Hamming(\Pi)}{d}, \tag{4}$$

where $metric(\Pi)$ denotes value of the chosen performance metric, $no - features(\Pi)$ is the number of features used by all classifiers from Π and $av - Hamming(\Pi)$ stands for the average $Hamming$ distance between the words represented individuals in Π. Procedure to count the criterion is shown at Algorithm 1.

Algorithm 1. Criterion count

1: **Input:** pool of individual classifiers Π, training set $\mathcal{TS}$
2: **Parameters:** α, $beta$
3: **Output:** value of criterion (eq. 4) for Π
4:
5: $counter \leftarrow 0$
6: $nobits \leftarrow 0$
7: $word \leftarrow [00..0]$
8: **for** $i \leftarrow 1$ to $K - 1$ **do**
9:    **for** $j \leftarrow i$ to K **do**
10:       $counter \leftarrow counter + 1$
11:       $nobits \leftarrow nobits +$ number of bits of $[b_i^1, b_i^2, \ldots, b_i^d]$ XOR $[b_j^1, b_j^2, \ldots, b_j^d]$
12:    **end for**
13:    $word \leftarrow word OR \left[b_i^1, b_i^2, \ldots, b_i^d \right]$
14: **end for**
15: $word \leftarrow word OR \left[b_K^1, b_K^2, \ldots, b_K^d \right]$
16: $no - features \leftarrow$ number of bits in $word * \frac{1}{d}$
17: $av - Hamming \leftarrow \frac{nobits}{counter*d}$
18: $BAC \leftarrow$ balanced accuracy of Π calculated on $\mathcal{TS}$
19: $criterion \leftarrow BAC - \alpha * no - used - features + \beta * av - Hamming - dist$
20: **return** $criterion$

We employ the genetic algorithm for finding the optimal ensemble line-up. The pseudocode of GES (*Genetic Ensemble Selection*) procedure is presented in Algorithm 2. Let's shortly describe its main steps.

Algorithm 2. Feature Selection Genetic Ensemble Classifier procedure

---

1: **Input:**
2: **Parameters:** e - number of generations p - population size ms - mutation size mp - mutation probability cs - crossing size cp - crossing probability n - elite size
3: **Output:**
4: $population \leftarrow$ create p random subspaces of features
5: **for** $i \leftarrow 0$ **to** e **do**
6:     $mindexes \leftarrow$ select ms individuals indexes list with mp probability
7:     **for all** m in $mindexes$ **do**
8:         $population[m] \leftarrow$ invert random features bits of $population[m]$
9:     **end for**
10:
11:     $cindexes \leftarrow$ select cs individuals indexes list with cp probability
12:     **for all** c in $cindexes$ **do**
13:         $p1 \leftarrow population[c]$
14:         $j \leftarrow$ generates a random number in the range of 0 to $population$ length
15:         $p2 \leftarrow population[j]$
16:         $c1, c2 \leftarrow$ exchange randomly selected features between $p1$ and $p2$
17:         $population[c] \leftarrow c1$
18:         $population[j] \leftarrow c2$
19:     **end for**
20:
21:     $scores \leftarrow$ new list
22:     **for** $j \leftarrow 0$ **to** $population$ length **do**
23:         $quality \leftarrow$ calculate quality for $population[j]$
24:         insert $quality$ into $scores[j]$
25:     **end for**
26:
27:     $population \leftarrow population$ sort by $scores$
28:     $elite \leftarrow$ select first n individuals from $population$
29:     $population \leftarrow elite +$ select $(p - n)$ individuals from $population$
30: **end for**

Initialisation Procedure. Algorithm 2 starts with the randomly generated population of the ensembles $Population = \{\Pi_1, \Pi_2, ..., \Pi_S\}$. Its size is an arbitrarily chosen by an input parameter. Basically, the larger the population exists, the more comprehensive optimisation is performed.

Evaluate Population Procedure. Individuals in the population are evaluated by criterion 4 calculated on the basis of algorithm (Eq. 1) using samples stored in the training set. Obtained results are used to select the elite (the best evaluated ensemble), being immune to mutation or replacement by crossover.

Genetic Operators. The mutation shall inject some randomness into chromosomes. We used a simple bit mutation (flipping random bits in a binary word). The crossover operator exchanges random member classifiers between two randomly chosen parent individuals to form child chromosomes, using standard two-point crossover operator [2].

Selection and Reproduction. To maintain the diversity of the population, which is essential for ensuring the ability to explore the space of possible solution of the problem effectively, the selection process has to be realised in probabilistic manner i.e., not only are the best chromosomes chosen, but also the probability of selection is straight proportional to their evaluation according to the criterion 4. To meet this assumption, a standard ranking selection procedure was implemented, where $S - 1$. Generated individuals and the elite individual are joined together and form offspring population.

3   Experimental Evaluation

Conducting appropriate experiments and analysis of the results should allow us to verify the idea and check the quality of the classification for the proposed method. This section describes the set-up of the conducted experiments, presents their results and analysis. During the experiments we would like to verify the following research hypotheses:

- Does the proposed approach improve the classification quality?
- Do the adjusted regularisation methods have a positive effect on the operation of the classifier?
- Is the random subspace an adequate solution for the feature selection for this classifier proposal?

3.1   Set-Up

All experiments were carried out using benchmark binary-class imbalanced data sets, which are posted and publicly available on the KEEL repository [1]. Selected data sets have a different imbalance ratios (from balanced problems to 1:33 proportion) and the various number of features. The tests were carried out on the basis of 5-fold stratified cross validation. This approach allows us a reliable check of the proposed method. The *Friedman ranking test* [5] was used for assessing the ranks of classifiers over all examined benchmarks. Checks whether the assigned rank differs significantly from assigning to each classifier an average rank. In the case if the hypothesis of ranking equality is rejected, the *Shaffer post hoc test* [7] is used to find out which of the tested methods are distinctive among an N × N comparison. The post hoc procedure was based on the significance level $\alpha = 0.05$.

The implementation was done using according to guidelines of *scikit-learn* library API [13]. *Logistic Regression* and *Gaussian Naive Bayes* were used as the base classifiers and the tests were carried out for the comparative analysis of different approaches of GES building classifier ensembles on them:

- GES—base *Genetic Ensemble Selection* classifier without regularization,
- GES-A—*Genetic Ensemble Selection* with *alpha*-regularization,
- GES-B—*Genetic Ensemble Selection* with both *alpha*- and *beta*-regularization.

Each comparison was supplemented with a bare-base method trained on the whole set of the features and a *Random Subspace* (RS) approach. All 10 approaches have been tested and evaluated using three metrics *F-score*, *Geometric Mean Score* and *Balanced accuracy score* [3], where metric used for evaluation is always used also as a part of optimization criterion Q. The entire implementation and code allowing to repeat all the experiments has been placed in a publicly available repository[1].

3.2   Results

The first experiment was aiming to identify the best hyperparameters for the genetic algorithm implemented within the GES. Parameters that have been tested include the probability of crossing individuals and the probability of mutation.

Conducted research allowed to indicate the best parameters value in the range from 0% to 10% for crossing and in the range from 0% to 2% for mutation. Figure 1 shows the visualisation of the obtained learning curves for two exemplary datasets. The intensity of the green tells us about the average value of the learning curve in the course. The height of the dot on the right and the result above each square is the quality on the test set of the already learned model. In addition, the intensity of the red of this dot tells whether it is among the worst or the best results. After analysis of all the results, the optimal parameters were selected as 2.5% for crossover and 1% for mutation. These hyperparameter settings were later used to carry out further experiments.

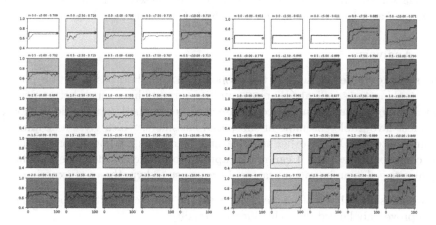

Fig. 1. Crossing and mutation probability influence on BAC.

---

[1] https://github.com/w4k2/genetic-ensemble-selection.

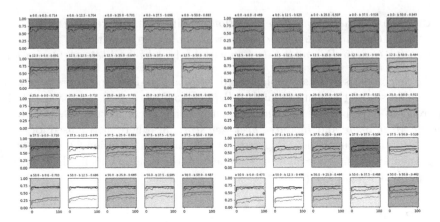

Fig. 2. Example of alpha and beta influence on BAC.

The next experiment was to find the best regularization parameters α and β. Parameters were tested in the range from 0 to 1 with and step of 0.125. Figure 2 shows the visualisations of the results obtained on exemplary datasets. The intensity of the yellow tells us about the average value of the learning curve in the course. The height of the dot on the right and the result above each square is the quality on the test set of the already learned model. In addition, the intensity of the red of this dot tells whether it is among the worst or the best results. The analysis of the obtained results allows to find best parameters.

The main experiment was aimed at conducting research on data sets after optimising the parameters of the proposed method. Tables 1, 2 and 3 shows the obtained results. Each table has been divided into two parts, for both employed base classifiers. The ranking tests outcome is presented in Table 4 for Logistic Regression as a base classifier and Table 5 for Gaussian Naive Bayes.

3.3   Analysis

The scores obtained show that the proposed GES method has a certain potential. Ranking tests indicate that GES and its variants have much better performance than the random subspace or base classifier method. GES is able to choose the most favourable ensemble of subspaced classifiers, which gives a significant advantage to RS. On the other side, learning using this method is computationally complex process and it can not be applied to problems where the classifier's learning speed is an important factor. In response to hypotheses of experiments:

- Results shows that features selection with the genetic algorithm approach can improve imbalanced data classification quality.
- Regularisation methods have no significant impact on the classifier's operation. This is indicated by good performance and lack of statistical difference between base GES and its variants—GES-A and GES-B where regularisation is reduced to one metric or totally excluded.

Table 1. Overview of the results of the main experiment for *F-score*.

dataset	Logistic Regression					Gaussian Naive Bayes				
	LR 1	RS 2	GES 3	GES-A 4	GES-B 5	GNB a	RS b	GES c	GES-A d	GES-B e
australian	0.762	0.728	0.808	0.794	0.800	0.719	0.732	0.834	0.852	0.844
	-	-	2	2	2	-	-	a,b	a,b	a,b
glass-0-1-2-3-vs-4-5-6	0.775	0.385	0.800	0.770	0.695	0.709	0.612	0.719	0.656	0.728
	2	-	2	2	2	-	-	-	-	-
glass-0-1-4-6-vs-2	0.000	0.215	0.147	0.214	0.267	0.219	0.096	0.252	0.267	0.242
	-	1	1	1	1	-	-	-	-	-
glass-0-1-5-vs-2	0.000	0.179	0.197	0.194	0.205	0.218	0.014	0.288	0.340	0.271
	-	1	1	-	1	b	-	b	b	-
glass-0-1-6-vs-2	0.000	0.199	0.181	0.172	0.168	0.199	0.088	0.244	0.220	0.273
	-	1	1	1	1	-	-	-	-	-
glass-0-1-6-vs-5	0.149	0.000	0.115	0.015	0.000	0.760	0.180	0.760	0.648	0.648
	-	-	-	-	-	b	-	b	b	b
glass-0-4-vs-5	0.507	0.040	0.711	0.333	0.360	0.960	0.268	0.773	0.960	0.727
	2	-	2	-	-	b	-	b	b	b
glass-0-6-vs-5	0.333	0.000	0.144	0.180	0.160	0.893	0.349	0.893	0.867	0.960
	-	-	-	-	-	b	-	b	b	b
glass0	0.516	0.000	0.699	0.701	0.660	0.642	0.627	0.628	0.625	0.632
	2	-	2	1,2	2	-	-	-	-	-
glass1	0.243	0.000	0.583	0.524	0.563	0.604	0.490	0.634	0.626	0.630
	2	-	1,2	1,2	1,2	-	-	-	-	-
glass2	0.000	0.165	0.152	0.068	0.158	0.189	0.168	0.216	0.189	0.190
	-	1	-	-	1	-	-	-	-	-
glass4	0.167	0.200	0.441	0.428	0.541	0.237	0.000	0.257	0.300	0.257
	-	-	-	-	-	b	-	-	b	-
glass5	0.149	0.200	0.333	0.200	0.348	0.768	0.140	0.634	0.457	0.768
	-	-	-	-	-	b	-	b	-	b
glass6	0.759	0.238	0.811	0.832	0.778	0.772	0.809	0.775	0.738	0.800
	2	-	2	2	2	-	-	-	-	-
heart	0.822	0.207	0.796	0.787	0.775	0.802	0.769	0.816	0.812	0.837
	2	-	2	2	2	-	-	-	-	b
hepatitis	0.919	0.912	0.912	0.903	0.899	0.719	0.615	0.889	0.840	0.795
	-	-	-	-	-	-	b	-	-	-
page-blocks-1-3-vs-4	0.533	0.421	0.548	0.550	0.568	0.493	0.347	0.419	0.419	0.419
	-	-	-	-	-	-	-	-	-	-
pima	0.629	0.000	0.634	0.646	0.635	0.621	0.501	0.600	0.621	0.616
	2	-	2	2	2	b	-	b	b	b
shuttle-c0-vs-c4	0.996	0.031	0.996	0.996	0.996	0.980	0.968	0.988	0.984	0.984
	2	-	2	2	2	-	-	-	-	-
shuttle-c2-vs-c4	1.000	0.960	0.960	1.000	0.960	0.813	0.800	0.800	0.800	0.867
	-	-	-	-	-	-	-	-	-	-
vowel0	0.540	0.113	0.637	0.553	0.573	0.709	0.717	0.736	0.716	0.726
	2	-	2	2	2	-	-	-	-	-
wisconsin	0.949	0.048	0.909	0.891	0.868	0.943	0.941	0.943	0.937	0.933
	2	-	2	2	2	-	-	-	-	-
yeast-0-2-5-6-vs-3-7-8-9	0.056	0.000	0.067	0.024	0.065	0.262	0.026	0.410	0.384	0.296
	2	-	-	-	-	b	-	b	b	b
yeast-0-2-5-7-9-vs-3-6-8	0.315	0.000	0.683	0.347	0.055	0.201	0.174	0.752	0.677	0.748
	2,5	-	2,5	-	-	b	-	a,b	a,b	a,b
yeast-0-3-5-9-vs-7-8	0.036	0.000	0.114	0.114	0.000	0.269	0.148	0.377	0.269	0.345
	-	-	-	-	-	-	-	b	-	-
yeast-0-5-6-7-9-vs-4	0.000	0.000	0.000	0.067	0.000	0.175	0.067	0.521	0.480	0.531
	-	-	-	-	-	b	-	a,b	a,b	a,b
yeast-2-vs-4	0.236	0.000	0.186	0.151	0.359	0.297	0.508	0.770	0.713	0.774
	2	-	-	-	2	-	-	a	a	a
yeast-2-vs-8	0.000	0.000	0.000	0.000	0.000	0.254	0.249	0.658	0.658	0.658
	-	-	-	-	-	-	-	a	a	a
yeast1	0.329	0.374	0.567	0.563	0.571	0.457	0.502	0.529	0.519	0.549
	-	-	1,2	1,2	1,2	-	-	a	a	a
yeast3	0.153	0.225	0.756	0.729	0.758	0.236	0.548	0.778	0.778	0.767
	-	-	1,2	1,2	1,2	-	a	a,b	a,b	a,b
yeast5	0.000	0.000	0.000	0.000	0.000	0.154	0.321	0.586	0.664	0.669
	-	-	-	-	-	-	a	a,b	a,b	a,b
Ranking	3.2742	4.1452	2.2258	2.7097	2.6452	3.4194	4.5161	2.0806	2.8226	2.1613

Table 2. Overview of the results of the main experiment for *Gmean* score.

dataset	LR 1	RS 2	GES 3	GES-A 4	GES-B 5	GNB a	RS b	GES c	GES-A d	GES-B e
		Logistic Regression					Gaussian Naive Bayes			
australian	0.788 2	0.711 -	0.812 2	0.827 2	0.815 2	0.752 -	0.763 -	0.884 a,b	0.865 a,b	0.864 a,b
glass-0-1-2-3-vs-4-5-6	0.833 2	0.000 -	0.891 2	0.878 2	0.926 2	0.792 -	0.685 -	0.789 -	0.780 -	0.775 -
glass-0-1-4-6-vs-2	0.000 -	0.338 -	0.519 1	0.471 1	0.481 1	0.495 -	0.282 -	0.603 -	0.614 -	0.619 -
glass-0-1-5-vs-2	0.000 -	0.000 -	0.591 1,2	0.592 1,2	0.596 1,2	0.510 b	0.051 -	0.611 b	0.559 b	0.638 b
glass-0-1-6-vs-2	0.000 -	0.286 -	0.421 1	0.458 1	0.537 1	0.481 -	0.258 -	0.494 -	0.578 -	0.463 -
glass-0-1-6-vs-5	0.259 -	0.000 -	0.555 2	0.411 -	0.417 2	0.930 b	0.691 -	0.788 -	0.935 b	0.735 -
glass-0-4-vs-5	0.624 2	0.000 -	0.893 2,5	0.394 -	0.337 -	0.994 b	0.584 -	0.822 -	0.994 b	0.994 b
glass-0-6-vs-5	0.461 2	0.000 -	0.362 -	0.318 -	0.435 2	0.936 b	0.715 -	0.936 b	0.936 b	0.936 b
glass0	0.619 2	0.000 -	0.739 2	0.749 2	0.757 2	0.632 -	0.605 -	0.624 -	0.633 -	0.624 -
glass1	0.377 2	0.000 -	0.564 1,2	0.589 1,2	0.514 2	0.543 -	0.529 -	0.553 -	0.505 -	0.548 -
glass2	0.000 -	0.385 1	0.397 1	0.395 1	0.536 1	0.568 -	0.459 -	0.562 -	0.529 -	0.514 -
glass4	0.228 -	0.231 -	0.811 1,2	0.595 -	0.663 -	0.341 b	0.000 -	0.341 b	0.342 b	0.523 b
glass5	0.274 -	0.200 -	0.639 -	0.677 -	0.768 1,2	0.929 b	0.672 -	0.788 -	0.744 -	0.885 b
glass6	0.855 2	0.000 -	0.879 2	0.898 2	0.896 2	0.862 -	0.856 -	0.864 -	0.890 -	0.844 -
heart	0.841 2	0.306 -	0.828 2	0.804 2	0.836 2	0.822 -	0.792 -	0.825 -	0.820 -	0.838 -
hepatitis	0.634 -	0.844 -	0.628 -	0.748 -	0.737 -	0.678 -	0.623 -	0.778 -	0.797 b	0.718 -
page-blocks-1-3-vs-4	0.688 -	0.643 -	0.921 -	0.902 -	0.908 -	0.653 -	0.543 -	0.581 -	0.547 -	0.548 -
pima	0.703 2	0.000 -	0.722 2	0.718 2	0.703 2	0.701 b	0.595 -	0.699 b	0.685 b	0.695 b
shuttle-c0-vs-c4	0.996 2	0.081 -	0.996 2	0.996 2	0.996 2	0.991 -	0.994 -	0.992 -	0.996 -	0.996 -
shuttle-c2-vs-c4	1.000 -	0.996 -	0.996 -	0.996 -	0.996 -	0.983 -	0.800 -	0.800 -	0.937 -	0.937 -
vowel0	0.719 2	0.161 -	0.703 2	0.695 2	0.758 2	0.873 -	0.814 -	0.829 -	0.821 -	0.803 -
wisconsin	0.959 2	0.120 -	0.885 2	0.907 2	0.888 2	0.962 -	0.958 -	0.954 -	0.965 -	0.957 -
yeast-0-2-5-6-vs-3-7-8-9	0.135 2	0.000 -	0.127 -	0.063 -	0.125 -	0.364 b	0.013 -	0.456 b	0.439 b	0.363 b
yeast-0-2-5-7-9-vs-3-6-8	0.399 2	0.000 -	0.700 2,4,5	0.108 -	0.108 -	0.393 b	0.172 -	0.905 a,b	0.885 a,b	0.846 a,b
yeast-0-3-5-9-vs-7-8	0.063 -	0.000 -	0.126 -	0.089 -	0.000 -	0.346 -	0.138 -	0.661 a,b	0.577 a,b	0.602 a,b
yeast-0-5-6-7-9-vs-4	0.000 -	0.000 -	0.000 -	0.000 -	0.000 -	0.182 -	0.112 -	0.754 a,b	0.727 a,b	0.751 a,b
yeast-2-vs-4	0.332 2,4	0.000 -	0.216 -	0.000 -	0.354 -	0.432 -	0.611 -	0.850 a	0.882 a,b	0.864 a
yeast-2-vs-8	0.000 -	0.000 -	0.000 -	0.000 -	0.000 -	0.564 -	0.513 -	0.745 -	0.723 -	0.749 -
yeast1	0.456 -	0.501 -	0.693 1,2	0.692 1,2	0.692 1,2	0.206 -	0.621 a	0.669 a	0.670 a	0.679 a
yeast3	0.282 -	0.384 -	0.897 1,2	0.896 1,2	0.907 1,2	0.446 -	0.680 a	0.894 a,b	0.912 a,b	0.905 a,b
yeast5	0.000 -	0.000 -	0.000 -	0.000 -	0.000 -	0.811 -	0.927 a	0.941 a	0.966 a	0.964 a
Ranking	3.2742	4.3871	2.3548	2.7258	2.2581	3.0323	4.5968	2.4516	2.3065	2.6129

Table 3. Overview of the results of the main experiment for *Balanced accuracy score*

dataset	Logistic Regression					Gaussian Naive Bayes				
	LR 1	RS 2	GES 3	GES-A 4	GES-B 5	GNB a	RS b	GES c	GES-A d	GES-B e
australian	0.791	0.727	0.812	0.823	0.818	0.770	0.781	0.869	0.863	0.853
	2	-	2	2	2	-	-	a,b	a,b	a,b
glass-0-1-2-3-vs-4-5-6	0.852	0.500	0.904	0.920	0.911	0.822	0.770	0.851	0.810	0.835
	2	-	2	2	2	-	-	-	-	-
glass-0-1-4-6-vs-2	0.500	0.611	0.544	0.619	0.642	0.561	0.477	0.665	0.643	0.640
	-	-	-	-	1	-	-	-	-	-
glass-0-1-5-vs-2	0.500	0.500	0.551	0.537	0.560	0.549	0.446	0.686	0.681	0.627
	-	-	-	-	-	-	-	b	b	-
glass-0-1-6-vs-2	0.500	0.583	0.607	0.636	0.563	0.547	0.448	0.588	0.657	0.583
	-	-	1	-	-	-	-	-	b	-
glass-0-1-6-vs-5	0.584	0.500	0.591	0.650	0.603	0.939	0.746	0.889	0.884	0.893
	-	-	2	-	2	b	-	-	-	-
glass-0-4-vs-5	0.750	0.500	0.894	0.706	0.832	0.994	0.679	0.944	0.994	0.994
	2	-	2	-	2	b	-	b	b	b
glass-0-6-vs-5	0.679	0.500	0.614	0.643	0.520	0.945	0.770	0.995	0.945	0.940
	-	-	-	-	-	-	-	b	b	b
glass0	0.657	0.500	0.763	0.776	0.769	0.701	0.684	0.705	0.701	0.705
	2	-	2	2	2	-	-	-	-	-
glass1	0.478	0.493	0.583	0.603	0.571	0.634	0.574	0.574	0.583	0.581
	-	-	-	2	2	-	-	-	-	-
glass2	0.500	0.565	0.567	0.544	0.534	0.604	0.551	0.672	0.632	0.682
	-	-	-	-	-	-	-	-	-	-
glass4	0.494	0.567	0.839	0.829	0.874	0.508	0.435	0.515	0.512	0.477
	-	-	1,2	1,2	1,2	-	-	-	-	-
glass5	0.594	0.600	0.790	0.793	0.752	0.938	0.729	0.888	0.949	0.949
	-	-	-	-	-	b	-	-	b	b
glass6	0.876	0.500	0.926	0.879	0.912	0.876	0.878	0.878	0.895	0.903
	2	-	2	2	2	-	-	-	-	-
heart	0.843	0.552	0.810	0.828	0.815	0.823	0.795	0.804	0.815	0.823
	2	-	2	2	2	-	-	-	-	-
hepatitis	0.745	0.849	0.745	0.760	0.759	0.722	0.689	0.810	0.796	0.853
	-	-	-	-	-	-	-	-	-	a,b
page-blocks-1-3-vs-4	0.790	0.748	0.932	0.904	0.909	0.762	0.692	0.714	0.697	0.695
	-	-	-	-	-	-	-	-	-	-
pima	0.722	0.500	0.725	0.727	0.723	0.713	0.653	0.706	0.708	0.707
	2	-	2	2	2	b	-	-	b	b
shuttle-c0-vs-c4	0.996	0.508	0.996	0.996	0.996	0.991	0.994	0.992	0.996	0.996
	2	-	2	2	2	-	-	-	-	-
shuttle-c2-vs-c4	1.000	0.996	0.996	0.946	0.996	0.984	0.900	0.900	0.946	0.946
	-	-	-	-	-	-	-	-	-	-
vowel0	0.772	0.533	0.767	0.765	0.799	0.881	0.830	0.835	0.874	0.829
	2	-	2	2	2	-	-	-	-	-
wisconsin	0.960	0.513	0.919	0.915	0.907	0.962	0.958	0.955	0.965	0.959
	2	-	2	2	2	-	-	-	-	-
yeast-0-2-5-6-vs-3-7-8-9	0.513	0.500	0.534	0.503	0.526	0.549	0.472	0.605	0.529	0.604
	-	-	-	-	-	-	-	-	-	-
yeast-0-2-5-7-9-vs-3-6-8	0.602	0.500	0.547	0.509	0.562	0.560	0.486	0.900	0.856	0.851
	2,4	-	2,4	-	-	b	-	a,b	a,b	a,b
yeast-0-3-5-9-vs-7-8	0.508	0.500	0.539	0.538	0.538	0.573	0.513	0.604	0.631	0.645
	-	-	-	-	-	-	-	-	b	b
yeast-0-5-6-7-9-vs-4	0.500	0.500	0.500	0.500	0.507	0.499	0.501	0.775	0.775	0.777
	-	-	-	-	-	-	-	a,b	a,b	a,b
yeast-2-vs-4	0.570	0.500	0.530	0.560	0.550	0.598	0.697	0.869	0.897	0.886
	2	-	-	-	-	-	-	a	a,b	a,b
yeast-2-vs-8	0.500	0.500	0.500	0.500	0.500	0.667	0.644	0.773	0.773	0.773
	-	-	-	-	-	-	-	-	-	-
yeast1	0.585	0.609	0.690	0.703	0.688	0.518	0.656	0.665	0.660	0.661
	-	-	1	1	1	-	a	a	a	a
yeast3	0.542	0.575	0.911	0.897	0.908	0.597	0.729	0.905	0.889	0.903
	-	-	1,2	1,2	1,2	-	a	a,b	a,b	a,b
yeast5	0.500	0.500	0.500	0.500	0.500	0.829	0.930	0.956	0.956	0.933
	-	-	-	-	-	-	a	a	a	a
Ranking	3.4516	4.2258	2.4355	2.4355	2.4516	3.2742	4.5968	2.4194	2.3387	2.371

Table 4. Shaffer test ($\alpha = 0.05$) for comparison between the proposed methods (GMean). Shaffer's procedure rejects those hypotheses that have an unadjusted p-value ≤ 0.005. Symbol '✗' stands for classifiers without significant differences, '✓' for situation in which the method on the left is superior.

Algorithms	BAC		F-score		GMean	
	$p-values$		$p-values$		$p-values$	
RS vs. GES	0.000008	✓	0.000002	✓	0	✓
RS vs. GES-A	0.000008	✓	0.000351	✓	0.000035	✓
RS vs. GES-B	0.00001	✓	0.000188	✓	0	✓
LR vs. GES	0.011402	✓	0.009042	✓	0.022069	✓
LR vs. GES-A	0.011402	✓	0.159833	✗	0.172104	✗
LR vs. GES-B	0.012775	✓	0.117284	✗	0.011402	✓
GES vs. GES-B	0.967965	✗	0.2964	✗	0.809582	✗
GES-A vs. GES-B	0.967965	✗	0.872374	✗	0.244153	✗
GES vs. GES-A	1	✗	0.228269	✗	0.355641	✗

Table 5. Shaffer test ($\alpha = 0.05$) for comparison between the proposed methods (GMean). Shaffer's procedure rejects those hypotheses that have an unadjusted p-value ≤ 0.005. Symbol '✗' stands for classifiers without significant differences, '✓' for situation in which the method on the left is superior.

Algorithms	BAC		F-score		GMean	
	$p-values$		$p-values$		$p-values$	
RS vs. GES	0	✓	0	✓	0	✓
RS vs. GES-A	0	✓	0.000025	✓	0	✓
RS vs. GES-B	0	✓	0	✓	0.000001	✓
GNB vs. GES-A	0.019841	✓	0.000858	✓	0.148235	✗
GNB vs. GES-B	0.024512	✓	0.137291	✗	0.070724	✗
GNB vs. GES	0.033293	✓	0.001733	✓	0.2964	✗
GES vs. GES-A	0.840851	✗	0.064689	✗	0.717764	✗
GES vs. GES-B	0.904101	✗	0.840851	✗	0.687971	✗
GES-A vs. GES-B	0.935981	✗	0.09964	✗	0.445429	✗

– Classification based only on *Random Subspace* gives quite poor performance. Employment of *Random Subspace* to select features along with the optimised selection allows a significant improvement compared to the classification on the full feature space.

4   Conclusions and Future Directions

Presented work focuses on an innovative ensemble method (GES) dedicated for imbalanced data classification. It is based on the genetic algorithm aiming to optimise selection of feature combinations. Conducted experimental evaluation shows the superiority of GES compared to both the basic random selection of features as well as to the base classifiers used to build ensembles learned on a full feature space.

An important aspect of analysis was the inclusion of regularisation component in fitness function, aiming to ensure the diversity of the classifiers committee. Experiments have shown that introducing it does not give statistically significant improvement in results, so in order to further develop of this method, a different regularisation proposal should be developed. New regularisation should give a greater impact on the classifier operation and better performance of the classification. The proposed method can be considered a useful and effective tool in the classification of data with varying degrees of imbalance.

Acknowledgments. This work was supported by the Polish National Science Centre under the grant No. 2017/27/B/ST6/01325 as well as by the statutory funds of the Department of Systems and Computer Networks, Faculty of Electronics, Wroclaw University of Science and Technology.

References

1. Alcalá-Fdez, J., et al.: Keel data-mining software tool: data set repository, integration of algorithms and experimental analysis framework. J. Mult. Valued Log. Soft Comput. **17**, 255–287 (2011)
2. Back, T., Fogel, D., Michalewicz, Z.: Handbook of Evolutionary Computation. Oxford University Press, New York (1997)
3. Branco, P., Torgo, L., Ribeiro, R.P.: Relevance-based evaluation metrics for multiclass imbalanced domains. In: Kim, J., Shim, K., Cao, L., Lee, J.-G., Lin, X., Moon, Y.-S. (eds.) PAKDD 2017. LNCS (LNAI), vol. 10234, pp. 698–710. Springer, Cham (2017). https://doi.org/10.1007/978-3-319-57454-7_54
4. Canuto, A.M., Nascimento, D.S.: A genetic-based approach to features selection for ensembles using a hybrid and adaptive fitness function. In: The 2012 international joint conference on neural networks (IJCNN), pp. 1–8. IEEE (2012)
5. Demšar, J.: Statistical comparisons of classifiers over multiple data sets. J. Mach. Learn. Res. **7**, 1–30 (2006)
6. Du, L., Xu, Y., Jin, L.: Feature selection for imbalanced datasets based on improved genetic algorithm. In: Decision Making and Soft Computing: Proceedings of the 11th International FLINS Conference, pp. 119–124. World Scientific (2014)
7. García, S., Fernández, A., Luengo, J., Herrera, F.: Advanced nonparametric tests for multiple comparisons in the design of experiments in computational intelligence and data mining: experimental analysis of power. Inf. Sci. **180**(10), 2044–2064 (2010)
8. Haque, M.N., Noman, N., Berretta, R., Moscato, P.: Heterogeneous ensemble combination search using genetic algorithm for class imbalanced data classification. PloS One **11**(1), e0146116 (2016)

9. Koziarski, M., Krawczyk, B., Woźniak, M.: The deterministic subspace method for constructing classifier ensembles. Pattern Anal. Appl. **20**(4), 981–990 (2017)
10. Krawczyk, B.: Learning from imbalanced data: open challenges and future directions. Prog. Artif. Intell. **5**(4), 221–232 (2016). https://doi.org/10.1007/s13748-016-0094-0
11. Ksieniewicz, P., Woźniak, M.: Imbalanced data classification based on feature selection techniques. In: Yin, H., Camacho, D., Novais, P., Tallón-Ballesteros, A.J. (eds.) IDEAL 2018. LNCS, vol. 11315, pp. 296–303. Springer, Cham (2018). https://doi.org/10.1007/978-3-030-03496-2_33
12. Lee, H.M., Chen, C.M., Chen, J.M., Jou, Y.L.: An efficient fuzzy classifier with feature selection based on fuzzy entropy. IEEE Trans. Syst. Man Cybern. Part B (Cybern.) **31**(3), 426–432 (2001)
13. Pedregosa, F., et al.: Scikit-learn: machine learning in Python. J. Mach. Learn. Res. **12**, 2825–2830 (2011)
14. Wozniak, M., Graña, M., Corchado, E.: A survey of multiple classifier systems as hybrid systems. Inf. Fusion **16**, 3–17 (2014)

The Feasibility of Deep Learning Use for Adversarial Model Extraction in the Cybersecurity Domain

Michał Choraś[1,2], Marek Pawlicki[1,2(✉)], and Rafał Kozik[1,2]

[1] ITTI Sp. z o.o., Poznań, Poland
[2] UTP University of Science and Technology, Bydgoszcz, Poland
marek.pawlicki@utp.edu.pl

Abstract. Machine learning algorithms found their way into a surprisingly wide range of applications, providing utility and allowing for insights gathered from data in a way never before possible. Those tools, however, have not been developed with security in mind. A deployed algorithm can meet a multitude of risks in the real world. This work explores one of those risks - the feasibility of an exploratory attack geared towards stealing an algorithm used in the cybersecurity domain. The process we have used is thoroughly explained and the results are promising.

Keywords: Pattern recognition · Adversarial machine learning · ANN

1 Introduction

Recent advances in machine learning (ML) and the surge in computational power have opened the way to the proliferation of Artificial Intelligence (AI) in all walks of life. The insights inferred from gathered data with the use of ML techniques have radically transformed, for example, the fields of health care and finance, along with uses in security systems [1].

Machine Learning has the uncanny ability to gather the interdependencies among features in large datasets. A range of methods emerged, like the Support Vector Machines (SVM), Clustering, Neural Networks, etc. The ML procedure usually follows the same outline - one where a training phase is followed by a deployment phase, where classification or regression is performed.

The training phase is where the algorithm 'fits' a model to the provided data - usually a large set. ML often achieves surprisingly accurate results in a wide range of applications [2].

With the real-world applications of AI came the realization that its security requires immediate attention. Malicious users, called 'Adversaries' in the AI world, can skillfully influence the inputs fed to the AI algorithms in a way that changes the classification or regression results [3]. Regardless of the machine learning's ubiquity, the awareness of the security threats and ML susceptibility to adversarial attacks is fairly uncommon [1]. Currently, numerous vulnerabilities

© Springer Nature Switzerland AG 2019
H. Yin et al. (Eds.): IDEAL 2019, LNCS 11872, pp. 353–360, 2019.
https://doi.org/10.1007/978-3-030-33617-2_36

have been exposed by researchers, most notably speech recognition, autonomous vehicles, and overall deep learning [3].

The research on securing machine learning is ongoing, with a range of different approaches found in recent literature. The puzzle of a truly immune system, however, still remains unsolved. The solutions already developed are yet to be proven in real-life applications as well [4]. The existing fixes introduced contemporary research include approaches like training the algorithms with the inclusion of perturbed examples, concepts like distillation or using a Generative Adversarial Network. They are reported to work for particular types of attacks, though they do not provide security against all types of strikes. Those solutions can also lead to underperforming ML solutions [3].

This paper is structured as follows: Firstly, the definitions of model extraction, black- and white-box attacks and the general domain of adversarial attacks on machine learning are disclosed. Secondly, the experimental setup is illustrated, along with the technologies and datasets used. This is followed by the description of the experimental process and finally, the results of the experiments are given.

2   Overview of Adversarial Attacks

Concisely put, exploratory attacks are a way to form a functional equivalent of a deployed ML algorithm, to extract the decision boundary, the setup of the algorithm, its properties and the information about the training dataset [5].

The danger an adversary poses is determined by the information available to it. The level of access the malevolent user has determined the range of attacks they can employ. This is referenced to as Adversarial Capabilities [1,3]. The level of acquaintance the adversary possesses of the machine learning algorithm architecture influences their behaviour. That level can be broadly categorized as white- and black box.

Black box attacks are undertaken with no prior familiarity with the model or any of its parts. This can be performed by carefully providing and observing the input/output pairs of the algorithm under attack, due to the transferability among many ML algorithms [1]. This category of intrusions can be then divided into a list of sub-groups:

- Non-Adaptive Black-Box Attack: In order to craft adversarial examples on a local copy of the target model, the adversary steals the classifier. In this category of attacks the adversary only gets access to the model's training data distribution, and uses it to train a local copy of the model with a chosen procedure [3].
- Adaptive Black-Box Attack: in case of the attacker not having any knowledge of the algorithm whatsoever, he can then proceed to assault the classifier with an attack similar to chosen-plaintext attack from the field of cryptography, where the adversary polls the target model with carefully selected querries [3,6].
- Strict Black-Box Attack: in case the adversary can gather the input-output pairs of the classifier but does not have the ability to poll the model with

adaptable inputs. This kind of attack can succeed when a large number of pairs is collected [3].

In a black-box attack, the adversary initiates the assault on an ML classifier without knowledge of the training data. To initiate a black-box attack the agent sends a set of samples to the algorithm and receives the output labels. Then, using the input-output pairs a deep learning (based on a multi-layer neural network) classifier is trained, building a functional copy of the original classifier.

The deep neural network is chosen for its astonishing modeling capacity in pattern recognition. It displays a strong ability to fit the data [7]. Moreover, [5] illustrates that a deep neural network can be used to build a functionally equivalent classifier to a Naive Bayes or an SVM classifier. In addition to that, [8] illustrates how to use a deep neural network to steal a real-life classifier through polling its API (Application Programming Interface) with the use of a free license allowance and a dataset scraped off the internet. The thing of importance in case of black-box attacks is that the adversarial objective is to train a local version of a classifier [3].

On the other hand, the white-box case can be evaluated from a number of angles. The architecture of the model can vary significantly with a myriad of hyper-parameter setups. The knowledge of the specific values can be used to find vulnerabilities which can then be abused by the attacker. A well-informed antagonist can alter the input data to steer the algorithm into spaces where its performance is weak [1]. In a white-box attack, the adversary is assumed to have total knowledge of the inner workings of an ML algorithm, that means the type of ML used, parameters like the number of hidden layers or the number of neurons, as well as the hyper-parameter setups, like the optimizer. The parameters of a deployed model are also assumed to be revealed to him. This knowledge is then used to find the areas where the classifier is open to attack. The knowledge of model weights can be translated into an immensely powerful attack [3].

As for the attack types these can be categorised in the following groups of methods:

- Model Inversion,
- Membership Inference attack,
- Model Extraction via APIs,
- Information Inference [3].

Model Extraction or model stealing disregards the confidentiality of ML, allowing the adversary to create a 'surrogate model' for the attacked ML algorithm [9].

3   Experiments

The aim of this paper is to evaluate if it is possible to steal a classifier in the cybersecurity domain by probing an established ML algorithm with a batch of data and training a deep neural network (DNN) on the observed responses. This kind of attack could be a first step to launching more sophisticated adversarial attacks, like poisoning attacks or evasion attacks. Having a local version of a classifier is, therefore, a valuable commodity for a malevolent user. It is important to mention that this work is preliminary research and at this stage, the stolen algorithm is not a live, operational system, but an artificial neural network trained on one of the benchmark cybersecurity datasets. The details of the setup are disclosed in a later section. To make the experiment as real-life as possible under the current circumstances, a set of assumptions was made.

- In a real-life cybersecurity situation, the only observable response would be the restriction of access for an agent displaying a given set of behaviours. This is in a way similar to the oracle attack known from cryptography [6]. Therefore the extracted model will only be able to perform binary classification.
- To make it possible to compare the extracted model with the original algorithm will be a binary classifier as well, only making the distinction between normal and anomalous traffic.
- As the research progresses, attempts will be made to steal a classifier in a completely black-box manner, at this stage however some initial knowledge of the algorithm is assumed. More precisely, the work has been conducted using the NSL-KDD dataset.

3.1 Original Classifier

The original classifier was a multi-layer artificial neural network (ANN) trained on the NSL-KDD [10] dataset.

A multi-layer ANN refers to a network with multiple computational layers, known as the hidden layers. The computations in those layers are not perceivable from the users' point of view. The data are fed forward from the input layer throughout the consecutive layers until they reach the output layer.

This is known as the feed-forward neural network [11]. It is a custom dictated with the ANN performance that the number of nodes in the layer following the input layer is usually smaller as compared to the input. This gives an adversary a clue as to the design of the network. The number of neurons and hidden layers is dictated by the performance of the network [7]. The utilisation of hidden layers with a number of neurons smaller than those of inputs contributes to a generalisation in representation, which might assist in the network's performance on new data [11].

3.2 Dataset Description

NSL-KDD is a dataset well known in the cybersecurity community, as it is an improved benchmark replacing the previous industry-standard - KDD'99. Even though the dataset is unfortunately an old one, and it still displays some of the undesirable traits, the shortage of open IDS datasets makes NSL-KDD the dataset of choice for cybersecurity-related research [10].

The set is designed to be suitable for machine learning, thanks to its size which does not impose researches with the need to randomly subsample the dataset. The results of the research are more comparable thanks to this property.

3.3   TensorFlow and Keras

In our work we utilise TensorFlow, an open-source library, shared with the community by the fellow machine learning scientists and engineers working for the Google Brain team. The library provides immense support for machine and deep learning with amazing performance [12] Keras, an interface to TensorFlow and a range of other machine learning libraries, grants the developer ease and fast pace of experimentation through a modular design [13].

3.4   Model Extraction

We have used a multilayer perceptron as the model extraction architecture, with 4 hidden layers of 512 neurons each and the Rectified Linear Unit activation function. This is motivated by the fact that a deep learning classifier achieved superior results for model extraction of an SVM and a Naive Bayes classifier in [5]. As depicted in [5], the model extraction algorithm procedure can be summarised as follows:

Stealing an Algorithm:

1. polling the classifier with input data
2. observing the labels returned by the classifier
3. using input/label data to train a deep learning classifier and optimize its hyperparameters.

   The pipeline of the process is also depicted in Fig. 1.

3.5   Experimental Process

To properly test the feasibility of an algorithm trained on the labels inferred from another algorithm we have prepared the experiment as follows:

   Firstly, the dataset was transformed into a binary classification problem. To achieve this the classes contained in the set were converted to 'attack(1)' or 'benign(0)' respectively. The details of the conversion are found in Table 1. After the conversion the dataset contains 58628 'attack' records and 67342 'benign' records, that constitutes roughly 46.5% attacks and 53,5% benign datapoints.

   Secondly, the binary NSL-KDD dataset was split into three parts:

- Set A - used to train the original algorithm
- Set B - used to poll the original classifier and receive classification labels, and then to train the new DNN Classifier
- Set C - used to test the DNN and compare the results with the original classifier (Figs. 2 and 3)

Thirdly, the original algorithm is polled with the features from Set B, the responses are recorded, paired with their respective polls and formed into a new dataset. This new dataset is then utilised to train a DNN.

   Finally, we use the remainder of the dataset (Set C) to test the DNN and also feed it to the original classifier to compare the results.

DNN Architecture. According to [11], a shallow neural network of one layer and sufficient number of neurons has, in theory, the ability to fit to any function. The same source suggests going for depth is also a viable alternative (although not without its own problems). Following in the footsteps of [5], where authors show they are capable of stealing a Naive Bayes and an SVM classifier using a DNN, the authors have used a network of 4 dense layers, 512 neurons each and a Rectified Linear Unit activation function. The architecture was chosen so as to be more complex than the original classifier, taking an educated guess based on the typical number of features in cybersecurity datasets.

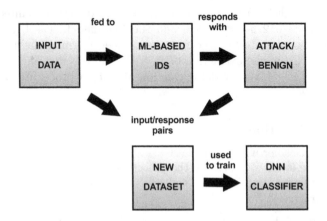

Fig. 1. Model extraction pipeline

Table 1. Mapping of NSL-KDD classes to attack (1) and benign (0)

Classes in NSL-KDD	Classification in this paper
'ipsweep', 'multihop', 'rootkit', 'warezclient',	attack (1)
'guess_passwd', 'phf', 'nmap',	attack (1)
'teardrop', 'neptune', 'loadmodule', 'imap', 'portsweep'	attack (1)
'pod', 'ftp_write', 'warezmaster',	attack (1)
'back', 'land', 'smurf', 'satan', 'buffer_overflow'	attack (1)
'normal'	benign (0)

Fig. 2. Original Classifier's confusion matrix on subset C

	0	1
0	20108	95
1	120	17468

Fig. 3. Extracted Classifier's confusion matrix on subset C

4   Results and Future Work

In Fig. 2 the classification results of the original algorithm can be found. The '0' refers to benign traffic, the '1' refers to detected attacks. As seen in the confusion matrix, the original classifier only committed 76 false positives (FP) and 78 false negatives (FN) in a total of 37791 classifications. This amounts to an accuracy of 99.59%.

In Fig. 3 one can notice that the extracted model only gave 120 FP and 95 FN's in 37791 classifications, which gives it an accuracy of 99.43%, just on par with the original classifier.

4.1   Conclusions

The results show that in a lab setting it is possible to achieve comparable results for a model created by polling a classifier and using the inferred labels as a training dataset for a deep neural network.

4.2   Future Work

This paper is a stepping stone for further research into adversarial attacks on machine learning. In the near future, we are planning to explore the real-life scenarios for model extraction in the cybersecurity domain, as well as poisoning and evasion attacks. Furthermore, we hope to find ways to secure machine learning against different sorts of adversarial attacks, working as a part of the SAFAIR (Secure And FAIR machine learning for citizen) program of the European SPARTA initiative.

Acknowledgments. This work is funded under the SPARTA project, which has received funding from the European Union's Horizon 2020 research and innovation programme under grant agreement No 830892.

References

1. Papernot, N., McDaniel, P., Sinha, A., Wellman, M.P.: SoK: security and privacy in machine learning. In: 2018 IEEE European Symposium on Security and Privacy (EuroS P), pp. 399–414, April 2018
2. Ateniese, G., Felici, G., Mancini, L.V., Spognardi, A., Villani, A., Vitali, D.: Hacking smart machines with smarter ones: how to extract meaningful data from machine learning classifiers. CoRR, abs/1306.4447 (2013)

3. Chakraborty, A., Alam, M., Dey, V., Chattopadhyay, A., Mukhopadhyay, D.: Adversarial attacks and defences: a survey. CoRR, abs/1810.00069 (2018)

4. Liao, X., Ding, L., Wang, Y.: Secure machine learning, a brief overview. In: 2011 Fifth International Conference on Secure Software Integration and Reliability Improvement - Companion, pp. 26–29, June 2011

5. Shi, Y., Sagduyu, Y., Grushin, A.: How to steal a machine learning classifier with deep learning. In: 2017 IEEE International Symposium on Technologies for Homeland Security (HST), pp. 1–5, April 2017

6. Cachin, C., Camenisch, J.L. (eds.): EUROCRYPT 2004. LNCS, vol. 3027. Springer, Heidelberg (2004). https://doi.org/10.1007/b97182

7. da Silva, I.N., Hernane Spatti, D., Andrade Flauzino, R., Liboni, L.H.B., dos Reis Alves, S.F.: Artificial Neural Networks A Practical Course. Springer, Cham (2017). https://doi.org/10.1007/978-3-319-43162-8

8. Shi, Y., Sagduyu, Y.E., Davaslioglu, K., Li, J.H.: Generative adversarial networks for black-box API attacks with limited training data. CoRR, abs/1901.09113 (2019)

9. Quiring, E., Arp, D., Rieck, K.: Forgotten siblings: unifying attacks on machine learning and digital watermarking. In: 2018 IEEE European Symposium on Security and Privacy (EuroS P), pp. 488–502, April 2018

10. NSL-KDD dataset

11. Aggarwal, C.C.: Neural Networks and Deep Learning A Textbook (2018)

12. Abadi, M., et al.: TensorFlow: large-scale machine learning on heterogeneous systems (2015). Software tensorflow.org

13. Chollet, F., et al.: Keras (2015). https://github.com/fchollet/keras

Author Index